CHEMICAL ENGINEERING

Solutions to the Problems in Chemical Engineering Volume 1

Coulson & Richardson's Chemical Engineering

Chemical Engineering, Volume 1, Sixth edition
Fluid Flow, Heat Transfer and Mass Transfer
J. M. Coulson and J. F. Richardson
with J. R. Backhurst and J. H. Harker

Chemical Engineering, Volume 2, Fourth edition
Particle Technology and Separation Processes
J. M. Coulson and J. F. Richardson
with J. R. Backhurst and J. H. Harker

Chemical Engineering, Volume 3, Third edition
Chemical & Biochemical Reactors & Process Control
Edited by J. F. Richardson and D. G. Peacock

Solutions to the Problems in Volume 1, First edition
J. R. Backhurst and J. H. Harker
with J. F. Richardson

Chemical Engineering, Volume 5, Second edition
Solutions to the Problems in Volumes 2 and 3
J. R. Backhurst and J. H. Harker

Chemical Engineering, Volume 6, Third edition
Chemical Engineering Design
R. K. Sinnott

Coulson & Richardson's

CHEMICAL ENGINEERING

J. M. COULSON and J. F. RICHARDSON

Solutions to the Problems in Chemical Engineering
Volume 1

By

J. R. BACKHURST and J. H. HARKER

University of Newcastle upon Tyne

With

J. F. RICHARDSON

University of Wales Swansea

OXFORD AUCKLAND BOSTON JOHANNESBURG MELBOURNE NEW DELHI

Butterworth-Heinemann
Linacre House, Jordan Hill, Oxford OX2 8DP
225 Wildwood Avenue, Woburn, MA 01801-2041
A division of Reed Educational and Professional Publishing Ltd

A member of the Reed Elsevier plc group

First published 2001

Transferred to digital printing 2004

British Library Cataloguing in Publication Data
A catalogue record for this book is available from the British Library

Library of Congress Cataloguing in Publication Data
A catalogue record for this book is available from the Library of Congress

ISBN 0 7506 4950 X

Typeset by Laser Words, Madras, India

FOR EVERY TITLE THAT WE PUBLISH, BUTTERWORTH-HEINEMANN
WILL PAY FOR BTCV TO PLANT AND CARE FOR A TREE.

Contents

Preface

Each of the volumes of the Chemical Engineering Series includes numerical examples to illustrate the application of the theory presented in the text. In addition, at the end of each volume, there is a selection of problems which the reader is invited to solve in order to consolidate his (or her) understanding of the principles and to gain a better appreciation of the order of magnitude of the quantities involved.

Many readers who do not have ready access to assistance have expressed the desire for solutions manuals to be available. This book, which is a successor to the old Volume 4, is an attempt to satisfy this demand as far as the problems in Volume 1 are concerned. It should be appreciated that most engineering problems do not have unique solutions, and they can also often be solved using a variety of different approaches. If therefore the reader arrives at a different answer from that in the book, it does not necessarily mean that it is wrong.

This edition of the solutions manual relates to the sixth edition of Volume 1 and incorporates many new problems. There may therefore be some mismatch with earlier editions and, as the volumes are being continually revised, they can easily get out-of-step with each other.

None of the authors claims to be infallible, and it is inevitable that errors will occur from time to time. These will become apparent to readers who use the book. We have been very grateful in the past to those who have pointed out mistakes which have then been corrected in later editions. It is hoped that the present generation of readers will prove to be equally helpful!

J. F. R.

SECTION 1

Units and Dimensions

PROBLEM 1.1

98% sulphuric acid of viscosity 0.025 N s/m^2 and density 1840 kg/m^3 is pumped at 685 cm^3/s through a 25 mm line. Calculate the value of the Reynolds number.

Solution

Cross-sectional area of line $= (\pi/4)0.025^2 = 0.00049$ m^2.

Mean velocity of acid, $u = (685 \times 10^{-6})/0.00049 = 1.398$ m/s.

$\therefore$ Reynolds number, $Re = du\rho/\mu = (0.025 \times 1.398 \times 1840)/0.025 = \underline{\underline{2572}}$

PROBLEM 1.2

Compare the costs of electricity at 1 p per kWh and gas at 15 p per therm.

Solution

Each cost is calculated in p/MJ.

  1 kWh = 1 kW $\times$ 1 h = (1000 J/s)(3600 s) = 3,600,000 J or 3.6 MJ

  1 therm = 105.5 MJ

  $\therefore$ cost of electricity = 1 p/3.6 MJ or (1/3.6) = $\underline{\underline{0.28 \text{ p/MJ}}}$

    cost of gas = 15 p/105.5 MJ or (15/105.5) = $\underline{\underline{0.14 \text{ p/MJ}}}$

PROBLEM 1.3

A boiler plant raises 5.2 kg/s of steam at 1825 kN/m^2 pressure, using coal of calorific value 27.2 MJ/kg. If the boiler efficiency is 75%, how much coal is consumed per day? If the steam is used to generate electricity, what is the power generation in kilowatts assuming a 20% conversion efficiency of the turbines and generators?

Solution

From the steam tables, in Appendix A2, Volume 1, total enthalpy of steam at 1825 kN/m^2 = 2798 kJ/kg.

$$\therefore \qquad \text{enthalpy of steam} = (5.2 \times 2798) = 14{,}550 \text{ kW}$$

Neglecting the enthalpy of the feed water, this must be derived from the coal. With an efficiency of 75%, the heat provided by the coal $= (14{,}550 \times 100)/75 = 19{,}400$ kW. For a calorific value of 27,200 kJ/kg, rate of coal consumption $= (19{,}400/27{,}200)$
$$= 0.713 \text{ kg/s}$$

or: $\qquad (0.713 \times 3600 \times 24)/1000 = \underline{\underline{61.6 \text{ Mg/day}}}$

20% of the enthalpy in the steam is converted to power or:

$$(14{,}550 \times 20)/100 = 2910 \text{ kW or } 2.91 \text{ MW say } \underline{\underline{3 \text{ MW}}}$$

PROBLEM 1.4

The power required by an agitator in a tank is a function of the following four variables:

(a) diameter of impeller,
(b) number of rotations of the impeller per unit time,
(c) viscosity of liquid,
(d) density of liquid.

From a dimensional analysis, obtain a relation between the power and the four variables.

The power consumption is found, experimentally, to be proportional to the square of the speed of rotation. By what factor would the power be expected to increase if the impeller diameter were doubled?

Solution

If the power $P = \phi(DN\rho\mu)$, then a typical form of the function is $P = kD^a N^b \rho^c \mu^d$, where k is a constant. The dimensions of each parameter in terms of $\mathbf{M}$, $\mathbf{L}$, and $\mathbf{T}$ are: power, $P = \mathbf{ML}^2/\mathbf{T}^3$, density, $\rho = \mathbf{M/L}^3$, diameter, $D = \mathbf{L}$, viscosity, $\mu = \mathbf{M/LT}$, and speed of rotation, $N = \mathbf{T}^{-1}$

Equating dimensions:

$\mathbf{M}:\quad 1 = c + d$
$\mathbf{L}:\quad 2 = a - 3c - d$
$\mathbf{T}: -3 = -b - d$

Solving in terms of d : $a = (5 - 2d)$, $b = (3 - d)$, $c = (1 - d)$

$$\therefore \qquad P = k\left(\frac{D^5}{D^{2d}} \frac{N^3}{N^d} \frac{\rho}{\rho^d} \mu^d \right)$$

or: $\qquad P/D^5 N^3 \rho = k(D^2 N\rho/\mu)^{-d}$

that is: $\qquad N_P = k\,Re^m$

Thus the power number is a function of the Reynolds number to the power m. In fact N_P is also a function of the Froude number, DN^2/g. The previous equation may be written as:

$$P/D^5N^3\rho = k(D^2N\rho/\mu)^m$$

Experimentally:
$$P \propto N^2$$

From the equation, $P \propto N^m N^3$, that is $m + 3 = 2$ and $m = -1$

Thus for the same fluid, that is the same viscosity and density:

$$(P_2/P_1)(D_1^5N_1^3/D_2^5N_2^3) = (D_1^2N_1/D_2^2N_2)^{-1} \text{ or: } (P_2/P_1) = (N_2^2D_2^3)/(N_1^2D_1^3)$$

In this case, $N_1 = N_2$ and $D_2 = 2D_1$.

$$\therefore \qquad\qquad (P_2/P_1) = 8D_1^3/D_1^3 = \underline{\underline{8}}$$

A similar solution may be obtained using the Recurring Set method as follows:

$$P = \phi(D, N, \rho, \mu), f(P, D, N, \rho, \mu) = 0$$

Using **M, L** and **T** as fundamentals, there are five variables and three fundamentals and therefore by Buckingham's π theorem, there will be two dimensionless groups.

Choosing D, N and ρ as the recurring set, dimensionally:

$$\left.\begin{array}{l} D \equiv \mathbf{L} \\ N \equiv \mathbf{T}^{-1} \\ \rho \equiv \mathbf{ML}^{-3} \end{array}\right] \text{ Thus: } \left[\begin{array}{l} \mathbf{L} \equiv D \\ \mathbf{T} \equiv N^{-1} \\ \mathbf{M} \equiv \rho\mathbf{L}^3 = \rho D^3 \end{array}\right.$$

First group, π_1, is $P(\mathbf{ML}^2\mathbf{T}^{-3})^{-1} \equiv P(\rho D^3 D^2 N^3)^{-1} \equiv \dfrac{P}{\rho D^5 N^3}$

Second group, π_2, is $\mu(\mathbf{ML}^{-1}\mathbf{T}^{-1})^{-1} \equiv \mu(\rho D^3 D^{-1} N)^{-1} \equiv \dfrac{\mu}{\rho D^2 N}$

Thus:
$$f\left(\frac{P}{\rho D^5 N^3}, \frac{\mu}{\rho D^2 N}\right) = 0$$

Although there is little to be gained by using this method for simple problems, there is considerable advantage when a large number of groups is involved.

PROBLEM 1.5

It is found experimentally that the terminal settling velocity u_0 of a spherical particle in a fluid is a function of the following quantities:

particle diameter, d; buoyant weight of particle (weight of particle − weight of displaced fluid), W; fluid density, ρ, and fluid viscosity, μ.

Obtain a relationship for u_0 using dimensional analysis.

Stokes established, from theoretical considerations, that for small particles which settle at very low velocities, the settling velocity is independent of the density of the fluid except in so far as this affects the buoyancy. Show that the settling velocity *must* then be inversely proportional to the viscosity of the fluid.

Solution

If: $u_0 = kd^a W^b \rho^c \mu^d$, then working in dimensions of **M, L** and **T**:

$$(\mathbf{L/T}) = k(\mathbf{L}^a(\mathbf{ML/T^2})^b(\mathbf{M/L^3})^c(\mathbf{M/LT})^d)$$

Equating dimensions:

M : $0 = b + c + d$
L : $1 = a + b - 3c - d$
T : $-1 = -2b - d$

Solving in terms of b:

$$a = -1, c = (b-1), \text{ and } d = (1 - 2b)$$

∴ $u_0 = k(1/d)(W^b)(\rho^b/\rho)(\mu/\mu^{2b})$ where k is a constant,

or: $u_0 = k(\mu/d\rho)(W\rho/\mu^2)^b$

Rearranging:

$$(du_0\rho/\mu) = k(W\rho/\mu^2)^b$$

where $(W\rho/\mu^2)$ is a function of a form of the Reynolds number.

For u_0 to be independent of ρ, b must equal unity and $\underline{\underline{u_0 = kW/d\mu}}$

Thus, for constant diameter and hence buoyant weight, the settling velocity is inversely proportional to the fluid viscosity.

PROBLEM 1.6

A drop of liquid spreads over a horizontal surface. What are the factors which will influence:

(a) the rate at which the liquid spreads, and
(b) the final shape of the drop?

Obtain dimensionless groups involving the physical variables in the two cases.

Solution

(a) The rate at which a drop spreads, say R m/s, will be influenced by: viscosity of the liquid, μ; volume of the drop, V expressed in terms of d, the drop diameter; density of the liquid, ρ; acceleration due to gravity, g and possibly, surface tension of the liquid,

σ. In this event: $R = \mathrm{f}(\mu, d, \rho, g, \sigma)$. The dimensions of each variable are: $R = \mathbf{L/T}$, $\mu = \mathbf{M/LT}$, $d = \mathbf{L}$, $\rho = \mathbf{M/L^3}$, $g = \mathbf{L/T^2}$, and $\sigma = \mathbf{M/T^2}$. There are 6 variables and 3 fundamentals and hence $(6 - 3) = 3$ dimensionless groups. Taking as the recurring set, d, ρ and g, then:

$$d \equiv \mathbf{L}, \qquad \mathbf{L} = d$$
$$\rho \equiv \mathbf{M/L^3} \quad \therefore \mathbf{M} = \rho \mathbf{L}^3 = \rho d^3$$
$$g \equiv \mathbf{L/T^2} \quad \therefore \mathbf{T}^2 = \mathbf{L}/g = d/g \text{ and } \mathbf{T} = d^{0.5}/g^{0.5}$$

Thus, dimensionless group 1: $R\mathbf{T/L} = Rd^{0.5}/dg^{0.5} = R/(dg)^{0.5}$

dimensionless group 2: $\mu \mathbf{LT/M} = \mu d(d^{0.5})/(g^{0.5}\rho d^3) = \mu/(g^{0.5}\rho d^{1.5})$

dimensionless group 3: $\sigma \mathbf{T}^2/\mathbf{M} = \sigma d/(g\rho d^3) = \sigma/(g\rho d^2)$

$\therefore$
$$R/(dg)^{0.5} = \mathrm{f}\left(\frac{\mu}{g^{0.5}\rho d^{1.5}}, \frac{\sigma}{g\rho d^2}\right)$$

or:
$$\underline{\underline{\frac{R^2}{dg} = \mathrm{f}\left(\frac{\mu^2}{g\rho^2 d^3}, \frac{\sigma}{g\rho d^2}\right)}}$$

(b) The final shape of the drop as indicated by its diameter, d, may be obtained by using the argument in (a) and putting $R = 0$. An alternative approach is to assume the final shape of the drop, that is the final diameter attained when the force due to surface tension is equal to that attributable to gravitational force. The variables involved here will be: volume of the drop, V; density of the liquid, ρ; acceleration due to gravity, g, and the surface tension of the liquid, σ. In this case: $d = f(V, \rho, g, \sigma)$. The dimensions of each variable are: $d = \mathbf{L}$, $V = \mathbf{L}^3$, $\rho = \mathbf{M/L^3}$, $g = \mathbf{L/T^2}$, $\sigma = \mathbf{M/T^2}$. There are 5 variables and 3 fundamentals and hence $(5 - 3) = 2$ dimensionless groups. Taking, as before, d, ρ and g as the recurring set, then:

$$d \equiv \mathbf{L}, \qquad \mathbf{L} = d$$
$$\rho \equiv \mathbf{M/L^3} \quad \therefore \mathbf{M} = \rho \mathbf{L}^3 = \rho d^3$$
$$g \equiv \mathbf{L/T^2} \quad \therefore \mathbf{T}^2 = \mathbf{L}/g = d/g \text{ and } \mathbf{T} = d^{0.5}/g^{0.5}$$

Dimensionless group 1: $V/\mathbf{L}^3 = V/d^3$

Dimensionless group 2: $\sigma \mathbf{T}^2/\mathbf{M} = \sigma d/(g\rho d^3) = \sigma/(g\rho d^2)$

and hence:
$$\underline{\underline{(d^3/V) = \mathrm{f}\left(\frac{\sigma}{g\rho d^2}\right)}}$$

PROBLEM 1.7

Liquid is flowing at a volumetric flowrate of Q per unit width down a vertical surface. Obtain from dimensional analysis the form of the relationship between flowrate and film thickness. If the flow is streamline, show that the volumetric flowrate is directly proportional to the density of the liquid.

Solution

The flowrate, Q, will be a function of the fluid density, ρ, and viscosity, μ, the film thickness, d, and the acceleration due to gravity, g,

or: $Q = f(\rho, g, \mu, d)$, or: $Q = K\rho^a g^b \mu^c d^d$ where K is a constant.

The dimensions of each variable are: $Q = \mathbf{L}^2/\mathbf{T}$, $\rho = \mathbf{M}/\mathbf{L}^3$, $g = \mathbf{L}/\mathbf{T}^2$, $\mu = \mathbf{M}/\mathbf{LT}$ and $d = \mathbf{L}$.

Equating dimensions:

$$\mathbf{M}: \quad 0 = a + c$$
$$\mathbf{L}: \quad 2 = -3a + b - c + d$$
$$\mathbf{T}: \quad -1 = -2b - c$$

from which, $c = 1 - 2b$, $a = -c = 2b - 1$, and $d = 2 + 3a - b + c$
$$= 2 + 6b - 3 - b + 1 - 2b = 3b$$

$\therefore$
$$Q = K(\rho^{2b-1} g^b \mu^{1-2b} d^{3b})$$

or:
$$\frac{Q\rho}{\mu} = K(\rho^2 g d^3/\mu^2)^b \text{ and } Q \propto \mu^{1-2b}.$$

For streamline flow, $Q \propto \mu^{-1}$

and:
$$-1 = 1 - 2b \text{ and } b = 1$$

$\therefore$
$$Q\rho/\mu = K(\rho^2 g d^3/\mu^2), Q = K(\rho g d^3/\mu)$$

and: $\underline{\underline{Q \text{ is directly proportional to the density, } \rho}}$

PROBLEM 1.8

Obtain, by dimensional analysis, a functional relationship for the heat transfer coefficient for forced convection at the inner wall of an annulus through which a cooling liquid is flowing.

Solution

Taking the heat transfer coefficient, h, as a function of the fluid velocity, density, viscosity, specific heat and thermal conductivity, u, ρ, μ, C_p and k, respectively, and of the inside and outside diameters of the annulus, d_i and d_0 respectively, then:

$$h = f(u, d_i, d_0, \rho, \mu, C_p, k)$$

The dimensions of each variable are: $h = \mathbf{H}/\mathbf{L}^2\mathbf{T}\theta$, $u = \mathbf{L}/\mathbf{T}$, $d_i = \mathbf{L}$, $d_0 = \mathbf{L}$, $\rho = \mathbf{M}/\mathbf{L}^3$, $\mu = \mathbf{M}/\mathbf{LT}$, $C_p = \mathbf{H}/\mathbf{M}\theta$, $k = \mathbf{H}/\mathbf{LT}\theta$. There are 8 variables and 5 fundamental dimensions and hence there will be $(8 - 5) = 3$ groups. $\mathbf{H}$ and θ always appear however as the group H/θ and in effect the fundamental dimensions are 4 ($\mathbf{M}$, $\mathbf{L}$, $\mathbf{T}$ and $\mathbf{H}/\theta$) and there will be $(8 - 4) = 4$ groups. For the recurring set, the variables d_i, μ, k and ρ will be chosen. Thus:

$$
\begin{aligned}
d_i &\equiv \mathbf{L}, & \mathbf{L} &= d_i \\
\rho &\equiv \mathbf{M}/\mathbf{L}^3 & \mathbf{M} &= \rho \mathbf{L}^3 = \rho d_i^3 \\
\mu &\equiv \mathbf{M}/\mathbf{L}\mathbf{T}, & \mathbf{T} &= \mathbf{M}/\mathbf{L}\mu = \rho d_i^3/d_i\mu = \rho d_i^2/\mu \\
k &\equiv (\mathbf{H}/\theta)/\mathbf{L}\mathbf{T}, & (\mathbf{H}/\theta) &= k\mathbf{L}\mathbf{T} = kd_i\rho d_i^2/\mu = k\rho d_i^3/\mu
\end{aligned}
$$

Dimensionless group 1: $hL^2T/(H/\theta) = hd_i^2\rho d_i^2/\mu(k\rho d_i^3/\mu) = hd_i/k$

Dimensionless group 2: $uT/L = u\rho d_i^2/\mu d_i = d_i u\rho/\mu$

Dimensionless group 3: $d_0/L = d_0/d_i$

Dimensionless group 4: $C_p M/(H/\theta) = C_p\rho d_i^3/k(\rho d_i^3/\mu) = C_p\mu/k$

$\therefore$ $\quad hd_i/k = \text{f}(d_i u\rho/\mu, C_p\mu/k, d_0/d_i)$ which is a form of equation 9.94.

PROBLEM 1.9

Obtain by dimensional analysis a functional relationship for the wall heat transfer coefficient for a fluid flowing through a straight pipe of circular cross-section. Assume that the effects of natural convection may be neglected in comparison with those of forced convection.

It is found by experiment that, when the flow is turbulent, increasing the flowrate by a factor of 2 always results in a 50% increase in the coefficient. How would a 50% increase in density of the fluid be expected to affect the coefficient, all other variables remaining constant?

Solution

For heat transfer for a fluid flowing through a circular pipe, the dimensional analysis is detailed in Section 9.4.2 and, for forced convection, the heat transfer coefficient at the wall is given by equations 9.64 and 9.58 which may be written as:

$$hd/k = f(du\rho/\mu, C_p\mu/k)$$

or:

$$hd/k = K(du\rho/\mu)^n (C_p\mu/k)^m$$

$\therefore$

$$h_2/h_1 = (u_2/u_1)^n.$$

Increasing the flowrate by a factor of 2 results in a 50% increase in the coefficient, or:

$$1.5 = 2.0^n \text{ and } n = (\ln 1.5/\ln 2.0) = 0.585.$$

Also:

$$h_2/h_1 = (\rho_2/\rho_1)^{0.585}$$

When $(\rho_2/\rho_1) = 1.50$, $h_2/h_1 = (1.50)^{0.585} = 1.27$ and the coefficient is increased by $\underline{\underline{27\%}}$

PROBLEM 1.10

A stream of droplets of liquid is formed rapidly at an orifice submerged in a second, immiscible liquid. What physical properties would be expected to influence the mean size of droplet formed? Using dimensional analysis obtain a functional relation between the variables.

Solution

The mean droplet size, d_p, will be influenced by: diameter of the orifice, d; velocity of the liquid, u; interfacial tension, σ; viscosity of the dispersed phase, μ; density of the dispersed phase, ρ_d; density of the continuous phase, ρ_c, and acceleration due to gravity, g. It would also be acceptable to use the term $(\rho_d - \rho_c)g$ to take account of gravitational forces and there may be some justification in also taking into account the viscosity of the continuous phase.

On this basis:
$$d_p = f(d, u, \sigma, \mu, \rho_d, \rho_c, g)$$

The dimensions of each variable are: $d_p = \mathbf{L}$, $d = \mathbf{L}$, $u = \mathbf{L/T}$, $\sigma = \mathbf{M/T^2}$, $\mu = \mathbf{M/LT}$, $\rho_d = \mathbf{M/L^3}$, $\rho_c = \mathbf{M/L^3}$, and $g = \mathbf{L/T^2}$. There are 7 variables and hence with 3 fundamental dimensions, there will be $(7 - 3) = 4$ dimensionless groups. The variables d, u and σ will be chosen as the recurring set and hence:

$$
\begin{aligned}
d &\equiv \mathbf{L}, & \mathbf{L} &= d \\
u &\equiv \mathbf{L/T}, & \mathbf{T} &= \mathbf{L}/u = d/u \\
\sigma &\equiv \mathbf{M/T^2}, & \mathbf{M} &= \sigma \mathbf{T^2} = \sigma d^2/u^2
\end{aligned}
$$

Thus, dimensionless group 1: $\mu \mathbf{LT/M} = \mu d(d/u)/(\sigma d^2/u^2) = \mu u/\sigma$

dimensionless group 2: $\rho_d \mathbf{L^3/M} = \rho_d d^3/(\sigma d^2/u^2) = \rho_d du^2/\sigma$

dimensionless group 3: $\rho_c \mathbf{L^3/M} = \rho_c d^3/(\sigma d^2/u^2) = \rho_c du^2/\sigma$

dimensionless group 4: $g\mathbf{T^2/L} = g(d^2/u^2)/d = gd/u^2$

and the function becomes: $d_p = f(\mu u/\sigma, \rho_d du^2/\sigma, \rho_c du^2/\sigma, gd/u^2)$

PROBLEM 1.11

Liquid flows under steady-state conditions along an open channel of fixed inclination to the horizontal. On what factors will the depth of liquid in the channel depend? Obtain a relationship between the variables using dimensional analysis.

Solution

The depth of liquid, d, will probably depend on: density and viscosity of the liquid, ρ and μ; acceleration due to gravity, g; volumetric flowrate per unit width of channel, Q, and the angle of inclination, θ,

or:
$$d = f(\rho, \mu, g, Q, \theta)$$

Excluding θ at this stage, there are 5 variables and with 3 fundamental dimensions there will be $(5 - 3) = 2$ dimensionless groups. The dimensions of each variable are: $d = \mathbf{L}$, $\rho = \mathbf{M/L^3}$, $\mu = \mathbf{M/LT}$, $g = \mathbf{L/T^2}$, $Q = \mathbf{L^2/T}$, and, choosing Q, ρ and g as the recurring set, then:

$$Q = \mathbf{L^2/T} \quad \mathbf{T} = L^2/Q$$
$$g = \mathbf{L/T^2} \quad \mathbf{L} = gT^2 = gL^4/Q^2, L^3 = Q^2/g, L = Q^{2/3}/g^{1/3}$$
$$\text{and } \mathbf{T} = Q^{4/3}/Qg^{2/3} = Q^{1/3}/g^{2/3}$$
$$\rho = \mathbf{M/L^3} \quad \mathbf{M} = L^3\rho = (Q^2/g)\rho = Q^2\rho/g$$

Thus, dimensionless group 1: $d/L = dg^{1/3}/Q^{2/3}$ or d^3g/Q^2

dimensionless group 2: $\mu LT/M = \mu(Q^{2/3}/g^{1/3})(Q^{1/3}/g^{2/3})/Q^2\rho g = \mu/Q\rho$

and the function becomes: $\underline{\underline{d^3g/Q^2 = f(\mu/Q\rho, \theta)}}$

PROBLEM 1.12

Liquid flows down an inclined surface as a film. On what variables will the thickness of the liquid film depend? Obtain the relevant dimensionless groups. It may be assumed that the surface is sufficiently wide for edge effects to be negligible.

Solution

This is essentially the same as Problem 1.11, though here the approach used is that of equating indices.

If, as before:
$$d = K(\rho^a, \mu^b, g^c, Q^d, \theta^e)$$

then, excluding θ at this stage, the dimensions of each variable are: $d = \mathbf{L}$, $\rho = \mathbf{M/L^3}$, $\mu = \mathbf{M/LT}$, $g = \mathbf{L/T^2}$, $Q = \mathbf{L^2/T}$.
Equating dimensions:

$$\mathbf{M}: 0 = a + b \tag{i}$$

$$\mathbf{L}: 1 = -3a - b + c + 2d \tag{ii}$$

$$\mathbf{T}: 0 = -b - 2c - d \tag{iii}$$

Solving in terms of b and c then:

from (i) $\quad a = -b$

from (iii) $\quad d = -b - 2c$

and in (ii) $\quad 1 = 3b - b + c - 2b - 4c$ or: $c = -1/3$ $\quad \therefore \quad d = 2/3 - b$

Thus:
$$d = K(\rho^{-b} \cdot \mu^b \cdot g^{-1/3} \cdot Q^{2/3-b})$$

$$dg^{1/3}/Q^{2/3} = K(\mu/\rho Q)^b$$

and:
$$\underline{\underline{d^3 g/Q^2 = K(\mu/\rho Q)^b (\theta)^e}} \text{ as before.}$$

PROBLEM 1.13

A glass particle settles under the action of gravity in a liquid. Upon what variables would the terminal velocity of the particle be expected to depend? Obtain a relevant dimensionless grouping of the variables. The falling velocity is found to be proportional to the square of the particle diameter when other variables are kept constant. What will be the effect of doubling the viscosity of the liquid? What does this suggest regarding the nature of the flow?

Solution

See Volume 1, Example 1.3

PROBLEM 1.14

Heat is transferred from condensing steam to a vertical surface and the resistance to heat transfer is attributable to the thermal resistance of the condensate layer on the surface.

What variables are expected to affect the film thickness at a point?

Obtain the relevant dimensionless groups.

For streamline flow it is found that the film thickness is proportional to the one third power of the volumetric flowrate per unit width. Show that the heat transfer coefficient is expected to be inversely proportional to the one third power of viscosity.

Solution

For a film of liquid flowing down a vertical surface, the variables influencing the film thickness δ, include: viscosity of the liquid (water), μ; density of the liquid, ρ; the flow per unit width of surface, Q, and the acceleration due to gravity, g. Thus: $\delta = f(\mu, \rho, Q, g)$. The dimensions of each variable are: $\delta = \mathbf{L}$, $\mu = \mathbf{M/LT}$, $\rho = \mathbf{M/L^3}$, $Q = \mathbf{L^2/T}$, and $g = \mathbf{L/T^2}$. Thus, with 5 variables and 3 fundamental dimensions, $(5 - 3) = 2$ dimensionless groups are expected. Taking μ, ρ and g as the recurring set, then:

$$\mu \equiv \mathbf{M/LT}, \quad \mathbf{M} = \mu \mathbf{LT}$$
$$\rho \equiv \mathbf{M/L^3}, \quad \mathbf{M} = \rho \mathbf{L^3} \qquad \therefore \rho \mathbf{L^3} = \mu \mathbf{LT}, \quad \mathbf{T} = \rho \mathbf{L^2}/\mu$$
$$g \equiv \mathbf{L/T^2} = \mu^2 \mathbf{L}/\rho^2 \mathbf{L^4} = \mu^2/\rho^2 \mathbf{L^3} \quad \therefore \mathbf{L^3} = \mu^2/\rho^2 g \text{ and } \mathbf{L} = \mu^{2/3}/(\rho^{2/3} g^{1/3})$$

$$\therefore \qquad \mathbf{T} = \rho(\mu^2/\rho^2 g)^{2/3}/\mu = \mu^{1/3}/(\rho^{1/3} g^{2/3})$$

and:
$$\mathbf{M} = \mu(\mu^2/\rho^2 g)^{1/3}(\mu^{1/3}/(\rho^{1/3} g^{2/3})) = \mu^2/(\rho g)$$

Thus, dimensionless group 1: $Q\mathbf{T}/\mathbf{L}^2 = Q(\mu^{1/3}/(\rho^{1/3}g^{2/3}))/(\mu^{4/3}/(\rho^{4/3}g^{2/3})) = Q\rho/\mu$

dimensionless group 2: $\delta\mathbf{L} = \delta\mu^{2/3}/(\rho^{2/3}g^{1/3})$ or, cubing $= \delta^3\rho^2g/\mu^2$

and: $$\underline{\underline{(\delta^3\rho^2g/\mu^2) = f(Q\rho/\mu)}}$$

This may be written as: $(\delta^3\rho^2g/\mu^2) = K(Q\rho/\mu)^n$

For streamline flow, $\delta \propto Q^{1/3}$ or $n = 1$

and hence: $(\delta^3\rho^2g/\mu^2) = KQ\rho/\mu, \delta^3 = KQ\mu/(\rho g)$ and $\delta = (KQ\mu/\rho g)^{1/3}$

As the resistance to heat transfer is attributable to the thermal resistance of the condensate layer which in turn is a function of the film thickness, then: $h \propto k/\delta$ where k is the thermal conductivity of the film and since $\delta \propto \mu^{1/3}$, $h \propto k/\mu^{1/3}$, that is the coefficient is inversely proportional to the one third power of the liquid viscosity.

PROBLEM 1.15

A spherical particle settles in a liquid contained in a narrow vessel. Upon what variables would you expect the falling velocity of the particle to depend? Obtain the relevant dimensionless groups.

For particles of a given density settling in a vessel of large diameter, the settling velocity is found to be inversely proportional to the viscosity of the liquid. How would this depend on particle size?

Solution

This problem is very similar to Problem 1.13, although, in this case, the liquid through which the particle settles is contained in a narrow vessel. This introduces another variable, D, the vessel diameter and hence the settling velocity of the particle is given by: $u = f(d, \rho, \mu, D, \rho_s, g)$. The dimensions of each variable are: $u = \mathbf{L/T}$, $d = \mathbf{L}$, $\rho = \mathbf{M/L}^3$, $\mu = \mathbf{M/LT}$, $D = \mathbf{L}$, $\rho_s = \mathbf{M/L}^3$, and $g = \mathbf{L/T}^2$. With 7 variables and 3 fundamental dimensions, there will be $(7 - 3) = 4$ dimensionless groups. Taking d, ρ and μ as the recurring set, then:

$d \equiv \mathbf{L}, \qquad \mathbf{L} = d$
$\rho \equiv \mathbf{M/L}^3, \quad \mathbf{M} = \rho\mathbf{L}^3 = \rho d^3$
$\mu \equiv \mathbf{M/LT}, \quad \mathbf{T} = \mathbf{M/L}\mu = \rho d^3/d\mu = \rho d^2/\mu$

Thus: dimensionless group 1: $u\mathbf{T}/\mathbf{L} = u\rho d^2/(\mu d) = du\rho/\mu$

dimensionless group 2: $D/\mathbf{L} = D/d$

dimensionless group 3: $\rho_s\mathbf{L}^3/\mathbf{M} = \rho_s d^3/(\rho d^3) = \rho_s/\rho$

and dimensionless group 4: $g\mathbf{T}^2/\mathbf{L} = g\rho^2d^4/(\mu^2d) = g\rho^2d^3/\mu^2$

Thus: $$\underline{\underline{(du\rho/\mu) = f((D/d)(\rho_s/\rho)(g\rho^2d^3/\mu^2))}}$$

In particular, $(du\rho/\mu) = K(g\rho^2d^3/\mu^2)^n$ where K is a constant.

For particles settling in a vessel of large diameter, $u \propto 1/\mu$. But $(u/\mu) \propto (1/\mu^2)^n$ and, when $n = 1$, $n \propto 1/\mu$. In this case:

$$(du\rho/\mu) = K(g\rho^2 d^3/\mu^2)$$

or:
$$du \propto d^3 \text{ and } u \propto d^2$$

Thus the settling velocity is proportional to the square of the particle size.

PROBLEM 1.16

A liquid is in steady state flow in an open trough of rectangular cross-section inclined at an angle θ to the horizontal. On what variables would you expect the mass flow per unit time to depend? Obtain the dimensionless groups which are applicable to this problem.

Solution

This problem is similar to Problems 1.11 and 1.12 although, here, the width of the trough and the depth of liquid are to be taken into account. In this case, the mass flow of liquid per unit time, G will depend on: fluid density, ρ; fluid viscosity, μ; depth of liquid, h; width of the trough, a; acceleration due to gravity, g and the angle to the horizontal, θ. Thus: $G = \mathrm{f}(\rho, \mu, h, a, g, \theta)$. The dimensions of each variable are: $G = \mathbf{M/T}$, $\rho = \mathbf{M/L^3}$, $\mu = \mathbf{M/LT}$, $h = \mathbf{L}$, $a = \mathbf{L}$, $g = \mathbf{L/T^2}$ and neglecting θ at this stage, with 6 variables with dimensions and 3 fundamental dimensions, there will be $(6 - 3) = 3$ dimensionless groups. Taking h, ρ and μ as the recurring set then:

$$h \equiv \mathbf{L}, \qquad \mathbf{L} = h$$
$$\rho \equiv \mathbf{M/L^3}, \quad \mathbf{M} = \rho \mathbf{L}^3 = \rho h^3$$
$$\mu \equiv \mathbf{M/LT}, \quad \mathbf{T} = \mathbf{M}/\mathbf{L}\mu = \rho h^3/(h\mu) = \rho h^2/\mu$$

Thus: dimensionless group 1: $G\mathbf{T}/\mathbf{M} = G\rho h^2/(\mu \rho h^3) = G/\mu h$

dimensionless group 2: $a/\mathbf{L} = a/h$

dimensionless group 3: $g\mathbf{T}^2/\mathbf{L} = g\rho^2 h^4/(\mu^2 h) = g\rho^2 h^3/\mu^2$

and:
$$(G/\mu h) = \mathrm{f}((a/h)(g\rho^2 h^3/\mu^2))$$

PROBLEM 1.17

The resistance force on a spherical particle settling in a fluid is given by Stokes' Law. Obtain an expression for the terminal falling velocity of the particle. It is convenient to express experimental results in the form of a dimensionless group which may be plotted against a Reynolds group with respect to the particle. Suggest a suitable form for this dimensionless group.

Force on particle from Stokes' Law $= 3\pi\mu du$; where μ is the fluid viscosity, d is the particle diameter and u is the velocity of the particle relative to the fluid.

What will be the terminal falling velocity of a particle of diameter 10 μm and of density 1600 kg/m^3 settling in a liquid of density 1000 kg/m^3 and of viscosity 0.001 Ns/m^2?

If Stokes' Law applies for particle Reynolds numbers up to 0.2, what is the diameter of the largest particle whose behaviour is governed by Stokes' Law for this solid and liquid?

Solution

The accelerating force due to gravity = (mass of particle − mass of liquid displaced)g. For a particle of radius r, volume = $4\pi r^3/3$, or, in terms of diameter, d, volume = $4\pi(d^3/2^3)/3 = \pi d^3/6$. Mass of particle = $\pi d^3\rho_s/6$, where ρ_s is the density of the solid. Mass of liquid displaced = $\pi d^3\rho/6$, where ρ is the density of the liquid, and accelerating force due to gravity = $(\pi d^3\rho_s/6 - \pi d^3\rho/6)g = (\pi d^3/6)(\rho_s - \rho)g$.

At steady state, that is when the terminal velocity is attained, the accelerating force due to gravity must equal the drag force on the particle F, or: $(\pi d^3/6)(\rho_s - \rho)g = 3\pi\mu d u_0$ where u_0 is the terminal velocity of the particle.

Thus:
$$u_0 = (d^2 g/18\mu)(\rho_s - \rho) \tag{i}$$

It is assumed that the resistance per unit projected area of the particle, R', is a function of particle diameter, d; liquid density, ρ; liquid viscosity, μ, and particle velocity, u or $R' = f(d, \rho, \mu, u)$. The dimensions of each variable are $R' = \mathbf{M/LT^2}$, $d = \mathbf{L}$, $\rho = \mathbf{M/L^3}$, $\mu = \mathbf{M/LT}$ and $u = \mathbf{L/T}$. With 5 variables and 3 fundamental dimensions, there will be $(5 - 3) = 2$ dimensionless groups. Taking d, ρ and u as the recurring set, then:

$$d \equiv \mathbf{L}, \qquad \mathbf{L} = d$$
$$\rho \equiv \mathbf{M/L^3}, \quad \mathbf{M} = \rho\mathbf{L}^3 = \rho d^3$$
$$u \equiv \mathbf{L/T}, \qquad \mathbf{T} = \mathbf{L}/u = d/u$$

Thus: dimensionless group 1: $R'\mathbf{LT^2/M} = R'd(d^2/u^2)/(\rho d^3) = R'/\rho u^2$

dimensionless group 2: $\mu\mathbf{LT/M} = \mu d(d/u)/(\rho d^3) = \mu/(du\rho)$

and:
$$R'/\rho u^2 = f(\mu/du\rho)$$

or:
$$R'/\rho u^2 = K(du\rho/\mu)^n = K\,Re^n \tag{ii}$$

In this way the experimental data should be plotted as the group $(R/\rho u^2)$ against Re.

For this particular example, $d = 10$ μm $= (10 \times 10^{-6}) = 10^{-5}$ m; $\rho_s = 1600$ kg/m^3; $\rho = 1000$ kg/m^3 and $\mu = 0.001$ Ns/m^2.

Thus, in equation (i): $u_0 = ((10^{-5})^2 \times 9.81/(18 \times 0.001))(1600 - 1000)$

$$= 3.27 \times 10^{-5} \text{ m/s or } \underline{\underline{0.033 \text{ mm/s}}}$$

When $Re = 0.2$, $du\rho/\mu = 0.2$ or when the terminal velocity is reached:

$$du_0 = 0.2\mu/\rho = (0.2 \times 0.001)/1000 = 2 \times 10^{-7}$$

or:
$$u_0 = (2 \times 10^{-7})/d$$

In equation (i): $u_0 = (d^2 g/18\mu)(\rho_s - \rho)$

$$(2 \times 10^{-7})/d = (d^2 \times 9.81/(18 \times 0.001))(1600 - 1000)$$

$\therefore$ $d^3 = 6.12 \times 10^{-13}$

and: $d = 8.5 \times 10^{-5}$ m or $\underline{\underline{85 \ \mu m}}$

PROBLEM 1.18

A sphere, initially at a constant temperature, is immersed in a liquid whose temperature is maintained constant. The time t taken for the temperature of the centre of the sphere to reach a given temperature θ_c is a function of the following variables:

Diameter of sphere, d
Thermal conductivity of sphere, k
Density of sphere, ρ
Specific heat capacity of sphere, C_p
Temperature of fluid in which it is immersed, θ_s.

Obtain relevant dimensionless groups for this problem.

Solution

In this case, $t = f(d, k, \rho, C_p, \theta_c, \theta_s)$. The dimensions of each variable are: $t = \mathbf{T}$, $d = \mathbf{L}$, $k = \mathbf{ML/T\theta}$, $C_p = \mathbf{L^2/T^2\theta}$, $\theta_c = \mathbf{\theta}$, $\theta_s = \mathbf{\theta}$. There are 7 variables and hence with 4 fundamental dimensions, there will be $(7 - 4) = 3$ dimensionless groups. Taking d, ρ, C_p and θ_c as the recurring set, then:

$$d \equiv \mathbf{L}, \qquad \mathbf{L} = d,$$
$$\rho \equiv \mathbf{M/L^3}, \quad \mathbf{M} = \rho \mathbf{L}^3 = \rho d^3$$
$$\theta_c \equiv \mathbf{\theta}, \qquad \mathbf{\theta} = \theta_c$$
$$C_p \equiv \mathbf{L^2/T^2\theta} \quad C_p = d^2/\mathbf{T}^2\theta_c \text{ and } \mathbf{T}^2 = d^2/C_p\theta_c \text{ or: } \mathbf{T} = d/C_p^{0.5}\theta_c^{0.5}$$

Thus: dimensionless group 1: $t/\mathbf{T} = tC_p^{0.5}\theta_c^{0.5}/d$

dimensionless group 2: $k\mathbf{T\theta}/\mathbf{ML} = k(d/C_p^{0.5}\theta_c^{0.5})\theta_c/(\rho d^3)d = k\theta_c^{0.5}/C_p^{0.5}\rho d^3$

dimensionless group 3: $\theta_s/\theta = \theta_s/\theta_c$

PROBLEM 1.19

Upon what variables would the rate of filtration of a suspension of fine solid particles be expected to depend? Consider the flow through unit area of filter medium and express the variables in the form of dimensionless groups.

It is found that the filtration rate is doubled if the pressure difference is doubled. What would be the effect of raising the temperature of filtration from 293 to 313 K?

The viscosity of the liquid is given by:

$$\mu = \mu_0(1 - 0.015(T - 273))$$

where μ is the viscosity at a temperature T K and μ_0 is the viscosity at 273 K.

Solution

The volume flow of filtrate per unit area, u m^3/m^{2}s, will depend on the fluid density, ρ; fluid viscosity, μ; particle size, d; pressure difference across the bed, ΔP, and the voidage of the cake, e or: $v = f(\rho, \mu, d, \Delta P, e)$. The dimensions of each of these variables are $u = \mathbf{L/T}$, $\rho = \mathbf{M/L^3}$, $\mu = \mathbf{M/LT}$, $d = \mathbf{L}$, $\Delta P = \mathbf{M/LT^2}$ and $e =$ dimensionless. There are 6 variables and 3 fundamental dimensions and hence $(6 - 3) = 3$ dimensionless groups. Taking, d, ρ and μ as the recurring set, then:

$$d \equiv \mathbf{L}, \qquad \mathbf{L} = d$$
$$\rho \equiv \mathbf{M/L^3}, \quad \mathbf{M} = \rho\mathbf{L^3} = \rho d^3$$
$$\mu \equiv \mathbf{M/LT}, \quad \mathbf{T} = \mathbf{M/L}\mu = \rho d^3/(d\mu) = \rho d^2/\mu$$

Thus: dimensionless group 1: $u\mathbf{T/L} = u\rho d^2/(\mu d) = du\rho/\mu$

dimensionless group 2: $\Delta P\mathbf{LT^2/M} = \Delta Pd(\rho d^2/\mu)^2/\rho d^3 = \Delta P\rho d^2/\mu^2$

and the function is: $\qquad \underline{\underline{\Delta P\rho d^2/\mu^2 = f(du\rho/\mu)}}$

This may be written as: $\qquad \Delta P\rho d^2/\mu^2 = K(du\rho/\mu)^n$

Since the filtration rate is doubled when the pressure difference is doubled, then:

$$u \propto \Delta P \text{ and } n = 1, \Delta P\rho d^2/\mu^2 = Kdu\rho/\mu$$

and: $\qquad u = (1/K)\Delta Pd/\mu$, or $u \propto 1/\mu$

$(\mu_{293}/\mu_{313}) = \mu_0(1 - 0.015(293 - 273))/\mu_0(1 - 0.015(313 - 273)) = (0.7/0.4) = 1.75$

$\therefore \qquad\qquad (v_{313}/v_{293}) = (\mu_{293}/\mu_{313}) = 1.75$

and the filtration rate will increase by 75%.

SECTION 2

Flow of Fluids — Energy and Momentum Relationships

PROBLEM 2.1

Calculate the ideal available energy produced by the discharge to atmosphere through a nozzle of air stored in a cylinder of capacity 0.1 m³ at a pressure of 5 MN/m². The initial temperature of the air is 290 K and the ratio of the specific heats is 1.4.

Solution

From equation 2.1: $dU = \delta q - \delta W$. For an adiabatic process: $\delta q = 0$ and $dU = -\delta W$, and for an isentropic process: $dU = C_v dT = -\delta W$ from equation 2.25.

As $\gamma = C_p/C_v$ and $C_p = C_v + \mathbf{R}$ (from equation 2.27), $C_v = \mathbf{R}/(\gamma - 1)$

$\therefore \qquad\qquad W = -C_v \Delta T = -\mathbf{R}\Delta T/(\gamma - 1) = (\mathbf{R}T_1 - \mathbf{R}T_2)/(\gamma - 1)$

and: $\qquad \mathbf{R}T_1 = P_1 v_1$ and $\mathbf{R}T_2 = P_2 v_2$ and hence: $W = (P_1 v_1 - P_2 v_2)/(\gamma - 1)$

$P_1 v_1^{\gamma} = P_2 v_2^{\gamma}$ and substituting for v_2 gives:

$$W = [(P_1 v_1)/(\gamma - 1)](1 - (P_2/P_1)^{(\gamma-1)/\gamma})$$

and: $\qquad\qquad \Delta U = -W = [(P_1 v_1)/(\gamma - 1)][(P_2/P_1)^{(\gamma-1)/\gamma} - 1]$

In this problem:

$$P_1 = 5 \text{ MN/m}^2, P_2 = 0.1013 \text{ MN/m}^2, T_1 = 290 \text{ K, and } \gamma = 1.4.$$

The specific volume, $v_1 = (22.4/29)(290/273)(0.1013/5) = 0.0166 \text{ m}^3/\text{kg}$.

$\therefore \qquad -W = [(5 \times 10^6 \times 0.0166)/0.4][(0.1013/5)^{0.4/1.4} - 1] = -0.139 \times 10^6 \text{ J/kg}$

Mass of gas $= (0.1/0.0166) = 6.02 \text{ kg}$

$\therefore \qquad\qquad \Delta U = -(0.139 \times 10^6 \times 6.20) = -0.84 \times 10^6 \text{ J or } \underline{\underline{-840 \text{ kJ}}}$

PROBLEM 2.2

Obtain expressions for the variation of: (a) internal energy with change of volume, (b) internal energy with change of pressure, and (c) enthalpy with change of pressure, all at constant temperature, for a gas whose equation of state is given by van der Waals' law.

Solution

See Volume 1, Example 2.2.

PROBLEM 2.3

Calculate the energy stored in 1000 cm^3 of gas at 80 MN/m^2 at 290 K using STP as the datum.

Solution

The key to this solution lies in the fact that the operation involved is an irreversible expansion. Taking C_v as constant between T_1 and T_2, $\Delta U = -W = nC_v(T_2 - T_1)$ where n is the kmol of gas and T_2 and T_1 are the final and initial temperatures, then for a constant pressure process, the work done, assuming the ideal gas laws apply, is given by:

$$W = P_2(V_2 - V_1) = P_2[(n\mathbf{R}T_2/P_2) - (n\mathbf{R}T_1/P_1)]$$

Equating these expressions for W gives: $-C_v(T_2 - T_1) = P_2\left(\dfrac{\mathbf{R}T_2}{P_2} - \dfrac{\mathbf{R}T_1}{P_1}\right)$

In this example:

$P_1 = 80000$ kN/m^2, $P_2 = 101.3$ kN/m^2, $V_1 = (1 \times 10^{-3})$ m^3, $\mathbf{R} = 8.314$ kJ/kmol K, and

$T_1 = 290$ K

Hence: $-C_v(T_2 - 290) = 101.3\mathbf{R}[(T_2/101.3) - (290/80,000)]$

By definition, $\gamma = C_p/C_v$ and $C_p = \rho_v + \mathbf{R}$ (from equation 2.27) or: $C_v = \mathbf{R}/(\gamma - 1)$

Substituting: $\hspace{4cm} T_2 = 174.15$ K.

$$PV = n\mathbf{R}T \text{ and } n = (80000 \times 10^{-3})/(8.314 \times 290) = 0.033 \text{ kmol}$$

$\therefore \ \Delta U = -W = C_v n(T_2 - T_1) = (1.5 \times 8.314 \times 0.033)(174.15 - 290) = \underline{\underline{-47.7 \text{ kJ}}}$

PROBLEM 2.4

Compressed gas is distributed from a works in cylinders which are filled to a pressure P by connecting them to a large reservoir of gas which remains at a steady pressure P and temperature T. If the small cylinders are initially at a temperature T and pressure P_0, what is the final temperature of the gas in the cylinders if heat losses can be neglected and if the compression can be regarded as reversible? Assume that the ideal gas laws are applicable.

Solution

From equation 2.1, $dU = \delta q - \delta W$. For an adiabatic operation, $q = 0$ and $\delta q = 0$ and $\delta W = Pdv$ or $dU = -Pdv$. The change in internal energy for any process involving an ideal gas is given by equation 2.25:

$$C_v dT = -P dv = dU.$$

$$P = \mathbf{R}T/Mv \text{ and hence: } dT/T = (-\mathbf{R}/MC_v)(dv/v)$$

By definition: $\gamma = C_p/C_v$ and $C_p = C_v + \mathbf{R}/M$ (from equation 2.27)

$$\therefore \qquad \mathbf{R}/MC_v = \gamma - 1 \text{ and } dT/T = -(\gamma - 1)(dv/v)$$

Integrating between conditions 1 and 2 gives:

$$\ln(T_2/T_1) = -(\gamma - 1)\ln(v_2/v_1) \text{ or } (T_2/T_1) = (v_2/v_1)^{\gamma-1}$$

$$P_1 v_1/T_1 = P_2 v_2/T_2 \text{ and hence } v_1/v_2 = (P_2/P_1)(T_1/T_2)$$

and: $$(T_2/T_1) = (P_2/P_1)^{(\gamma-1)/\gamma}$$

Using the symbols given, the final temperature, $\underline{\underline{T_2 = T(P/P_0)^{(\gamma-1)/\gamma}}}$

Flow in Pipes and Channels

PROBLEM 3.1

Calculate the hydraulic mean diameter of the annular space between a 40 mm and a 50 mm tube.

Solution

The hydraulic mean diameter, d_m, is defined as four times the cross-sectional area divided by the wetted perimeter. Equation 3.69 gives the value d_m for an annulus of outer radius r and inner radius r_i as:

$$d_m = 4\pi(r^2 - r_i^2)/2\pi(r + r_i) = 2(r - r_i) = (d - d_i)$$

If $r = 25$ mm and $r_i = 20$ mm, then:

$$d_m = 2(25 - 20) = \underline{\underline{10 \text{ mm}}}$$

PROBLEM 3.2

0.015 m³/s of acetic acid is pumped through a 75 mm diameter horizontal pipe 70 m long. What is the pressure drop in the pipe?
Viscosity of acid $= 2.5$ mNs/m², density of acid $= 1060$ kg/m³, and roughness of pipe surface $= 6 \times 10^{-5}$ m.

Solution

Cross-sectional area of pipe $= (\pi/4)(0.075)^2 = 0.0044$ m².

Velocity of acid in the pipe, $u = (0.015/0.0044) = 3.4$ m/s.

Reynolds number $= \rho u d/\mu = (1060 \times 3.4 \times 0.07)/(2.5 \times 10^{-3}) = 1.08 \times 10^5$

Pipe roughness $e = 6 \times 10^{-5}$ m and $e/d = (6 \times 10^{-5})/0.075 = 0.0008$

The pressure drop is calculated from equation 3.18 as: $-\Delta P_f = 4(R/\rho u^2)(l/d)(\rho u^2)$

From Fig. 3.7, when $Re = 1.08 \times 10^5$ and $e/d = 0.0008$, $R/\rho u^2 = 0.0025$.

Substituting: $-\Delta P_f = (4 \times 0.0025)(70/0.075)(1060 \times 3.4^2)$
$$= 114{,}367 \text{ N/m}^2 \text{ or: } \underline{\underline{114.4 \text{ kN/m}^2}}$$

PROBLEM 3.3

A cylindrical tank, 5 m in diameter, discharges through a mild steel pipe 90 m long and 230 mm diameter connected to the base of the tank. Find the time taken for the water level in the tank to drop from 3 m to 1 m above the bottom. The viscosity of water is 1 mNs/m^2.

Solution

If at any time the depth of water in the tank is h and levels 1 and 2 are the liquid levels in the tank and the pipe outlet respectively, then the energy balance equation states that:

$$\Delta(u^2/2) + g\Delta z + v(P_2 - P_1) + F = 0$$

In this example, $P_1 = P_2 =$ atmospheric pressure and $v(P_2 - P_1) = 0$. Also $u_1/u_2 = (0.23/5)^2 = 0.0021$ so that u_1 may be neglected. The energy balance equation then becomes:

$$u^2/2 - hg + 4(R/\rho u^2)(l/d)u^2 = 0$$

The last term is obtained from equation 3.19 and $\Delta z = -h$.
Substituting the known data:

$$u^2/2 - 9.81h + 4(R/\rho u^2)(90/0.23)u^2 = 0$$

or: $$u^2 - 19.62h + 3130(R/\rho u^2)u^2 = 0$$

from which: $$u = 4.43\sqrt{h}/\sqrt{[1 + 3130(R/\rho u^2)]}$$

In falling from a height h to $h - \mathrm{d}h$,

the quantity of water discharged $= (\pi/4)5^2(-\mathrm{d}h) = 19.63\mathrm{d}h$ m^3.

Volumetric flow rate $= (\pi/4)(0.23)^2 u = 0.0415u = 0.184\sqrt{h}/\sqrt{[1 + 3130(R/\rho u^2)]}$, and the time taken for the level to fall from h to $h - \mathrm{d}h$ is:

$$\frac{-19.63}{0.184}\frac{\mathrm{d}h}{\sqrt{h}}\sqrt{[1 + 3130(R/\rho u^2)]} = -106.7h^{-0.5}\sqrt{[1 + 3130(R/\rho u^2)]}\,\mathrm{d}h$$

∴ the time taken for the level to fall from 3 m to 1 m is:

$$t = -106.7\sqrt{[1 + 3130(R/\rho u^2)]}\int_3^1 h^{-0.5}\mathrm{d}h$$

$R/\rho u^2$ depends upon the Reynolds number which will fall as the level in the tank falls and upon the roughness of the pipe e which is not specified in this example. The pressure drop along the pipe $= h\rho g = 4Rl/d$ N/m^2 and $R = h\rho g d/4l$.
From equation 3.23:

$$(R/\rho u^2)\,Re^2 = Rd^2\rho/\mu^2 = h\rho^2 g d^3/4l\mu^2$$

$$= (h \times 1000^2 \times 9.81 \times 0.23^3)/(4 \times 90 \times 10^{-6}) = 3.315 \times 10^8 h$$

Thus as h varies from 3 m to 1 m, $(R/\rho u^2)(Re)^2$ varies from (9.95×10^8) to $(3.315 \times 10^8.)$

If $R/\rho u^2$ is taken as 0.002, Re will vary from (7.05×10^5) to (4.07×10^5). From Fig. 3.7 this corresponds to a range of e/d of between 0.004 and 0.005 or a roughness of between 0.92 and 1.15 mm, which is too high for a commercial pipe.

If e is taken as 0.05 mm, $e/d = 0.0002$, and, for Reynolds numbers near 10^6, $R/\rho u^2 = 0.00175$. Substituting $R/\rho u^2 = 0.00175$ and integrating gives a time of 398 s for the level to fall from 3 m to 1 m. If $R/\rho u^2 = 0.00175$, Re varies from (7.5×10^5) to (4.35×10^5), and from Fig. 3.7, $e/d = 0.00015$, which is near enough to the assumed value. Thus the time for the level to fall is approximately 400 s.

PROBLEM 3.4

Two storage tanks A and B containing a petroleum product discharge through pipes each 0.3 m in diameter and 1.5 km long to a junction at D. From D the product is carried by a 0.5 m diameter pipe to a third storage tank C, 0.8 km away. The surface of the liquid in A is initially 10 m above that in C and the liquid level in B is 7 m higher than that in A. Calculate the initial rate of discharge of the liquid if the pipes are of mild steel. The density of the petroleum product is 870 kg/m^3 and the viscosity is 0.7 mNs/m^2.

Solution

See Volume 1, Example 3.4

PROBLEM 3.5

Find the drop in pressure due to friction in a glazed porcelain pipe 300 m long and 150 mm diameter when water is flowing at the rate of 0.05 m^3/s.

Solution

For a glazed porcelain pipe, $e = 0.0015$ mm, $e/d = (0.0015/150) = 0.00001$.

Cross-sectional area of pipe $= (\pi/4)(0.15)^2 = 0.0176$ m^2.

Velocity of water in pipe, $u = (0.05/0.0176) = 2.83$ m/s.

Reynolds number $= \rho u d/\mu = (1000 \times 2.83 \times 0.15)/10^{-3} = 4.25 \times 10^5$

From Fig. 3.7, $R/\rho u^2 = 0.0017$.

The pressure drop is given by equation 3.18: $-\Delta P_f = 4(R/\rho u^2)(l/d)(\rho u^2)$

or: $4 \times 0.0017(300/0.15)(1000 \times 2.83^2) = 108,900$ N/m^2 or 1 MN/m^2

PROBLEM 3.6

Two tanks, the bottoms of which are at the same level, are connected with one another by a horizontal pipe 75 mm diameter and 300 m long. The pipe is bell-mouthed at each end so that losses on entry and exit are negligible. One tank is 7 m diameter and contains water to a depth of 7 m. The other tank is 5 m diameter and contains water to a depth of 3 m.

If the tanks are connected to each other by means of the pipe, how long will it take before the water level in the larger tank has fallen to 6 m? Assume the pipe to be of aged mild steel.

Solution

The system is shown in Fig. 3a. If at any time t the depth of water in the larger tank is h and the depth in the smaller tank is H, a relationship between h and H may be found.

$$\text{Area of larger tank} = (\pi/4)7^2 = 38.48 \text{ m}^2,$$

$$\text{area of smaller tank} = (\pi/4)5^2 = 19.63 \text{ m}^2.$$

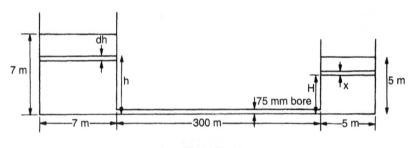

Figure 3a.

When the level in the large tank falls to h, the volume discharged $= (7 - h) \times 38.48$ m^3. The level in the small tank will rise by a height x, given by:

$$x = 38.48(7 - h)/19.63 = (13.72 - 1.95h)$$

$$H = (x + 3) = (16.72 - 1.95h)$$

The energy balance equation is: $\Delta u^2/2 + g\Delta z + v(P_1 - P_2) = F$

$\Delta u^2/2$ may be neglected, and $P_1 = P_2 =$ atmospheric pressure, so that:

$$g\Delta z = F = g\Delta z + 4(R/\rho u^2)(l/d)u^2, \quad \Delta z = (h - H) = (2.95h - 16.72)$$

or: $$(2.95h - 16.72)g = 4(R/\rho u^2)(l/d)u^2$$

and: $$u = \sqrt{[(2.95h - 16.72)g/4(R/\rho u^2)(l/d)]}$$

As the level falls from h to $h - dh$ in time dt, the volume discharged $= 38.48(dh)$ m^3.

Hence: time, $dt = \dfrac{-38.48\, dh}{(\pi/4)(0.075)^2 \sqrt{[(2.95h - 16.72)g/4(R/\rho u^2)(l/d)]}}$

or: $$dt = \frac{-2780\, dh\sqrt{[4(R/\rho u^2)(l/d)]}}{\sqrt{(2.95h - 16.72)}}$$

If $R/\rho u^2$ is taken as 0.002, then:

$$\int dt = -15{,}740 \int_7^6 \frac{dh}{\sqrt{(2.95h - 16.72)}} \quad \text{and} \quad t = 10590 \text{ s}$$

Average volumetric flowrate $= 38.48(7 - 6)/10590 = 0.00364 \text{ m}^3/\text{s}$

Cross-sectional area of pipe $= 0.00442 \text{ m}^2$.

Average velocity in the pipe $= (0.00364/0.00442) = 0.82 \text{ m/s}$.

Reynolds number $= (1000 \times 0.82 \times 0.75)/10^{-3} = 6.2 \times 10^4$.

From Fig. 3.7, if $e = 0.05$ mm, $e/d = 0.00067$ and $R/\rho u^2 = 0.0025$, which is near enough to the assumed value of 0.002 for a first estimate.

Thus the time for the level to fall is approximately <u>10590 s</u> (2.94 h).

PROBLEM 3.7

Two immiscible fluids **A** and **B**, of viscosities μ_A and μ_B, flow under streamline conditions between two horizontal parallel planes of width b, situated a distance $2a$ apart (where a is much less than b), as two distinct parallel layers one above the other, each of depth a. Show that the volumetric rate of flow of **A** is:

$$\left(\frac{-\Delta P a^3 b}{12\mu_A l}\right)\left(\frac{7\mu_A + \mu_B}{\mu_A + \mu_B}\right)$$

where, $-\Delta P$ is the pressure drop over a length l in the direction of flow.

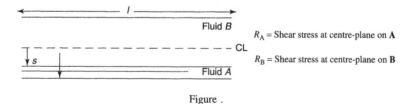

Figure .

Solution

Considering a force balance on the fluid lying within a distance s from the centre plane, then:

For A:
$$-\Delta P(sb) = bl\mu_A\left(\frac{du_s}{ds}\right)_A + R_A l$$

where R_A is the shear stress at the centre plane,

or:
$$-d\mu_s = \frac{-\Delta P}{\mu_A l} s \, ds - \frac{R_A}{\mu_A} ds$$

Integrating:
$$(-u_s)_A = \frac{-\Delta P}{\mu_A l}\frac{s^2}{2} - \frac{R_A}{\mu_A}s + k_1$$

Similarly for B:
$$(-u_s)_B = \frac{-\Delta P}{\mu_A l}\frac{s^2}{2} - \frac{R_B}{\mu_B}s + k_2$$

where R_B is the shear stress at the centre plane on B.

Noting that: $\qquad R_A = -R_B$

At $s = a$: $\qquad (u_s)_A = 0 \quad (u_s)_B = 0$

On the centre plane: At $s = 0$

$$(u_s)_A = (u_s)_B$$

$$A = -\frac{(-\Delta P)}{\mu_A l}\frac{a^2}{2} + \frac{R_A}{\mu_A}a$$

$$B = -\frac{(-\Delta P)}{\mu_B l}\frac{a^2}{2} + \frac{R_B}{\mu_B}a$$

Thus: $\qquad u_{sA} = \frac{-\Delta P}{2\mu_A l}\{a^2 - s^2\} - \frac{R_A}{\mu_A}(a - s)$

and: $\qquad u_{sB} = \frac{-\Delta P}{2\mu_B l}\{a^2 - s^2\} + \frac{R_A}{\mu_B}(a - s) \quad (\text{since } R_A = -R_B)$

Centre line velocity, $\qquad (u_A)_{CL} = \frac{-\Delta P a^2}{2\mu_A l} + \frac{R_A a}{\mu_A}$

and: $\qquad (u_B)_{CL} = \frac{-\Delta P a^2}{2\mu_B l} + \frac{R_A a}{\mu_B}$

Equating: $\qquad \frac{-\Delta P_a^2}{2l}\left\{\frac{1}{\mu_A} - \frac{1}{\mu_B}\right\} = aR_A\left\{\frac{1}{\mu_A} + \frac{1}{\mu_B}\right\}$

$\therefore \qquad R_A = \frac{-\Delta P a}{2l}\left\{\frac{\mu_B - \mu_A}{\mu_B + \mu_A}\right\}$

Substituting:

$$(u_s)_A = \frac{-\Delta P}{2\mu_A l}(a^2 - s^2) - \frac{-\Delta P a}{2\mu_A l}(a - s)\frac{\mu_B - \mu_A}{\mu_B + \mu_A}$$

$$= \frac{-\Delta P}{2\mu_A l}\left\{(a^2 - s^2) + \frac{\mu_A - \mu_B}{\mu_A + \mu_B}a(a - s)\right\}$$

Total flowrate of **A** $= Q_A$ is given by:

$$Q_A = \int_0^a bu_s\,ds = \frac{-\Delta P b}{2\mu_A l}\left\{\int_0^a \left[(a^2 - s^2) + \left(\frac{\mu_A - \mu_B}{\mu_A + \mu_B}\right)(a^2 - as)\right]ds\right\}$$

$$= \frac{-\Delta P b}{2\mu_A l}\left[\left(a^2 s - \frac{s^3}{3}\right) + \left[\frac{\mu_A - \mu_B}{\mu_A + \mu_B}\right]\left(a^2 s - a\frac{s^2}{2}\right)\right]_0^a$$

$$= \frac{-\Delta P b}{2\mu_A l}\left[\frac{2a^3}{3} + \left[\frac{\mu_A - \mu_B}{\mu_A + \mu_B}\right]\frac{a^3}{2}\right]$$

$$= \frac{-\Delta P b a^3}{2\mu_A l}\left[\frac{2}{3} + \frac{\mu_A - \mu_B}{2(\mu_A + \mu_B)}\right]$$

$$= \frac{-\Delta Pba^3}{2\mu_A l} \left[\frac{4\mu_A + 4\mu_B + 3\mu_A - 3\mu_B}{6(\mu_A + \mu_B)} \right]$$

$$= \frac{-\Delta Pba^3}{12\mu_A l} \left[\frac{7\mu_A + \mu_B}{\mu_A + \mu_B} \right]$$

PROBLEM 3.8

Glycerol is pumped from storage tanks to rail cars through a single 50 mm diameter main 10 m long, which must be used for all grades of glycerol. After the line has been used for commercial material, how much pure glycerol must be pumped before the issuing liquid contains not more than 1% of the commercial material? The flow in the pipeline is streamline and the two grades of glycerol have identical densities and viscosities.

Solution

Making a force balance over an element distance r from the axis of a pipe whose radius is a, then:

$$-\Delta P\pi r^2 = -\mu(du/dr)2\pi rl$$

where u is the velocity at distance r and l is the length of the pipe.

Hence: $$du = -(-\Delta P/2\mu l)r\,dr$$

and: $$u = -(-\Delta P/4\mu l)r^2 + \text{constant}$$

When $r = a$, $u = 0$, the constant $= (-\Delta P/4\mu l)a^2$ and hence $u = (-\Delta P/4\mu l)(a^2 - r^2)$.

At a distance r from the axis, the time taken for the fluid to flow through a length l is given by $4\mu l^2/-\Delta P(a^2 - r^2)$.

The volumetric rate of flow from $r = 0$ to $r = r$ is:

$$= \int_0^r (-\Delta P/4\mu l)(a^2 - r^2)2\pi r dr$$

$$= (-\Delta P\pi/8\mu l)(2a^2r^2 - r^4)$$

Volumetric flowrate over the whole pipe $= -\Delta P\pi a^4/8\mu l$
and the mean velocity $= -\Delta Pa^2/8\mu l$.

The required condition at the pipe outlet is:

$$\frac{(-\Delta P\pi/8\mu l)(2a^2r^2 - r^4)}{-\Delta P\pi a^4/8\mu l} = 0.99$$

from which $r = 0.95a$.

The time for fluid at this radius to flow through length $l = 4\mu l^2/-\Delta Pa^2(1 - 0.95^2) = (41\mu l^2)/(-\Delta Pa^2)$

Hence, the volume to be pumped $= (41\mu l^2/(-\Delta P)a^2)(\pi(-\Delta P)a^4/8\mu l)$
$$= 41\pi a^2/8 = 41\pi(0.25)^2/8 = \underline{\underline{0.10 \text{ m}^3}}$$

PROBLEM 3.9

A viscous fluid flows through a pipe with slightly porous walls so that there is a leakage of kP, where P is the local pressure measured above the discharge pressure and k is a constant. After a length l, the liquid is discharged into a tank.

If the internal diameter of the pipe is d and the volumetric rate of flow at the inlet is Q_0, show that the pressure drop in the pipe is given by $-\Delta P = (Q_0/\pi kd)a \tanh al$,

where: $$a = (128k\mu/d^3)^{0.5}$$

Assume a fully developed flow with $(R/\rho u^2) = 8\,Re^{-1}$.

Solution

Across a small element of the pipe, the change in liquid flow is:

$$-dQ = kP\pi d\,dl$$

and the change in velocity is:

$$du = -4kP\,dl/d$$

Also: $$R/\rho u^2 = 8\mu/ud\rho \quad \text{and:} \ R = 8\mu u/d \tag{i}$$

Making a force balance over the element:

$$-dP(\pi/4)d^2 = R\pi d\,dl = 8\mu u\pi\,dl$$

and: $$-dP = 32\mu u\,dl/d^2 \tag{ii}$$

From equations (i) and (ii):

$$-dP/du = -8\mu u/kPd$$

and: $$PdP = 8\mu u\,du/kd$$

Over the whole pipe:

$$(P_0^2/2 - P^2/2) = (8\mu/kd)[(u_0^2/2) - (u^2/2)]$$

and: $$u^2 = u_0^2 + (P^2 - P_0^2)(kd/8\mu)$$

Assuming zero outlet pressure as a datum, then substituting for u in equation (ii):

$$\int_{P_0}^{0} -dp \bigg/ \sqrt{[u_0^2 + (P^2 - P_0^2)(kd/8\mu)]} = 32\mu l/d^2$$

Thus: $$\sqrt{(8\mu/kd)}\,\sinh^{-1}\left[P_0 \bigg/ \sqrt{(8\mu u_0^2/kd) - P_0^2}\right] = 32\mu l/d^2$$

and: $$P_0 \bigg/ \sqrt{[8\mu u_0^2/kd) - P_0^2]} = \sinh\sqrt{(128\mu k/d^3)}l$$

Writing $\sqrt{(128\mu k/d^3)} = a$:

$$(8\mu u_0^2/kd) = [a^2(d^3/128\mu k)][8\mu Q_0^2 \times 16)/(kd\pi^2 d^4)] = Q_0^2/(k^2 d^2 \pi^2 a^2)$$

and:

$$P_0^2 = [(Q_0^2/(k^2 d^2 \pi^2 a^2) \sinh^2 al]/(1 + \sinh^2 al)$$

or:

$$-\Delta P = \underline{\underline{P_0 = (Q_0/kd\pi)a \tanh al}}$$

PROBLEM 3.10

A petroleum product of viscosity 0.5 m Ns/m^2 and density 700 kg/m^3 is pumped through a pipe of 0.15 m diameter to storage tanks situated 100 m away. The pressure drop along the pipe is 70 kN/m^2. The pipeline has to be repaired and it is necessary to pump the liquid by an alternative route consisting of 70 m of 200 mm pipe followed by 50 m of 100 mm pipe. If the existing pump is capable of developing a pressure of 300 kN/m^2, will it be suitable for use during the period required for the repairs? Take the roughness of the pipe surface as 0.05 mm.

Solution

This problem may be solved by using equation 3.23 and Fig. 3.8 to find the volumetric flowrate and then calculating the pressure drop through the alternative pipe system. From equation 3.23: $(R/\rho u^2) Re^2 = -\Delta P_f d^3 \rho/4l\mu^2$

$$= (70,000 \times 0.15^3 \times 700)/(4 \times 100 \times 0.5^2 \times 10^{-6}) = 1.65 \times 10^9$$

From Fig. 3.8, $Re = 8.8 \times 10^5 = (700 \times 10.15u)/(0.5 \times 10^{-3})$ and the velocity $u = 4.19$ m/s.

Cross-sectional area $= (\pi/4)0.15^2 = 0.0177$ m^2.

Volumetric flowrate $= (4.19 \times 0.0177) = 0.074$ m^3/s.

The velocity in the 0.2 m diameter pipe $= 0.074/(\pi/4)0.2^2 = 2.36$ m/s.

The velocity in the 0.1 m diameter pipe $= 9.44$ m/s.

Reynolds number in the 0.2 m pipe $= (700 \times 2.36 \times 0.2/0.5 \times 10^{-3}) = 6.6 \times 10^5$.

Reynolds number in the 0.1 m pipe $= (700 \times 9.44 \times 0.1/0.5 \times 10^{-3}) = 1.32 \times 10^6$.

The values of e/d for the 0.2 m and the 0.1 m pipes are 0.00025 and 0.0005 respectively.

From Fig. 3.7, $R/\rho u^2 = 0.0018$ and 0.002 respectively, and from equation 3.18:

$$-\Delta P_f = [4 \times 0.0018(70/0.2)(700 \times 2.36^2)] + [4 \times 0.002(50/0.1)(700 \times 9.44^2)]$$

$$= 255,600 \text{ N/m}^2 = \underline{\underline{255.6 \text{ kN/m}^2}}$$

In addition, there will be a small pressure drop at the junction of the two pipes although this has been neglected in this solution. <u>Thus the existing pump is satisfactory for this duty.</u>

PROBLEM 3.11

Explain the phenomenon of hydraulic jump which occurs during the flow of a liquid in an open channel.

A liquid discharges from a tank into an open channel under a gate so that the liquid is initially travelling at a velocity of 1.5 m/s and a depth of 75 mm. Calculate, from first principles, the corresponding velocity and depth after the jump.

Solution

See Volume 1, Example 3.9.

PROBLEM 3.12

What is a non-Newtonian fluid? Describe the principal types of behaviour exhibited by these fluids. The viscosity of a non-Newtonian fluid changes with the rate of shear according to the approximate relationship:

$$\mu_a = k \left(-\frac{du_x}{dr} \right)^{-0.5}$$

where μ_a is the viscosity, and du/dr is the velocity gradient normal to the direction of motion.

Show that the volumetric rate of streamline flow through a horizontal tube of radius a is:

$$\frac{\pi}{5} a^5 \left(\frac{-\Delta P}{2kl} \right)^2$$

where $-\Delta P$ is the pressure drop over a length l of the tube.

Solution

For a power-law fluid, the apparent viscosity is given by equation 3.123. Noting that the velocity gradient du_x/dr is negative then:

$$\mu_a = k \left(\left| \frac{du_x}{dy} \right| \right)^{n-1} = k \left(-\frac{du_x}{dr} \right)^{n-1}$$

Thus, in this problem: $n - 1 = -0.5$ and:

$$n - 1 = -0.5, \text{ giving } n = 0.5$$

The fluid is therefore shear-thinning.

Equation 3.136 gives the mean velocity in a pipe:

$$u = \left(\frac{-\Delta P}{4kl} \right)^{\frac{1}{n}} \left(\frac{n}{6n + 2} \right) d^{\frac{n+1}{n}} \qquad \text{(equation 3.136)}$$

Putting $n = 0.5$, the volumetric flowrate Q is given by:

$$Q = \left(\frac{-\Delta P}{4kl}\right)^2 \left(\frac{1}{10}\right) d^3 \left(\frac{\pi}{4}d^2\right)$$

Putting $d = 2a$, then:

$$\underline{\underline{Q = \left(\frac{\pi}{5}\right) a^5 \left(\frac{-\Delta P}{2kl}\right)^2}}$$

PROBLEM 3.13

Calculate the pressure drop when 3 kg/s of sulphuric acid flows through 60 m of 25 mm pipe ($\rho = 1840$ kg/m^3, $\mu = 0.025$ N s/m^2).

Solution

Reynolds number $= \rho u d / \mu = 4G/\pi\mu d = 4.30/(\pi \times 0.025 \times 0.025) = 6110$

If e is taken as 0.05 mm, then: $e/d = (0.05/25) = 0.002$.

From Fig. 3.7, $R/\rho u^2 = 0.0046$.

Acid velocity in pipe $= 3.0/[1840 \times (\pi/4)(0.025)^2] = 3.32$ m/s.

From equation 3.18, the pressure drop due to friction is given by:

$$-\Delta P = 4(R/\rho u^2)(l/d)\rho u^2$$

$$= 4 \times 0.0046(60/0.025)1840 \times 3.32^2 = 895{,}620 \text{ N/m}^2 \text{ or } \underline{\underline{900 \text{ kN/m}^2}}$$

PROBLEM 3.14

The relation between cost per unit length C of a pipeline installation and its diameter d is given by: $C = a + bd$ where a and b are independent of pipe size. Annual charges are a fraction β of the capital cost. Obtain an expression for the optimum pipe diameter on a minimum cost basis for a fluid of density ρ and viscosity μ flowing at a mass rate of G. Assume that the fluid is in turbulent flow and that the Blasius equation is applicable, that is the friction factor is proportional to the Reynolds number to the power of minus one quarter. Indicate clearly how the optimum diameter depends on flowrate and fluid properties.

Solution

The total annual cost of a pipeline consists of a capital charge plus the running costs. The chief element of the running cost is the power required to overcome the head loss which is given by:

$$h_f = 8(R/\rho u^2)(l/d)(u^2/2g) \qquad \text{(equation 3.20)}$$

If $R/\rho u^2 = 0.04/Re^{0.25}$, the head loss per unit length l is:

$$h_f/l = 8(0.04/Re^{0.25})(l/d)(u^2/2g) = 0.016(u^2/d)(\mu/\rho ud)^{0.25}$$

$$= 0.016u^{1.75}\mu^{0.25}/(\rho^{0.25}d^{1.25})$$

The velocity $u = G/\rho A = G/\rho(\pi/4)d^2 = 1.27G/\rho d^2$

$$\therefore \quad h_f/l = 0.016(1.27G/\rho d^2)^{1.75}\mu^{0.25}/(\rho^{0.25}d^{1.25}) = 0.024G^{1.75}\mu^{0.25}/(\rho^2 d^{4.75})$$

The power required for pumping if the pump efficiency is η is:

$$\mathbf{P} = Gg(0.024G^{1.75}\mu^{0.25}/\rho^2 d^{4.75})/\eta$$

If $\eta = 0.5$, $\mathbf{P} = 0.47G^{2.75}\mu^{0.25}/(\rho^2 d^{4.75})$ **W**

If c = power cost/W, the cost of pumping is given by: $0.47cG^{2.75}\mu^{0.25}/\rho^2 d^{4.75}$

The total annual cost is then $= (\beta a + \beta bd) + (\gamma G^{2.75}\mu^{0.25}\mu^{0.25}/\rho^2 d^{4.75})$ where $\gamma = 0.47c$

Differentiating the total cost with respect to the diameter gives:

$$dC/dd = \beta b - 4.75\gamma G^{2.75}\mu^{0.25}/\rho^2 d^{5.75}$$

For minimum cost, $dC/dd = 0$, $d^{5.75} = 4.75\gamma G^{2.75}\mu^{0.25}/\rho^2\beta b$ and $\underline{\underline{d = KG^{0.48}\mu^{0.43}/\rho^{0.35}}}$

where: $$K = (4.75\gamma\beta b)^{0.174}$$

PROBLEM 3.15

A heat exchanger is to consist of a number of tubes each 25 mm diameter and 5 m long arranged in parallel. The exchanger is to be used as a cooler with a rating of 4 MW and the temperature rise in the water feed to the tubes is to be 20 deg K.

If the pressure drop over the tubes is not to exceed 2 kN/m^2, calculate the minimum number of tubes that are required. Assume that the tube walls are smooth and that entrance and exit effects can be neglected. Viscosity of water = 1 mNs/m^2.

Solution

Heat load = (mass flow × specific heat × temperature rise), or: $4000 = (m \times 4.18 \times 20)$

and: $$m = 47.8 \text{ kg/s}$$

Pressure drop = 2 kN/m^2 = $2000/(1000 \times 9.81) = 0.204$ m of water.

From equation 3.23, $(R/\rho u^2)Re^2 = -\Delta P_f d^3\rho/4l\mu^2$
$$= (2000 \times 0.25^3 \times 1000)/(4 \times 5 \times 10^{-6}) = 1.56 \times 10^6$$

If the tubes are smooth, then from Fig. 3.8: $Re = 2.1 \times 10^4$.

$\therefore$ water velocity = $(2.1 \times 10^4 \times 10^{-3})/(1000 \times 0.025) = 0.84$ m/s.

Cross-sectional area of each tube = $(\pi/4)0.25^2 = 0.00049$ m^2.

Mass flow rate per tube $= (0.84 \times 0.00049) = 0.000412$ m^3/s $= 0.412$ kg/s

Hence the number of tubes required $= (47.8/0.412) = \underline{\underline{116 \text{ tubes}}}$

PROBLEM 3.16

Sulphuric acid is pumped at 3 kg/s through a 60 m length of smooth 25 mm pipe. Calculate the drop in pressure. If the pressure drop falls by one half, what will be the new flowrate? Density of acid $= 1840$ kg/m^3. Viscosity of acid $= 25$ mN s/m^2.

Solution

Cross-sectional area of pipe $= (\pi/4)0.025^2 = 0.00049$ m^2.

Volumetric flowrate of acid $= (3.0/1840) = 0.00163$ m^3/s.

Velocity of acid in the pipe $= (0.00163/0.00049) = 3.32$ m/s.

Reynolds number, $\rho u d/\mu = (1840 \times 3.32 \times 0.025/25 \times 10^{-3}) = 6120$

From Fig. 3.7 for a smooth pipe and $Re = 6120$, $R/\rho u^2 = 0.0043$.

The pressure drop is calculated from equation 3.18:

$$-\Delta P_f = 4(R/\rho u^2)(l/d)(\rho u^2)$$

$$= 4 \times 0.0043(60/0.025)(1840 \times 3.32^2) = 837,200 \text{ N/m}^2 \text{ or } \underline{840 \text{ kN/m}^2}$$

If the pressure drop falls to 418,600 N/m^2, equation 3.23 and Fig. 3.8 may be used to calculate the new flow.

From equation 3.23 : $(R/\rho u^2) Re^2 = -\Delta P_f d^3 \rho/4l\mu^2$

$$= (418,600 \times 0.025^3 \times 1840)/(4 \times 60 \times 25^2 \times 10^{-6})$$

$$= 8.02 \times 10^4$$

From Fig. 3.8: $Re = 3800$ and the new velocity is:

$$u' = (3800 \times 25 \times 10^{-3})/(1840 \times 0.025) = 2.06 \text{ m/s}$$

and the mass flowrate $= (2.06 \times 0.00049 \times 1840) = \underline{\underline{1.86 \text{ kg/s}}}$

PROBLEM 3.17

A Bingham plastic material is flowing under streamline conditions in a pipe of circular cross-section. What are the conditions for one half of the total flow to be within the central core across which the velocity profile is flat? The shear stress acting within the fluid, R_y, varies with velocity gradient du_x/dy according to the relation: $R_y - R_c = -k(du_x/dy)$ where R_c and k are constants for the material.

Solution

The shearing characteristics of non-Newtonian fluids are shown in Fig. 3.24 of Volume 1. This type of fluid remains rigid when the shear stress is less than the yield stress R_Y and flows like a Newtonian fluid when the shear stress exceeds R_Y. Examples of Bingham plastics are many fine suspensions and pastes including sewage sludge and toothpaste. The velocity profile in laminar flow is shown in Fig. 3c.

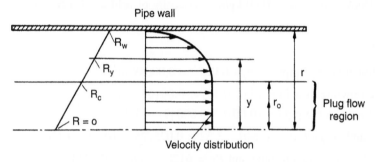

Figure 3c.

A force balance over the pipe assuming no slip at the walls gives: $-\Delta P \pi r^2 = R_w 2\pi rL$, and

$$-\Delta P/L = 2R_w/r \text{ where } R_w = \text{shear stress at the wall.} \tag{i}$$

A force balance over the annular core where $y > r_0$ gives:

$$-\Delta P \pi y^2 = 2\pi yLR_y$$

Hence:
$$R_y = yR_w/r \text{ and } y = rR_y/R_w \tag{ii}$$

when:
$$R_y = R_Y \text{ and } r_0 = rR_Y/R_w \tag{iii}$$

$$R_y - R_Y = -k(du_x/dy)$$

∴ from equation (ii):
$$-\frac{du_x}{dy} = \frac{R_y - R_Y}{k} = \frac{1}{k}\left(\frac{yR_w}{r} - R_Y\right) \tag{iv}$$

Integrating:
$$-ku_x = (y^2 R_w/2r) - R_Y y + C$$

When $y = r$, $u_x = 0$, $C = (-rR_w/2) + R_Y r$.

∴
$$ku_x = R_w\left(\frac{r}{2} - \frac{y^2}{2r}\right) - R_Y(r - y) \tag{v}$$

Substituting for y from equation (iii) gives:

$$ku_0 = R_w\left[\frac{r}{2} - \left(\frac{R_Y}{R_w}\right)\frac{r}{2}\right] - R_Y\left(r - \frac{R_Y}{R_w}r\right) \text{ and } ku_0 = \frac{r}{2R_w}(R_w - R_Y)^2 \tag{vi}$$

The total volumetric flowrate Q is obtained by integrating the equation for the velocity profile to give:

$$Q_{\text{total}} = \int_0^r \pi y^2 (-du_x/dy)\, dy$$

From equation (iv):
$$Q_{\text{total}} = \frac{1}{k} \int_0^r \pi y^2 \left(\frac{yR_w}{r} - R_Y \right) dy = \frac{\pi r^3}{k} \left(\frac{R_w}{4} - \frac{R_Y}{3} \right) \text{ m}^3/\text{s}$$

Over the central core, the volumetric flowrate Q_{core} is:

$$Q_{\text{core}} = \pi r_0^2 u_0 = \pi (rR_Y/R_w)^2 u_0 \text{ (from (3))}$$

From equation (vi):

$$Q_{\text{core}} = \pi (rR_Y/R_w)^2 (r/2kR_w)(R_w - R_Y)^2 = (\pi r^3 R_Y^2/2kR_w^3)(R_w - R_Y)^2$$

If half the total flow is to be within the central core, then:

$$Q_{\text{core}} = Q_{\text{total}}/2$$

$$(\pi r^3 R_Y^2/2kR_w^3)(R_w - R_Y)^2 = (\pi r^3/2k)(R_w/4 - R_Y/3)$$

and:
$$R_Y^2 (R_w - R_Y)^2 = R_w^3 \left(\frac{R_w}{4} - \frac{R_Y}{3} \right)$$

PROBLEM 3.18

Oil of viscosity 10 mN s/m^2 and density 950 kg/m^3 is pumped 8 km from an oil refinery to a distribution depot through a 75 mm diameter pipeline and is then despatched to customers at a rate of 500 tonne/day. Allowance must be made for periods of maintenance which may interrupt the supply from the refinery for up to 72 hours. If the maximum permissible pressure drop over the pipeline is 3450 kN/m^2, what is the shortest time in which the storage tanks can be completely recharged after a 72 hour shutdown? The roughness of the pipe surface is 0.05 mm.

Solution

From equation 3.23:
$$\frac{R}{\rho u^2} Re^2 = \frac{-\Delta P_f d^3 \rho}{4l\mu^2}$$

$-\Delta P_f = 3450 \text{ kN/m}^2 = 3.45 \times 10^6 \text{ N/m}^2$, $d = 0.075 \text{ m}$, $\rho = 950 \text{ kg/m}^3$, $l = 8000 \text{ m}$ and $\mu = 10 \text{ m Ns/m}^2 \equiv 0.01 \text{ Ns/m}^2$.

$$\therefore \quad \frac{R}{\rho u^2} Re^2 = (3.45 \times 10^6 \times 0.075^3 \times 950)/(4 \times 8000 \times 0.01^2) = 4.32 \times 10^5$$

$$e/d = (0.05/75) = 0.0007$$

From Fig. 3.8 with $(R/\rho u^2)\, Re^2 = (4.32 \times 10^5)$, $e/d = 0.0007$.

$$Re = 1.1 \times 10^4 = \rho du/\mu$$

$\therefore$ $u\rho = Re\,\mu/d = (1.1 \times 10^4 \times 0.01/0.075) = 1.47 \times 10^3$ kg/m^{2}s

$\therefore$ mass flowrate $= (1.47 \times 10^3 \times 10^{-3} \times 3600 \times 24 \times (\pi/4)0.075^2) = 561$ tonne/day

Depletion of storage in 72 h $= (561 \times 72/24) = 1683$ tonne

Maximum net gain in capacity in the system $= (561 - 500) = 61$ tonne/day and the time to recharge the tanks $= (1683/61) = \underline{\underline{27.6 \text{ days}}}$

PROBLEM 3.19

Water is pumped at 1.4 m^3/s from a tank at a treatment plant to a tank at a local works through two parallel pipes, 0.3 m and 0.6 m diameter respectively. What is the velocity in each pipe and, if a single pipe is used, what diameter will be needed if this flow of water is to be transported, the pressure drop being the same? Assume turbulent flow with the friction factor inversely proportional to the one quarter power of the Reynolds number.

Solution

The pressure drop through a pipe is given by equation 3.18:

$$-\Delta P = 4 \left(\frac{R}{\rho u^2} \right) \frac{l}{d} \rho u^2$$

In this case, $R/\rho u^2 = K\,Re^{-1/4}$ where K is a constant.

Hence: $-\Delta P = 4K \left(\frac{ud\rho}{\mu} \right)^{-1/4} \frac{l}{d} \rho u^2$

$$= K \frac{u^{1.75} l \rho^{0.75}}{d^{1.25} \mu^{0.25}} = K' u^{1.75}/d^{1.25}$$

For pipe 1 in which the velocity is u_1, $-\Delta P = K' u_1^{1.75}/0.3^{1.25}$ and the diameter is 0.3 m. Similarly for pipe 2, $-\Delta P = K' u_1^{1.75}/0.6^{1.25}$

Hence $(u_2/u_1)^{1.75} = (0.6/0.3)^{1.25} = 2.38$ and $u_2/u_1 = 1.64$

The total volumetric flowrate $= 1.4$ m^3/s $= \pi/4(d_1^2 u_1 + d_2^2 u_2)$ and substituting for d_1 and d_2 and $u_2 = 1.64u$, $u_1 = \underline{\underline{2.62 \text{ m/s}}}$ and $u_2 = \underline{\underline{4.30 \text{ m/s}}}$

If a single pipe of diameter d_3 is used for the same duty at the same pressure drop and the velocity is u_3, then:

$$(\pi/4)d_3^2 u_3 = 1.4 \text{ and } d_3^2 u_3 = 1.78 \tag{1}$$

and: $-\Delta P = K^1 u_3^{1.75}/d_3^{1.25}$

and: $(u_3/u_1)^{1.75} = (d_3/0.3)^{1.25}$

Since $u_1 = 2.62$ m/s, then:

$$0.185 u_3^{1.75} = 4.5 d_3^{1.25} \tag{2}$$

From equations (1) and (2), the required diameter, $d_3 = \underline{0.63 \text{ m}}$

PROBLEM 3.20

Oil of viscosity 10 mNs/m^2 and specific gravity 0.90, flows through 60 m of 100 mm diameter pipe and the pressure drop is 13.8 kN/m^2. What will be the pressure drop for a second oil of viscosity 30 mNs/m^2 and specific gravity 0.95 flowing at the same rate through the pipe? Assume the pipe wall to be smooth.

Solution

For the first oil, with a velocity in the pipe of u m/s then:

$$Re = u \times (0.90 \times 1000) \times (100/1000)/(10 \times 10^{-3}) = 9000u$$

$$\frac{R}{\rho u^2} Re^2 = \frac{-\Delta P d^3 \rho}{4l\mu^2}$$

$$= (13.8 \times 1000) \times 0.10^3 \times 900/(4 \times 60 \times 0.01^2) = 5.2 \times 10^5$$

From Fig. 3.8, when $(R/\rho u^2) Re^2 = 5.2 \times 10^5$ for a smooth pipe, $Re = 12000$.

Hence, velocity $u = (12,000/9000) = 1.33$ m/s.

For the second oil, the same velocity is used although the density and viscosity are now 950 kg/m^3 and 0.03 Ns/m^2.

Hence: $$Re = (1.33 \times 0.10 \times 950/0.03) = 4220$$

For a smooth pipe, Fig. 3.7 gives a friction factor, $R/\rho u^2 = 0.0048$ for this value of Re.

From Equation 3.18:

$$-\Delta P = 4(R/\rho u^2)(l/d)\rho u^2$$

$$= 4 \times 0.0048 \times (60/0.10) \times 950 \times 1.33^2$$

$$= 1.94 \times 10^4 \text{ N/m}^2 \equiv \underline{\underline{19.4 \text{ kN/m}^2}}$$

PROBLEM 3.21

Crude oil is pumped from a terminal to a refinery through a 0.3 m diameter pipeline. As a result of frictional heating, the temperature of the oil is 20 deg K higher at the refinery end than at the terminal end of the pipe and the viscosity has fallen to one half its original value. What is the ratio of the pressure gradient in the pipeline at the refinery end to that at the terminal end? Viscosity of oil at terminal = 90 mNs/m^2. Density of oil (approximately constant) = 960 kg/m^3. Flowrate of oil = 20,000 tonne/day.

Outline a method for calculating the temperature of the oil as a function of distance from the inlet for a given value of the heat transfer coefficient between the pipeline and the surroundings.

Solution

Oil flowrate = $(20,000 \times 1000)(24 \times 3600)$

$\qquad\qquad = 231.5$ kg/s or $(231/960) = 0.241$ m^3/s

Cross section area of pipe = $(\pi/4) \times 0.3^2 = 0.0707$ m^2

Oil velocity in pipe = $(0.241/0.0707) = 3.40$ m/s

Reynolds number at terminal = $(3.40 \times 0.3 \times 960/0.09) = 10,880$

Reynolds number at the refinery is twice this value or 21,760.

From equation 3.18: $\qquad\qquad -\Delta P = 4(R/\rho u^2)(l/d)(\rho u^2)$ $\qquad\qquad$ (equation 3.18)

and: $\qquad\qquad \dfrac{-(\Delta P/l)_{\text{refinery}}}{-(\Delta P/l)_{\text{terminal}}} = \dfrac{(R/\rho u^2)_{\text{refinery}}}{(R/\rho u^2)_{\text{terminal}}}$

which, from Fig. 3.7: $\qquad\qquad = (0.0030/0.00375) = \underline{\underline{0.80}}$

In a length of pipe dl:

$$-\mathrm{d}P = 4(R/\rho u^2)(\mathrm{d}l/d)\rho u^2 \text{ N/m}^2$$

Energy dissipated = $-\mathrm{d}PQ = (\pi/4)d^2u4(R/\rho u^2)(\mathrm{d}l/d)(\rho u^2)$ W where u is the velocity in the pipe.

The heat loss to the surroundings at a distance l from the inlet is $h(T - T_S)\pi\mathrm{d}l$ W where T_S is the temperature of the surroundings and T is the temperature of the fluid.

Heat gained by the fluid = $(\pi/4)d^2u\rho C_p\mathrm{d}T$ W where C_p (J/kg K) is the specific heat capacity of the fluid.

Thus an energy balance over the length of pipe dl gives:

$$(R/\rho u^2)\mathrm{d}\rho u^3\mathrm{d}l = h(T - T_s)\pi d\,\mathrm{d}l + (\pi/4)d^2u\rho C_p\,\mathrm{d}T$$

$(R/\rho u^2)$ varies with temperature as illustrated in the first part of this problem, and hence this equation may be written as:

$$A\,\mathrm{d}l = B\,\mathrm{d}l + C\,\mathrm{d}T$$

or: $\qquad\qquad \dfrac{C\mathrm{d}T}{A - B} = \mathrm{d}l$

(where A and B are both functions of temperature and C is a constant).
Integrating between l_1 and l_2, T_1 and T_2 gives:

$$\int_{l_1}^{l_2} \mathrm{d}l = \int_{T_1}^{T_2} \left[\frac{C\mathrm{d}T}{A - B} \right]$$

If T_1, T_s, h and ΔT are known (20 deg K in this problem), the integral may then be evaluated.

PROBLEM 3.22

Oil with a viscosity of 10 mNs/m^2 and density 900 kg/m^3 is flowing through a 500 mm diameter pipe 10 km long. The pressure difference between the two ends of the pipe is

10^6 N/m^2. What will the pressure drop be at the same flowrate if it is necessary to replace the pipe by one only 300 mm in diameter? Assume the pipe surface to be smooth.

Solution

$\mu = 0.01$ Ns/m^2, $\rho = 900$ kg/m^3, $d = 0.50$ m, $l = 10000$ m and $-\Delta P = 1 \times 10^6$ N/m^2.

$$(R/\rho u^2)\,Re^2 = (-\Delta P d^3 \rho)/(4l\mu^2) \qquad\qquad \text{(equation 3.23)}$$

$$= (1.10^6 \times 0.50^3 \times 900/(4 \times 10000 \times 0.01^2)) = 2.81 \times 10^7$$

From Fig. 3.8, $Re = ud\rho/\mu = 1.2 \times 10^5$

$$u = Re\,\mu/\rho d$$

$$= (1.2 \times 10^5 \times 0.01/900 \times 0.50) = 2.67 \text{ m/s}$$

If the diameter of the new pipe is 300 mm, the velocity is then:

$$= 2.67 \times (0.5/0.3)^2 = 7.42 \text{ m/s}$$

Reynolds number $= (7.42 \times 0.30 \times 900/0.01) = 2.0 \times 10^5$

From Fig. 3.7, $R/\rho u^2 = 0.0018$ and from equation 3.18:

$$-\Delta P = (4 \times 0.0018 \times (10000/0.3) \times 900 \times 7.42^2) = \underline{\underline{1.19 \times 10^7 \text{ N/m}^2}}$$

PROBLEM 3.23

Oil of density 950 kg/m^3 and viscosity 10^{-2} Ns/m^2 is to be pumped 10 km through a pipeline and the pressure drop must not exceed 2×10^5 N/m^2. What is the minimum diameter of pipe which will be suitable, if a flowrate of 50 tonne/h is to be maintained? Assume the pipe wall to be smooth. Use either a pipe friction chart *or* the Blasius equation $(R/\rho u^2 = 0.0396\ Re^{-1/4})$.

Solution

From equation 3.6 a force balance on the fluid in the pipe gives:

$$R = -\Delta P(d/4l)$$

or:

$$= 2 \times 10^5 (d/4 \times 10^4) = 5d$$

Velocity in the pipe $= G/\rho A$

$$= (50 \times 1000/3600)/(950 \times (\pi/4)d^2) = 0.186/d^2$$

Hence:

$$R/\rho u^2 = 5d/(950 \times (0.186/d^2)^2) = 15.21d^5$$

$$Re = ud\rho/\mu$$

$$= (0.186 \times d^2) \times d \times 950/(1 \times 10^{-2}) = 1.77 \times 10^3/d$$

The Blasius equation is: $R/\rho u^2 = 0.0396\,Re^{-0.25}$ (equation 3.11)

and hence: $15.21d^5 = 0.0396(d/1.77 \times 10^3)^{0.25}$

and: $\underline{d = 0.193 \text{ m}}$

In order to use the friction chart, Fig. 3.7, it is necessary to assume a value of $R/\rho u^2$, calculate d as above, check the resultant value of Re and calculate $R/\rho u^2$ and compare its value with the assumed value.

If $R/\rho u^2$ is assumed to be $= 0.0030$, then $15.21d^5 = 0.0030$ and $d = 0.182$ m

$\therefore$ $Re = (1.77 \times 10^3)/0.182 = 9750$

From Fig. 3.7, $R/\rho u^2 = 0.0037$ which does not agree with the original assumption.

If $R/\rho u^2$ is taken as 0.0024, d is calculated $= 0.175$ m, Re is 1.0×10^5 and $R/\rho u^2 = 0.0022$. This is near enough giving the minimum pipe diameter $= \underline{0.175 \text{ m}}$.

PROBLEM 3.24

On the assumption that the velocity profile in a fluid in turbulent flow is given by the Prandtl one-seventh power law, calculate the radius at which the flow between it and the centre is equal to that between it and the wall, for a pipe 100 mm in diameter.

Solution

See Volume 1, Example 3.5.

PROBLEM 3.25

A pipeline 0.5 m diameter and 1200 m long is used for transporting an oil of density 950 kg/m³ and of viscosity 0.01 Ns/m² at 0.4 m³/s. If the roughness of the pipe surface is 0.5 mm, what is the pressure drop? With the same pressure drop, what will be the flowrate of a second oil of density 980 kg/m³ and of viscosity 0.02 Ns/m²?

Solution

$\mu = 0.01$ Ns/m², $d = 0.5$ m and $A = (\pi/4)0.5^2 = 0.196$ m², $l = 1200$ m, $\rho = 950$ kg/m³ and: $u = (0.4/0.196) = 2.04$ m/s.

Reynolds number $= \rho u d/\mu = (950 \times 2.04 \times 0.5)/0.01 = 9.7 \times 10^4$

$e/d = (0.5/500) = 0.001$ and from Fig. 3.7, $R/\rho u^2 = 0.0027$

From equation 3.8, $-\Delta P = 4(R/\rho u^2)(l/d)\rho u^2$

$$= (4 \times 0.0027 \times 1200 \times 950 \times 2.04^2/0.5)$$

$$= \underline{\underline{1.03 \times 10^5 \text{ N/m}^2}}$$

$$(R/\rho u^2)\,Re^2 = -\Delta P d^3 \rho/4l\mu^2 \qquad\qquad\qquad \text{(equation 3.23)}$$

$$= (1.03 \times 10^5 \times 0.5^3 \times 980/4 \times 1200 \times 0.02^2) = 6.6 \times 10^6$$

From Fig. 3.8, $Re = 4.2 \times 10^4 = (980u \times 0.5/0.02)$

and: $u = 1.71$ m/s

$\therefore$ volumetric flowrate $= (1.71 \times 0.196) = \underline{\underline{0.34 \text{ m}^3/\text{s}}}$

PROBLEM 3.26

Water (density 1000 kg/m^3, viscosity 1 mNs/m^2) is pumped through a 50 mm diameter pipeline at 4 kg/s and the pressure drop is 1 MN/m^2. What will be the pressure drop for a solution of glycerol in water (density 1050 kg/m^3, viscosity 10 mNs/m^2) when pumped at the same rate? Assume the pipe to be smooth.

Solution

Cross-sectional area of pipe $= ((\pi/4) \times 0.05^2) = 0.00196$ m^2

Water velocity, $u = 4/(1000 \times 0.00196) = 2.04$ m/s.

Reynolds number, $Re = (2.04 \times 1000 \times 0.05/1 \times 10^{-3}) = 102,000$

From Fig. 3.7, $R/\rho u^2 = 0.0022$

From equation 3.18, $-\Delta P = (4 \times 0.0022 \times (l/0.05) \times 1000 \times 2.04^2) = 732l$

For glycerol/water flowing at the same velocity:

$$Re = (2.4 \times 1050 \times 0.05/1 \times 10^{-2}) = 10,700$$

From Fig. 3.7, $R/\rho u^2 = 0.0037$

and: $\Delta P = (4 \times 0.0037 \times (l/0.05) \times 1050 \times 2.04^2) = 1293l.$

$\therefore$ $-\Delta P_{\text{glycerol}}/ - \Delta P_{\text{water}} = (1293l/732/l) = 1.77$

and $-\Delta P_{\text{glycerol}} = (1.77 \times 1 \times 10^6) = \underline{\underline{1.77 \times 10^6 \text{ N/m}^2}}$

PROBLEM 3.27

A liquid is pumped in streamline flow through a pipe of diameter d. At what distance from the centre of the pipe will the fluid be flowing at the average velocity?

Solution

A force balance on an element of fluid of radius r gives:

$$-\Delta P \pi r^2 = -2\pi r l \mu \frac{du}{dr}$$

or:
$$-\frac{du}{dr} = -\frac{\Delta Pr}{2\mu l} \quad \text{or} \quad -u = \frac{\Delta Pr^2}{4\mu l} + \text{constant}$$

When $r = d/2 = a$, $u = 0$ and the constant $= -\left(-\dfrac{\Delta Pa^2}{4\mu l}\right)$

$$\therefore \qquad\qquad u = \frac{-\Delta P}{4\mu l}(a^2 - r^2)$$

The flow, dQ, through an annulus, radius r and thickness dr is given by:

$$dQ = 2\pi r \, dr \, u_r$$

$$= 2\pi r \, dr \left[\frac{-\Delta P}{4\mu l}(a^2 - r^2)\right]$$

and:
$$Q = \pi a^4 \left(\frac{-\Delta P}{8\mu l}\right)$$

The average velocity is: $u_{av} = Q/\pi a^2$

$$= \frac{-\Delta Pa^2}{8\mu l}$$

The radius at which $u = u_{av}$ is:

$$\frac{-\Delta Pa^2}{8\mu l} = \frac{-\Delta P}{4\mu l}(a^2 - r^2)$$

from which:
$$r^2 = a^2/2 = d^2/8 \text{ and } r = \underline{\underline{0.35d}}$$

PROBLEM 3.28

Cooling water supplied to a heat exchanger flows through 25 mm diameter tubes each 5 m long arranged in parallel. If the pressure drop over the heat exchanger is not to exceed 8000 N/m², how many tubes must be included for a total flowrate of water of 110 tonne/h? Density of water = 1000 kg/m³. Viscosity of water = 1 mNs/m². Assume pipes to be smooth-walled.

If ten per cent of the tubes became blocked, what would the new pressure drop be?

Solution

$$\frac{R}{\rho u^2} \times Re^2 = \frac{-\Delta P d^3 \rho}{4l\mu^2} \qquad\qquad\qquad \text{(equation 3.23)}$$

$$= (8000 \times (0.025)^3 \times 1000)/(4 \times 5 \times (1 \times 10^{-3})^2) = 6.25 \times 10^6$$

From Fig. 3.8: $Re = \rho u d / \mu = 5 \times 10^4$

$\therefore u = (5 \times 10^4 \times 1 \times 10^{-3})/(1000 \times 0.025) = 2.0$ m/s

Flowrate per tube $= (2.0 \times (\pi/4) \times 0.025^2) = 0.982 \times 10^4$ m^3/s

Total flowrate $= (110 \times 1000)/(1000 \times 3600) = 0.3056$ m^3/s

$\therefore$ Number of tubes required $= 0.3056/(0.982 \times 10^{-4}) = 31.1$ or 32 tubes

If 10 per cent of the tubes are blocked, velocity of fluid $= (2.0/0.9) = 2.22$ m/s

$Re = (5 \times 10^4)/0.9 = 5.5 \times 10^4$ and, from Fig. 3.7, $R/\rho u^2 = 0.00245$.

From equation 3.18, pressure drop is:

$$-\Delta P = (4 \times 0.00245 \times (5/0.025) \times 1000 \times 2.22^2) = 9650 \text{ N/m}^2, \text{ an increase of } 20.6\%$$

PROBLEM 3.29

The effective viscosity of a non-Newtonian fluid may be expressed by the relationship:

$$\mu_a = k'' \left(-\frac{du_x}{dr} \right)$$

where k'' is constant.

Show that the volumetric flowrate of this fluid in a horizontal pipe of radius a under isothermal laminar flow conditions with a pressure drop $-\Delta P/l$ per unit length is:

$$Q = \frac{2\pi}{7} a^{7/2} \left(\frac{-\Delta P}{2k''l} \right)^{1/2}$$

Solution

In Section 3.4.1 of Volume 1 it is shown that for any fluid, the shear stress, R_r, at a distance r from the centre of the pipe may be found from a force balance for an element of fluid of length l across which the pressure drop is $-\Delta P$ by:

$$-\Delta P \pi r^2 = 2\pi r l (-R_r) \text{ or } -R_r = \frac{r}{2} \left(\frac{-\Delta P}{l} \right) \qquad \text{(equation 3.7)}$$

The viscosity is related to the velocity of the fluid, u_x, and the shear stress, R_r, by:

$$R_r = \mu_a (-du_x/dr) \qquad \text{(from equation 3.4)}$$

If, for the non-Newtonian fluid, $\mu_a = k''(-du_x/dr)$

then:
$$R_r = k''(-du_x/dr)(-du_x/dr)$$
$$= k''(-du_x/dr)^2$$

Combining the two equations for R_r:

$$k''(-du_x/dr)^2 = \frac{r}{2} \left(\frac{-\Delta P}{l} \right)$$

or:
$$-\frac{du_x}{dr} = \left(\frac{-\Delta P}{2k''l}\right)^{1/2} r^{1/2}dr$$

∴
$$-u_x = \frac{2}{3}\left(\frac{-\Delta P}{2k''l}\right)^{1/2} r^{3/2} + \text{constant}.$$

When $r = a$ (at the wall), $u_x = 0$ and the constant $= -\left[\frac{2}{3}\left(\frac{-\Delta P}{2k''l}\right)^{1/2} a^{3/2}\right]$

and:
$$u_x = \left(\frac{-\Delta P}{2k''l}\right)^{1/2} \frac{2}{3}(a^{3/2} - r^{3/2})r$$

The volumetric flowrate Q is:

$$\int_0^Q dQ = \int_0^a 2\pi r u_x\, dr = 2\pi \left(\frac{-\Delta P}{2k''l}\right)^{1/2} \frac{2}{3} \int_0^a (ra^{3/2} - r^{5/2})\, dr$$

$$= 2\pi \left(\frac{-\Delta P}{2k''l}\right)^{1/2} \frac{2}{3}\left(\frac{3}{14}a^{7/12}\right) = \frac{2\pi}{7}\left(\frac{-\Delta P}{2k''l}\right)^{1/2} a^{7/2}$$

PROBLEM 3.30

Determine the yield stress of a Bingham fluid of density 2000 kg/m³ which will just flow out of an open-ended vertical tube of diameter 300 mm under the influence of its own weight.

Solution

The shear stress at the pipe wall, R_0, in a pipe of diameter d, is found by a force balance as given Volume 1, Section 3.4.1:

$$(-R_0)\pi\, dl = (-\Delta P)(\pi/4)d^2$$

or:
$$-R_0 = (-\Delta P)(d/4l) \qquad\qquad \text{(equation 3.6)}$$

If the fluid just flows from the vertical tube, then:

$$-\Delta P/l = \rho g$$

and:
$$-R_0 = \rho g(d/4) = (2000 \times 981 \times 0.3)/4 = \underline{\underline{1472 \text{ N/m}^2}}$$

PROBLEM 3.31

A fluid of density 1200 kg/m³ flows down an inclined plane at 15° to the horizontal. If the viscous behaviour is described by the relationship:

$$R_{yx} = -k\left(\frac{du_x}{dy}\right)^n$$

where $k = 4.0$ Ns$^{0.4}$/m^2 and $n = 0.4$, calculate the volumetric flowrate per unit width if the fluid film is 10 mm thick.

Solution

Flow with a free surface is discussed in Section 3.6 and the particular case of laminar flow down an inclined surface in Section 3.6.1. For a flow of liquid of depth δ, width w and density ρ down a surface inclined at an angle θ to the horizontal, a force balance in the x direction (parallel to the surface) may be written. The weight of fluid flowing down the plane at a distance y from the free surface is balanced by the shear stress at the plane. For unit width and unit height:

$$-R_{yx} = \rho g \sin \theta y$$

$$R_{yx} = -k(d_x/d_y)^n \text{ and } k(du_x/dy)^n = \rho g \sin \theta y$$

Substituting $k = 4.0$ Ns$^{0.4}$/m^2, $n = 0.4$, $\rho = 1200$ kg/m^3 and $\theta = 15°$:

$$4.0(du_x/dy)^{0.4} = (1200 \times 9.81 \times \sin 15°)y$$

or: $(du_x/dy = 762y)$ and $du_x/dy = 1.60 \times 10^7 y^{2.5}$

$$u_x = 4.57 \times 10^6 y^{3.5} + \text{constant}$$

When the film thickness $y = \delta = 0.01$ m, $u_x = 0$. Hence $0 = 0.457 + c$ and $c = -0.457$.

$\therefore$
$$u_x = 4.57 \times 10^6 y^{3.5} - 0.457$$

The volumetric flowrate down the surface is then:

$$\int_0^Q dQ = \int_0^w \int_{0.01}^0 u_x \, dw \, dy$$

or, for unit width: $Q/W = \int_{0.01}^0 (4.57 \times 10^6 y^{3.5} - 0.457) \, dy = \underline{\underline{0.00357 \text{ (m}^3/\text{s)}/\text{m}}}$

PROBLEM 3.32

A fluid with a finite yield stress is sheared between two concentric cylinders, 50 mm long. The inner cylinder is 30 mm diameter and the gap is 20 mm. The outer cylinder is held stationary while a torque is applied to the inner. The moment required just to produce motion is 0.01 Nm. Calculate the torque needed to ensure all the fluid is flowing under shear if the plastic viscosity is 0.1 Ns/m^2.

Solution

Concentric-cylinder viscometers are in widespread use. Figure 3d represents a partial section through such an instrument in which liquid is contained and sheared between the stationary inner and rotating outer cylinders. Either may be driven, but the flow regime

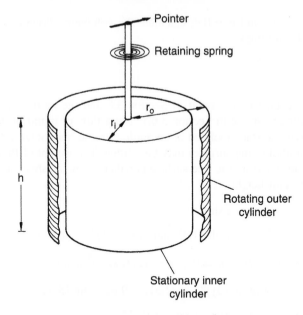

Figure 3d.   Partial section of concentric-cylinder viscometer

which is established with the outer rotating and the inner stationary is less disturbed by centrifugal forces. The couple transmitted through the fluid to the suspended stationary inner cylinder is resisted by a calibrated spring, the deflection of which allows calculation of the torque, T, and hence the inner wall shearing stress R_i is given by:

$$T = -R_i \, r_i \, 2\pi r_i h$$

This torque T originates from the outer cylinder which is driven at a uniform speed. On the inner surface of the outer cylinder the shear stress is R_o and:

$$T = -R_o r_o 2\pi r_o h$$

$$\therefore \qquad R_o = \frac{-T}{2\pi r_o^2 h}$$

and:

$$R_i = \frac{-T}{2\pi r_i^2 h}$$

For any intermediate radius r, the local shear stress is:

$$R_r = \frac{-T}{2\pi r^2 h} = R_o \left(\frac{r_o^2}{r^2}\right) = R_i \left(\frac{r_i^2}{r^2}\right)$$

In this example, $r_i = 0.015$ m, $r_2 = 0.035$ m, $h = 0.05$ m and $T = 0.01$ Nm which just produces motion at the surface of the inner cylinder.

Using these equations:

$$R_i = T/(2\pi r_i^2 h) = [0.01/(2\pi \times 0.015^2 \times 0.05)] = 141.5 \text{ N/m}^2$$

As motion just initiates under the action of this torque, this shear stress must equal the yield stress and:

$$R_Y = 141.5 \text{ N/m}^2$$

If all the fluid is to be in motion, the shear stress at the surface of the outer cylinder must be at least this value and the shear stress at the inner cylinder will be higher, and will be given by:

$$R_i = R_o(r_o/r_i)^2 = [141.5(0.035/0.015)^2] = 770 \text{ N/m}^2$$

The required torque is then:

$$T = R_i \times 2\pi r_i^2 h = (770 \times 2\pi \times 0.015^2 \times 0.05) = \underline{0.054 \text{ Nm}}$$

PROBLEM 3.33

Experiments with a capillary viscometer of length 100 mm and diameter 2 mm gave the following results:

Applied pressure (N/m^2)	Volumetric flowrate (m^3/s)
1×10^3	$1 \ \times 10^{-7}$
2×10^3	2.8×10^{-7}
5×10^3	1.1×10^{-6}
1×10^4	$3 \ \times 10^{-6}$
2×10^4	$9 \ \times 10^{-6}$
5×10^4	3.5×10^{-5}
1×10^5	$1 \ \times 10^{-4}$

Suggest a suitable model to describe the fluid properties.

Solution

Inspection of the data shows that the pressure difference increases less rapidly than the flowrate. Taking the first and the last entries in the table, it is seen that when the flowrate increases from 1×10^{-7} to 1×10^{-4} m^3/s, that is by a factor of 1000, the pressure difference increases from 1×10^3 to 1×10^5 N/m^2 that is by a factor of only 100. In this way, the fluid appears to be shear-thinning and the simplest model, the power-law model, will be tried.

From equation 3.136:

$$Q = (\pi/4)d^2 u = (-\Delta P/4kl)^{1/n}[n/(6n+2)](\pi/4)d^{(3n+1)/n}$$

Using the last set of data:

$$1.0 \times 10^{-4} = [(1 \times 10^5)/(4k \times 0.1)]^{1/n}(\pi/8)(n/(3n+1))(2 \times 10^{-3})^{(3n+1)/n}$$

or: $$Q = K(-\Delta P)^{1/n}$$

A plot of Q against $-\Delta P$ on logarithmic axes, shown in Figure 3e, gives a slope, $(1/n) = 1.5$ which is constant over the entire range of the experimental data.
This confirms the validity of the power-law model and, for this system:

$$n = 0.67$$

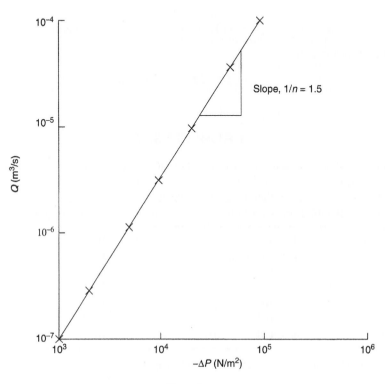

Figure 3e.

The value of the consistency coefficient k may be obtained by substituting $n = 0.67$ and the experimental data for any one set of data and, if desired, the constancy of this value may be confirmed by repeating this procedure for each set of the data.

For the last set of data:

$$Q = (\pi/4)d^2u$$

$$= (-\Delta P/4kl)^{1/n}[n/(6n + 2)](\pi/4)d^{(3n+1)/n} \quad \text{(from equation 3.136)}$$

Thus: $1 \times 10^{-4} = [(1 \times 10^5)/(4 \times 0.1k)]^{1.5}(1/9)(\pi/4)(002)^{4.5}$

and: $k = 0.183 \text{ Ns}^n\text{m}^2$

In S.I. units, the power-law equation is therefore:

$$R = 0.183(du_x/dy)^{0.67}$$

or: $\tau = 0.183\dot{\gamma}^{0.67}$

PROBLEM 3.34

Data obtained with a cone and plate viscometer, cone half-angle 89° cone radius 50 mm, were:

cone speed (Hz)	measured torque (Nm)
0.1	4.6×10^{-1}
0.5	7×10^{-1}
1	1.0
5	3.4
10	6.4
50	3.0×10

Suggest a suitable model to describe the fluid properties.

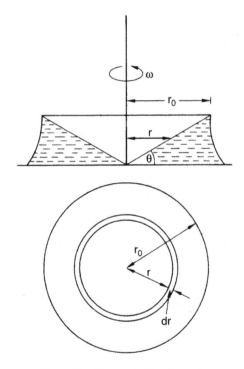

Figure 3f. Cone and plate viscometer

Solution

A cone and plate viscometer, such as the Ferranti–Shirley or the Weissenberg instruments, shears a fluid sample in the small angle (usually 4° or less) between a flat surface and a rotating cone whose apex just touches the surface. Figure 3f illustrates one such arrangement. This geometry has the advantage that the shear rate is everywhere uniform and equal to $\omega/\sin\theta$, since the local cone velocity is ωr and the separation between the solid surfaces at that radius is $r \sin \theta$ where ω is the angular velocity of rotation. The shear stress R_r acting on a small element or area dr wide will produce a couple $-2\pi r \, dr \, R_r r$ about the axis of rotation. With a uniform velocity gradient at all points in contact with the cone surface, the surface stress R will also be uniform, so that the suffix can be omitted and the total couple about the axis is:

$$-C = \int_0^{r_0} 2\pi r^2 R \, dr = \frac{2}{3}\pi r_0^3 R$$

The shear stress within the fluid can therefore be evaluated from this equation.

In this problem, $\theta = 1° = 0.0175$ rad and $r_0 = 0.05$ m

When the cone speed is 0.1 Hz, $\omega = 2\pi \times 0.1 = 0.628$

Hence the shear rate, $\omega/\sin \theta = (0.628/0.0175) = 36 \text{ s}^{-1}$

The shear stress is given by: $R = \dfrac{3c}{2\pi r_0^3}$

When $c = 4.6 \times 10^{-2}$ Nm, $R = (3 \times 4.6 \times 10^{-2}/(2\pi \times 0.05^3)) = 176 \text{ N/m}^2$

The remaining data may be treated in the same way to give:

Cone speed (Hz)	Shear rate (s^{-1})	Torque (Nm)	Shear stress (N/m^2)
0.1	36	0.46	1760
0.5	180	0.70	2670
1	360	1.0	3820
5	1800	3.4	13000
10	3600	6.4	24500
50	18000	30.0	114600

These data may be plotted on linear axes as shown in Fig. 3.24 or on logarithmic axes as in Fig. 3.26 given here as Figs 3g and 3h, respectively.

It will be seen from Fig. 3g that linear axes produce an excellent straight line with an intercept of 1500 N/m^2 and this indicates a Bingham plastic type of material whose characteristics are described by equation 3.122

$$|R_y| - R_y = \mu_p \left| \frac{du_x}{dy} \right| \qquad \text{(equation 3.122)}$$

From Fig. 3g, the slope is $\mu_p = 6.4$ Ns/m^2 and the graph confirms Bingham plastic behaviour.

PROBLEM 3.35

Tomato purée of density 1300 kg/m^3 is pumped through a 50 mm diameter factory pipeline at a flowrate of 0.00028 m^3/s. It is suggested that in order to double production:

(a) a similar line with pump should be put in parallel to the existing one, or

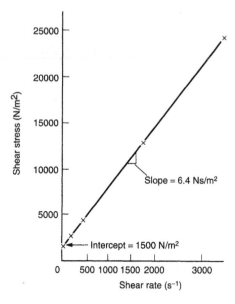

Figure 3g.

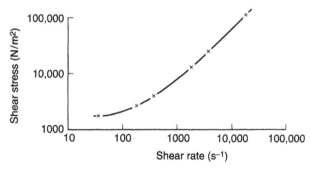

Figure 3h.

(b) a large pump should force the material through the present line, or
(c) a large pump should supply the liquid through a line of twice the cross-sectional area.

Given that the flow properties of the purée can be described by the Casson equation:

$$(-R_y)^{1/2} = (-R_Y)^{1/2} + \left(-\mu_c \frac{du_x}{dy}\right)^{1/2}$$

where R_Y is a yield stress, here 20 N/m^2, μ_c is a characteristic Casson plastic viscosity, 5 Ns/m^2, and du_x/d_y is the velocity gradient, evaluate the relative pressure drops of the three suggestions, assuming laminar flow throughout.

Solution

The Casson equation is a particular form of equation 3.122 which applies to a number of foodstuffs as well as tomato puree.

$$(-R_y)^{1/2} = (-R_Y)^{1/2} + \left(-\mu_c \frac{du_x}{dy}\right)^{1/2}$$

or:

$$\left(-\mu_c \frac{du_x}{dy}\right)^{1/2} = (-R_y)^{1/2} - (-R_Y)^{1/2}$$

and:

$$-\mu_c \frac{du_x}{dy} = (-R_Y) + (-R_y) - 2(R_y R_Y)^{1/2}$$

$$-\frac{du_x}{dy} = \frac{1}{\mu_c}\left[(-R_y) + (-R_Y) - (2R_y R_Y)^{1/2}\right]$$

The Rabinowitch–Mooney equation gives the total volumetric flowrate Q through the pipe as:

$$\frac{Q}{\pi a^3} = -\frac{1}{\mu_c R_w^3}\int_0^{R_w} (R_y)^2 f(R_y)\,dR_y \qquad \text{(equation 3.149)}$$

where a is the pipe radius and R_w is the stress at the wall. Substituting for $f(R_y)$:

$$\frac{Q}{\pi a^3} = \frac{1}{\mu_c R_w^3}\int_0^{R_w} (R_y^3 + R_y^2 R_Y + 2R_y^{5/2} R_Y^{1/2})\,dR_y$$

$$= \frac{1}{\mu_c R_w^3}\left[\frac{R_y^4}{4} + \frac{R_y^3 R_Y}{3} + \frac{4}{7}R_y^{7/2} R_Y^{1/2}\right]_0^{R_w}$$

$$= \frac{1}{\mu_c}\left[\frac{R_w}{4} + \frac{R_Y}{3} + \frac{4}{7}R_w^{1/2} R_Y^{1/2}\right]$$

In this problem, $R_Y = 20$ N/m^2, $\mu_c = 5$ Ns/m^2, $Q = 2.8 \times 10^{-4}$ m^3/s, $a = 0.025$ m and substituting these values, $R_w = 030.24$ N/m^2.

From equation 3.138, $R_w = (D/4)(-\Delta P/l) = 30.24$ N/m^2

and:

$$(-\Delta P/l) = \underline{\underline{2420\ (\text{N/m}^2)/\text{m}}}$$

For case (a), the pressure drop will remain *unchanged*.

For case (b), the flowrate $= 2Q$ and substituting $2Q$ for Q enables R_w to be recalculated as 98.0 N/m^2 and $(-\Delta P/l)$ to be determined as $\underline{\underline{7860\ (\text{N/m}^2)/\text{m}}}$.

For case (c), the flowrate $= 2Q$ and the pipe diameter $= a\sqrt{2}$. Again recalculation of R_w gives a value of 14.52 N/m^2 and $(-\Delta P/l) = \underline{\underline{821\ (\text{N/m}^2)/\text{m}}}$.

PROBLEM 3.36

The rheological properties of a particular suspension can be approximated reasonably well by either a "power law" or a "Bingham plastic" model over the shear rate range of 10 to 50 s^{-1}. If the consistency k is 10 Ns^n/m^2 and the flow behaviour index n is 0.2 in the power law model, what will be the approximate values of the yield stress and of the plastic viscosity in the Bingham plastic model?

What will be the pressure drop when the suspension is flowing under laminar conditions in a pipe 200 m long and 40 mm diameter, when the centre line velocity is 1 m/s, according to the power law model? Calculate the centre line velocity for this pressure drop for the Bingham plastic model and comment on the result.

Solution

See Volume 1, Example 3.10.

PROBLEM 3.37

Show how, by suitable selection of the index n, the power-law may be used to describe the behaviour of both shear-thinning and shear-thickening non-Newtonian fluids over a limited range of shear rates. What are the main objections to the use of the power law? Give some examples of different types of shear-thinning fluids.

A power-law fluid is flowing under laminar conditions through a pipe of circular cross-section. At what radial position is the fluid velocity equal to the mean velocity in the pipe? Where does this occur for a fluid with an n-value of 0.2?

Solution

Steady state shear-dependent behaviour is discussed in Volume 1, Section 3.7.1.

For a Newtonian fluid,
$$R = \mu \frac{du}{dy}$$
(equation 3.4)

For a non-Newtonian power law fluid, $R = k \left(\frac{du}{dy} \right)^n$ (equation 3.119)

$$= k \left(\frac{du}{dy} \right)^{n-1} \left(\frac{du}{dy} \right) = \mu_a \frac{du}{dy}$$

where the apparent viscosity $\mu_a = k \left(\frac{du}{dy} \right)^{n-1}$

When $n < 1$, shear-thinning behaviour is represented

$\quad\quad\quad n > 1$, shear-thickening behaviour is represented

$\quad\quad\quad n = 1$, the behaviour is Newtonian.

For shear-thinning fluids, $\mu_a \to \infty$ at zero shear stress and $\mu_a \to 0$ at infinite shear stress. Paint often exhibits shear thinning behaviour as its apparent viscosity is very high while in the can and when just applied to a wall but its apparent viscosity is very low as the brush applies it to the surface when it flows readily to give an even film. Toothpaste remains in its tube and on the brush when not subjected to shear but when sheared, as it is when the tube is squeezed, it flows readily through the nozzle to the brush.

For a fluid flowing in a pipe of radius a, length l with a central core of radius r, a force balance gives:

$$-\Delta P \pi r^2 = k \left[-\frac{du_r}{dr} \right]^n 2\pi r l$$

$$\frac{-\Delta P r}{2l} = k \left[-\frac{du_r}{dr} \right]^n$$

or:

$$-\frac{du_r}{dr} = \left[\frac{-\Delta P}{2kl} \right]^{1/n} r^{1/n}$$

Integrating:

$$-u_r = \left[\frac{-\Delta P}{2kl} \right]^{1/n} \frac{n}{n+1} r^{\frac{n+1}{n}} + C$$

When $r = a$, $u_r = 0$ and $C = -\left[\frac{-\Delta P}{2kl} \right]^{1/n} \frac{n}{n+1} a^{\frac{n+1}{n}}$

$$\therefore \qquad u_r = \left[\frac{-\Delta P}{2kl} \right]^{1/n} \frac{n}{n+1} \left[a^{\frac{n+1}{n}} - r^{\frac{n+1}{n}} \right]$$

The mean velocity is u given by the volumetric flow/area

or:

$$u = \frac{1}{\pi a^2} \int_0^Q dQ = \frac{1}{\pi a^2} \int_0^a 2\pi r \, dr \, u_x$$

$$\therefore \qquad u = \frac{1}{\pi a^2} \left(\frac{-\Delta P}{2kl} \right)^{1/n} \frac{n}{n+1} 2\pi \int_0^a \left[a^{\frac{n+1}{n}} r - r^{\frac{2n+1}{n}} \right] dr$$

$$\therefore \qquad u = \left(\frac{-\Delta P}{2kl} \right)^{1/n} \frac{n}{3n+1} a^{\frac{n+1}{n}}$$

When the mean velocity = average velocity, then:

$$\left(\frac{-\Delta P}{2kl} \right)^{1/n} \frac{n}{3n+1} a^{\frac{n+1}{n}} = \left(\frac{-\Delta P}{2kl} \right)^{1/n} \frac{n}{n+1} \left[a^{\frac{n+1}{n}} - r^{\frac{n+1}{n}} \right]$$

$$1 - \left(\frac{r}{a} \right)^{\frac{n+1}{n}} = \frac{n+1}{3n+1}$$

or:

$$\frac{r}{a} = \left(\frac{2n}{3n+1} \right)^{\frac{n}{n+1}}$$

When $n = 0.2$ then:

$$\frac{r}{a} = \left(\frac{0.4}{1.6} \right)^{\frac{0.2}{1.2}} = \underline{\underline{0.794}}$$

PROBLEM 3.38

A liquid whose rheology can be represented by the power-law model is flowing under streamline conditions through a pipe of 5 mm diameter. If the mean velocity of flow is 1 m/s and the velocity at the pipe axis is 1.2 m/s, what is the value of the power law index n?

Water, of viscosity 1 mNs/m^2 flowing through the pipe at the same mean velocity gives rise to a pressure drop of 10^4 N/m^2 compared with 10^5 N/m^2 for the non-Newtonian fluid. What is the consistency ("k" value) of the non-Newtonian fluid?

Solution

In problem 3.37, the mean velocity, u, is shown to be:

$$u = \left(\frac{-\Delta P}{2kl}\right)^{1/n} a^{\frac{n+1}{n}} \frac{n}{3n+1}$$

and the velocity at any distance y from the pipe axis is:

$$u_r = \left(\frac{-\Delta P}{2kl}\right)^{1/n} \frac{n}{n+1} \left[a^{\frac{n+1}{n}} - r^{\frac{n+1}{n}}\right]$$

The maximum velocity, u_{max}, will occur when $y = 0$ and:

$$u_{max} = \left(\frac{-\Delta P}{2kl}\right)^{1/n} \frac{n}{n+1} a^{\frac{n+1}{n}}$$

$$\therefore \quad \frac{u_{max}}{u} = \frac{1.2}{1.0} = \frac{3n+1}{n+1} \quad \text{and } n = \underline{\underline{0.111}}$$

As shown previously: $\quad u = \left(\frac{-\Delta P}{2kl}\right)^{1/n} a^{\frac{n+1}{n}} \frac{n}{3n+1}$

When $n = 0.111$ for the non-Newtonian fluid, $-\Delta P = 10^5$ N/m^2 and $u = 1$ m/s

$$\therefore \quad 1 = \left(\frac{10^5}{2kl}\right)^9 a^{10} \times 0.083$$

When $n = 1$ for water, $-\Delta P = 10^4$ N/m^2 and $u = 1$ m/s and $k = \mu$.

$$\therefore \quad 1 = \left(\frac{10^4}{2\mu l}\right) a^2 \times 0.25$$

or: $$1 = \left(\frac{10^4}{2\mu l}\right)^9 a^{18} \times 3.81 \times 10^{-6}$$

$$\therefore \quad \left(\frac{10^5}{2kl}\right)^9 a^{10} \times 0.083 = \left(\frac{10^4}{2\mu l}\right)^9 a^{18} \times 3.81 \times 10^{-6}$$

From which, when $a = 0.0025$ m and $\mu = 1 \times 10^{-3}$ Ns/m^2,

$$k = \underline{\underline{6.24 \text{ Ns}^n/\text{m}^2}}$$

PROBLEM 3.39

Two liquids of equal densities, the one Newtonian and the other a non-Newtonian "power-law" fluid, flow at equal volumetric rates down two wide vertical surfaces of the same widths. The non-Newtonian fluid has a power-law index of 0.5 and has the same apparent viscosity, in SI units, as the Newtonian fluid when its shear rate is 0.01 s^{-1}. Show that, for equal surface velocities of the two fluids, the film thickness for the Newtonian fluid is 1.125 times that of the non-Newtonian fluid.

Solution

For a power-law fluid:

$$R = k \left(\frac{du}{dy}\right)^n \qquad \text{(equation 3.121)}$$

$$= k \left(\frac{du}{dy}\right)^{n-1} \left(\frac{du_x}{dy}\right) = \mu_a (du_x/dy) \qquad \text{(equation 3.122)}$$

where μ_a is the apparent velocity $= k(du_x/dy)^{n-1}$
   For a Newtonian fluid:

$$R = \mu \left(\frac{du_x}{dy}\right) \qquad \text{(equation 3.3)}$$

When $n = 0.5$ and $(du_x/dy) = 0.01$, $\mu = \mu_a$ and:

$$\mu_a = \mu = k(du_x/dy)^{n-1} = k(0.01)^{-0.5} = 10 \, k = \mu \text{ and } k = 0.1\mu.$$

The equation of state of the power-law fluid is therefore:

$$R = 0.1\mu (du_x/dy)^{0.5}$$

For a fluid flowing down a vertical surface, length l and width w and film thickness S, at a distance y from the solid surface, a force balance gives:

$$(S - y)wl\rho g = Rwl = k(du_x/dy)^n wl$$

or:
$$\frac{du_x}{dy} = \left(\frac{\rho g}{k}\right)^{1/n} (S - y)^{1/n}$$

and:
$$u_x = \left(\frac{\rho g}{k}\right)^{1/n} (S - y)^{\frac{n+1}{n}} \left(-\frac{n}{n+1}\right) + \text{const.}$$

When $y = 0$, $u_x = 0$ and the constant $= \left(\frac{\rho g}{K}\right)^{1/n} S^{\frac{n+1}{n}} \frac{n}{n+1}$

and:
$$u_x = \left(\frac{\rho g}{k}\right)^{1/n} \frac{n}{n+1} \left[S^{\frac{n+1}{n}} - (S - y)^{\frac{n+1}{n}}\right]$$

At the free surface where $y = S$:

$$u_s = \left(\frac{\rho g}{k}\right)^{1/n} \left(\frac{n}{n+1}\right) S^{\frac{n+1}{n}} \tag{i}$$

The volumetric flowrate Q is given by:

$$Q = \int_0^s w\,dy \left(\frac{\rho g}{k}\right)^{1/n} \left(\frac{n}{n+1}\right) \left(S^{\frac{n+1}{n}} - s - y^{\frac{n+1}{n}}\right)$$

$$= w \left(\frac{\rho g}{k}\right)^{1/n} \left(\frac{n}{2n+1}\right) S^{\frac{2n+1}{n}} \tag{ii}$$

For the non-Newtonian fluid, $k = 0.1\mu$, $n = 0.5$ and equation (ii), when expressed in S.I. units, becomes:

$$= w \left(\frac{\rho g}{0.1\mu}\right)^2 \times 0.25\ s^4 = 25\ w \left(\frac{\rho g s^2}{\mu}\right)^2 \tag{iii}$$

For the Newtonian fluid, $n = 1$ and $k = \mu$ and substituting in equation (ii):

$$Q = w \left(\frac{\rho g}{\mu}\right) \times 0.33 S_N^3 \tag{iv}$$

where S_N is the thickness of the Newtonian film.
For equal flowrates, from equations (iii) and (iv):

$$25w \left(\frac{\rho g s^2}{\mu}\right)^2 = 0.33w \left(\frac{\rho g}{\mu}\right) S_N^3$$

or:

$$S_N^3 = 75 \left(\frac{\rho g}{\mu}\right) S^4 \tag{v}$$

For equal surface velocities, the term $(\rho g/K)$ in equation (i) can be substituted from equation (v) and:

For the non-Newtonian fluid: $u_s = \left(\frac{\rho g}{0.1\mu}\right)^2 \times 0.33 S^3$

$$= 100 \left(\frac{S_N^3}{75 S^4}\right)^2 \times 0.33 S^3$$

$$= 0.00592 S_N^6 / S^5$$

For the Newtonian fluid: $u_s = \left(\frac{\rho g}{\mu}\right) \times 0.5 S_N^2$

$$= \left(\frac{S_N^3}{75 S^4}\right) \times 0.5 S_N^2 = 0.0067 S_N^5 / S^4$$

and: $\qquad\qquad S_N/S = (0.00667/0.00592) = \underline{\underline{1.126}}$

PROBLEM 3.40

A fluid which exhibits non-Newtonian behaviour is flowing in a pipe of diameter 70 mm and the pressure drop over a 2 m length of pipe is 4×10^4 N/m². When the flowrate is doubled, the pressure drop increases by a factor of 1.5. A pitot tube is used to measure the velocity profile over the cross-section. Confirm that the information given below is consistent with the laminar flow of a *power-law* fluid.

Any equations used must be derived from the basic relation between shear stress R and shear rate:

$$R = k(\dot{\gamma})^n$$

radial distance from centre of pipe (s mm)	velocity (m/s)
0	0.80
10	0.77
20	0.62
30	0.27

Solution

For a power-law fluid:

At the initial flowrate: $4 \times 10^4 = ku^n$

With a flow of: $6 \times 10^4 = k(2u)^n$

Dividing: $1.5 = 2^n$

and hence: $n = 0.585$

For the power-law fluid:

$$R = k \left| \frac{du_x}{dy} \right|^n$$

A force balance on a fluid core of radius s in pipe of radius r gives:

$$R_s 2\pi s l = -\Delta P \pi s^2$$

or: $$R_s = k \left(-\frac{du_x}{ds} \right)^n = -\frac{\Delta P s}{2l}$$

$$-\frac{du_x}{ds} = \left(-\frac{\Delta P}{2kl} \right)^{1/n} s^{1/n}$$

Integrating: $$-u_x = \left(-\frac{\Delta P}{2ks} \right)^{1/n} \left(\frac{n}{n+1} \right) s^{\frac{n+1}{n}} + \text{constant}$$

When $s = r$, $u_x = 0$ (the no-slip condition):

and hence the constant
$$= -\left(\frac{-\Delta P}{2kl}\right)^{\frac{1}{n}}\left(\frac{n}{n+1}\right) r^{\frac{n+1}{n}}$$

Substituting:
$$u_x = \left(-\frac{\Delta P}{2kl}\right)^{1/n}\left(\frac{n}{n+1}\right)\left[r^{\frac{n+1}{n}} - s^{\frac{n+1}{n}}\right]$$

On the centre line:
$$u_{CL} = \left(\frac{-\Delta P}{2kl}\right)^{1/n}\left(\frac{n}{n+1}\right) r^{\frac{n+1}{n}}$$

and hence:
$$\frac{u_x}{u_{CL}} = 1 - \left(\frac{s}{r}\right)^{\frac{n+1}{n}}$$

at when $n = 0.585$:
$$= 1 - \left(\frac{s}{r}\right)^{2.71}$$

The following data are obtained for a pipe radius of $r = 35$ mm:

		experimental	
radius s (mm)	$\dfrac{u_x}{u_{CL}} = 1 - \left(\dfrac{s}{r}\right)^{2.71}$	u_x (m/s)	u_x/u_{CL}
0	1	0.80	1
10	0.966	0.77	0.96
20	0.781	0.62	0.77
30	0.341	0.27	0.34

Thus, the calculated and experimental values of u_x/u_{CL} agree within reasonable limits of experimental accuracy.

PROBLEM 3.41

A Bingham-plastic fluid (yield stress 14.35 N/m^2 and plastic viscosity 0.150 Ns/m^2) is flowing through a pipe of diameter 40 mm and length 200 m. Starting with the rheological equation, show that the relation between pressure gradient $-\Delta P/l$ and volumetric flowrate Q is:

$$Q = \frac{\pi(-\Delta P)r^4}{8l\mu_p}\left[1 - \frac{4}{3}X + \frac{1}{3}X^4\right]$$

where l is the pipe radius, μ_p is the plastic viscosity, and X is the ratio of the yield stress to the shear stress at the pipe wall.

Calculate the flowrate for this pipeline when the pressure drop is 600 kN/m^2. It may be assumed that the flow is laminar.

Solution

For a Bingham-plastic material, the shear stress R_s at radius s is given by:

$$R_s - R_Y = +\mu_p \left(-\frac{du_x}{ds}\right) \quad (R_s \gg R_Y)$$

$$\frac{du_x}{ds} = 0 \quad (R_s \ll R_Y)$$

The central unsheared core has radius $r_c = r(R_Y/R)$ (where r = pipe radius and R = wall shear stress) since the shear stress is proportional to the radius s.

In the annular region:

$$-\frac{du_x}{ds} = \frac{1}{\mu_p}(R - R_Y) = \frac{1}{\mu_p}\left(-\Delta P \frac{s}{2l} - R_Y\right) \quad \text{(from a force balance)}$$

$$-u_x = -\int du_x = \frac{1}{\mu_p}\left(-\Delta P \frac{s^2}{4l} - R_Y s\right) + \text{constant}$$

For the no-slip condition: $u_x = 0$, when $s = r$

Thus:
$$0 = \frac{1}{\mu_p}\left(-\Delta P \frac{r^2}{4l} - R_Y s\right) + \text{constant}$$

and:
$$u_s = \frac{1}{\mu_p}\left\{\frac{-\Delta P}{4l}(r^2 - s^2) - R_Y(r - s)\right\}$$

Substituting:
$$-\Delta P = \frac{2R}{l/r}$$

$$u_s = \frac{1}{\mu_p}\left\{\frac{R}{2r}(r^2 - s^2) - R_Y(r - s)\right\} \tag{i}$$

The volumetric flowrate through elemental annulus, $dQ_A = u_s 2\pi s ds$

Thus:
$$Q_A = \int_{r_c}^{r} \frac{1}{\mu_p}\left\{\frac{R}{2r}(r^2 - s^2) - R_Y(r - s)\right\} 2\pi s ds$$

$$= \frac{2\pi}{\mu_p}R\left[\frac{1}{2r}\left(\frac{r^2 s^2}{2} - \frac{s^4}{4}\right) - \frac{R_Y}{R}\left(\frac{rs^2}{2} - \frac{s^3}{3}\right)\right]_{r_c}^{r}$$

Writing $\dfrac{R_Y}{R} = X$ and $r_c = r\dfrac{R_Y}{R}$, then :

$$Q_A = \frac{2\pi}{\mu_p}R\left\{\frac{1}{2r}\left(\frac{r^4}{2} - \frac{r^4}{4}\right) - X\left(\frac{r^3}{2} - \frac{r^3}{3}\right) - \frac{1}{2r}\left(\frac{X^2 r^4}{2} - \frac{X^4 r^4}{4}\right)\right.$$

$$\left. + X\left(\frac{r^3 X^2}{2} - \frac{r^3 X^3}{3}\right)\right\}$$

$$= \frac{2\pi R}{\mu_p}r^3\left\{\frac{1}{8} - \frac{1}{6}X - \frac{1}{4}X^2 + \frac{1}{8}X^4 + \frac{1}{2}X^3 - \frac{1}{3}X^4\right\}$$

$$= \frac{\pi R r^3}{4\mu_p}\left\{1 - 4/3X - 2X^2 + 4X^3 - \frac{5}{3}X^4\right\} \qquad \text{(ii)}$$

In the core region

Substituting: $s = r_c = (R_Y/R)r = Xr$ in equation (i) for the core velocity u_c gives:

$$u_c = \frac{1}{\mu_p}\left\{\frac{R}{2r}(r^2 - X^2 r^2) - R_Y(r - X_r)\right\}$$

$$= \frac{R_r}{\mu_p}\left\{\frac{1}{2}(1 - X^2) - X(1 - X)\right\} - \frac{R_r}{4\mu_p}\{2(1 - X^2) - 4X + 4X^2)\}$$

$$= \frac{R_r}{4\mu_p}\left\{2 - 4X + 2X^2\right\}$$

The flowrate through the core is: $u_c\pi r_c^2 = u_c\pi X^2 r^2 = Q_c$

Thus: $$Q_c = \frac{Rr}{4\mu_p}\pi X^2 r^2\{2 - 4X + 2X^2\}$$

$$= \frac{Rr^3\pi}{4\mu_p}\{2X^2 - 4X^3 + 2X^4\}$$

The total flowrate is: $(Q_A + Q_c) = Q$

and: $$Q = \frac{\pi R r^3}{4\mu_p}\left\{1 - \tfrac{4}{3}X + \tfrac{1}{3}X^4\right\}$$

Putting $$R = \frac{-\Delta P r}{2l} \quad \text{then :}$$

$$\underline{Q = \frac{\pi(-\Delta P)r^4}{8l\mu_p}\left\{1 - \tfrac{4}{3}X + \tfrac{1}{3}X^4\right\}}$$

When: $-\Delta P = 6 \times 10^5 \text{ N/m}^2, l = 200 \text{ m} \quad d = 40 \text{ mm and } r = 0.02 \text{ m.}$

Then: $$R = -\Delta P\frac{r}{2l} = 6 \times \frac{0.02}{400} \times 10^5 = 30 \text{ N/m}^2$$

$$\mu_p = 0.150 \text{ Ns/m}^2$$

$$R_Y = 14.35 \text{ N/m}^2$$

and: $$X = \frac{R_Y}{R} = \frac{14.35}{30} = 0.478$$

Thus: $$Q = \frac{(\pi)(6 \times 10^5)(0.02)^4}{8 \times 200 \times 0.150}\left\{1 - \frac{4}{3} \times 0.478 + \frac{1}{3}(0.478)^3\right\}$$

$$= \underline{\underline{0.000503 \text{ m}^3/\text{s}}}$$

SECTION 4

Flow of Compressible Fluids

PROBLEM 4.1

A gas, having a molecular weight of 13 kg/kmol and a kinematic viscosity of 0.25 cm^2/s, flows through a pipe 0.25 m internal diameter and 5 km long at the rate of 0.4 m^3/s and is delivered at atmospheric pressure. Calculate the pressure required to maintain this rate of flow under isothermal conditions. The volume occupied by 1 kmol at 273 K and 101.3 kN/m^2 is 22.4 m^3. What would be the effect on the required pressure if the gas were to be delivered at a height of 150 m (i) above, and (ii) below its point of entry into the pipe?

Solution

From equation 4.57 and, as a first approximation, omitting the kinetic energy term:

$$(P_2 - P_1)/v_m + 4(R/\rho u^2)(l/d)(G/A)^2 = 0$$

At atmospheric pressure and 289 K, the density $= (13/22.4)(273/289) = 0.542$ kg/m^3

Mass flowrate of gas, $G = (0.4 \times 0.542) = 0.217$ kg/s.

Cross-sectional area, $A = (\pi/4)(0.25)^2 = 0.0491$ m^2.

Gas velocity, $u = (0.4/0.0491) = 8.146$ m/s

∴ $\qquad G/A = (0.217/0.0491) = 4.413$ $kg/m^2 s$

Reynolds number, $\quad Re = \rho du/\mu$

$$= (0.25 \times 8.146/0.25 \times 10^{-4}) = 8.146 \times 10^4$$

From Fig. 3.7, for $e/d = 0.002$, $R/\rho u^2 = 0.0031$

$$v_2 = (1/0.542) = 1.845 \text{ m}^3/\text{kg}$$

$$v_1 = (22.4/13)(298/273)(101.3/P_1) = 190.5/P_1 \text{ m}^3/\text{kg}$$

and: $\qquad v_m = (0.923P_1 + 95.25)/P_1 \text{ m}^3/\text{kg}$

Substituting in equation 4.57:

$$P_1(P_1 - 101.3)10^3/(0.923P_1 + 95.25) = 4(0.0031)(5000/0.25)(4.726)^2$$

and: $\qquad P_1 = \underline{\underline{111.1 \text{ kN/m}^2}}$

The kinetic energy term $= (G/A)^2 \ln(P_1/P_2) = (4.413)^2 \ln(111.1/101.3)$
$$= 1.81 \text{ kg}^2/\text{m}^4\text{s}^2$$

This is negligible in comparison with the other terms which equal $5539 \text{ kg}^2/\text{m}^4\text{s}^2$ so that the initial approximation is justified. If the pipe is not horizontal, the term $g\ dz$ in equation 4.49 must be included in the calculation. If equation 4.49 is divided by v^2, this term on integration becomes $g\Delta z/v_m^2$.

$\therefore$
$$v_m = (0.923 \times 111.1 + 95.25)/111.1 = 1.781 \text{ m}^3/\text{kg}$$
$$v_{\text{air}} = (24.0/29) = 0.827 \text{ m}^3/\text{kg}$$

As the gas is less dense than air, v_m is replaced by $(v_{\text{air}} - v_m) = -0.954 \text{ m}^3/\text{kg}$.

$\therefore$
$$g\Delta z/v_m^2 = (9.81 \times 150/0.954^2) = 1616 \text{ N/m}^2 \text{ or } 0.16 \text{ kN/m}^2$$

(i) If the delivery point is 150 m above the entry level, then since gas is less dense,

$$P_1 = (111.1 - 0.16) = \underline{\underline{110.94 \text{ kN/m}^2}}$$

(ii) If the delivery point is 150 m below the entry level then,

$$P_1 = (111.1 + 0.16) = \underline{\underline{111.26 \text{ kN/m}^2}}$$

PROBLEM 4.2

Nitrogen at 12 MN/m^2 pressure is fed through a 25 mm diameter mild steel pipe to a synthetic ammonia plant at the rate of 1.25 kg/s. What will be the pressure drop over a 30 m length of pipe for isothermal flow of the gas at 298 K? Absolute roughness of the pipe surface $= 0.005$ mm. Kilogram molecular volume $= 22.4 \text{ m}^3$. Viscosity of nitrogen $= 0.02$ mN s/m^2.

Solution

Molecular weight of nitrogen $= 28$ kg/kmol.

Assuming a mean pressure in the pipe of 10 MN/m^2, the specific volume, v_m at 10 MN/m^2 and 298 K is:

$$v_m = (22.4/28)(101.3/10 \times 10^3)(298/273) = 0.00885 \text{ m}^3/\text{kmol}$$

Reynolds number, $\rho u d/\mu = d(G/A)/\mu$).

$A = (\pi/4)(0.025)^2 = 4.91 \times 10^{-3} \text{ m}^2$.

$\therefore (G/A) = (1.25/4.91 \times 10^{-3}) = 2540 \text{ kg/m}^2\text{s}$

and: $\qquad\qquad Re = (0.025 \times 2540/0.02 \times 10^{-3}) = 3.18 \times 10^6$

From Fig. 3.7, for $Re = 3.18 \times 10^6$ and $e/d = (0.005/25) = 0.0002$,

$$R/\rho u^2 = 0.0017$$

In equation 4.57 and neglecting the first term:

$$(P_2 - P_1)/v_m + 4(R/\rho u^2)(l/d)(G/A)^2 = 0$$

or:

$$P_1 - P_2 = 4v_m(R/\rho u^2)(l/d)(G/A)^2$$

$$= 4 \times 0.00885(0.0017)(30/0.025)(2540)^2$$

$$= 466{,}000 \text{ N/m}^2 \text{ or } 0.466 \text{ MN/m}^2$$

This is small in comparison with $P_1 = 12 \text{ MN/m}^2$, and the average pressure of 10 MN/m^2 is seen to be too low. A mean pressure of 11.75 kN/m^2 is therefore selected and the calculation repeated to give a pressure drop of 0.39 MN/m^2. The mean pressure is then $(12 + 11.61)/2 = 11.8 \text{ MN/m}^2$ which is close enough to the assumed value.

It remains to check if the assumption that the kinetic energy term is negligible is justified.

Kinetic energy term $= (G/A)^2 \ln(P_1/P_2) = (2540)^2 \ln(12/11.61) = 2.13 \times 10^5 \text{kg2/m4s2}$

The term $(P_1 - P_2)/v_m$, where v_m is the specific volume at the mean pressure of $11.75 \text{ MN/m}^2 = (0.39 \times 10^6)/0.00753 = 5.18 \times 10^7 \text{ kg}^2/\text{m}^4\text{s}$.

Hence the omission of the kinetic energy term is justified

and the pressure drop $= \underline{\underline{0.39 \text{ MN/m}^2}}$

PROBLEM 4.3

Hydrogen is pumped from a reservoir at 2 MN/m^2 pressure through a clean horizontal mild steel pipe 50 mm diameter and 500 m long. The downstream pressure is also 2 MN/m^2 and the pressure of this gas is raised to 2.6 MN/m^2 by a pump at the upstream end of the pipe. The conditions of flow are isothermal and the temperature of the gas is 293 K. What is the flowrate and what is the effective rate of working of the pump? Viscosity of hydrogen $= 0.009 \text{ mN s/m}^2$ at 293 K.

Solution

Neglecting the kinetic energy term in equation 4.55, then:

$$(P_2^2 - P_1^2)/2P_1v_1 + 4(R/\rho u^2)(l/d)(G/A)^2 = 0$$

where $P_1 = 2.6 \text{ MN/m}^2$ and $P_2 = 2.0 \text{ MN/m}^2$.

Thus: $v_1 = (22.4/2)(293/273)(0.1013/2.6) = 0.468 \text{ m}^3/\text{kg}$

If $Re = 10^7$ and $e/d = 0.001$, from Fig. 3.7, $R/\rho u^2 = 0.0023$.

Substituting:

$$(2.0^2 - 2.6^2)10^{12}/(2 \times 2.6 \times 10^6 \times 0.468) + 4(0.0023)(500/0.05)(G/A)^2 = 0$$

from which $G/A = 111$ kg/m^{2}s.

$\therefore$ $\qquad Re = d(G/A)/\mu = (0.05 \times 111/(0.009 \times 10^{-3}) = 6.2 \times 10^5$

Thus the chosen value of Re was too high. If Re is taken as 6.0×10^5 and the problem reworked, $G/A = 108$ kg/m^{2}s and $Re = 6.03 \times 10^5$ which is in good agreement.

$$A = (\pi/4)(0.05)^2 = 0.00197 \text{ m}^2$$

and: $\qquad\qquad\qquad G = 108 \times 0.00197 = \underline{\underline{0.213 \text{ kg/s}}}$

The power requirement is given by equation 8.71 as $(1/\eta)GP_1v_1 \ln(P_1/P_2)$
If a 60% efficiency is assumed, then the power requirement is:

$$= (1/0.6) \times 0.213 \times 2.6 \times 10^6 \times 0.468 \ln(2.6/2)$$
$$= (1.13 \times 10^5) \text{ W or } \underline{\underline{113 \text{ kW}}}$$

PROBLEM 4.4

In a synthetic ammonia plant the hydrogen is fed through a 50 mm steel pipe to the converters. The pressure drop over the 30 m length of pipe is 500 kN/m^2, the pressure at the downstream end being 7.5 MN/m^2. What power is required in order to overcome friction losses in the pipe? Assume isothermal expansion of the gas at 298 K. What error is introduced by assuming the gas to be an incompressible fluid of density equal to that at the mean pressure in the pipe? $\mu = 0.02$ mNs/m^2.

Solution

If the downstream pressure $= 7.5$ MN/m^2 and the pressure drop due to friction $= 500$ kN/m^2, the upstream pressure $= 8.0$ MN/m^2 and the mean pressure $= 7.75$ MN/m^2. The mean specific volume is: $v_m = (22.4/2)(298/273)(0.1013/7.75) = 0.16$ m^3/kg

and: $\qquad\qquad v_1 = (22.4/2)(298/273)(0.1013/8.0) = 0.15$ m^3/kg

It is necessary to assume a value of $R/\rho u^2$, calculate G/A and the Reynolds number and check that the value of e/d is reasonable. If the gas is assumed to be an incompressible fluid of density equal to the mean pressure in the pipe and $R/\rho u^2 = 0.003$, the pressure drop due to friction $= 500$ kN/m^2 is:

$\therefore$ $\qquad\qquad (500 \times 10^3/0.16) = 4(0.003)(30/0.05)(G/A)^2$

and $\qquad\qquad\qquad G/A = 658$ kg/m^{2}s.

$$Re = d(G/A)/\mu = (0.05 \times 658/0.02 \times 10^{-3}) = 1.65 \times 10^6$$

From Fig. 3.7 this corresponds to a value of e/d of approximately 0.002, which is reasonable for a steel pipe.

For compressible flow:

$$(G/A)^2 \ln(P_1/P_2) + (P_2^2 - P_1^2)/2P_1v_1 + 4(R/\rho u^2)(l/d)(G/A)^2 = 0 \quad \text{(equation 4.55)}$$

Substituting:

$$(G/A)^2 \ln(8.0/7.5) + (7.5^2 - 8.0^2)10^{12}/(2 \times 8.0 \times 10^6 \times 0.15)$$
$$+ 4(0.003)(30/0.05)(G/A)^2 = 0$$

from which: $G/A = 667$ kg/m^{2}s and $G = 667 \times (\pi/4)(0.05)^2 = 1.31$ kg/s

Little error is made by the simplifying assumption in this particular case.

The power requirement is given by equation 8.71:

$$= (1/\eta)GP_1v_1 \ln(P_1/P_2)$$

If the compressor efficiency $= 60\%$,

$$\text{power requirement} = (1/0.6) \times 1.31 \times 8.0 \times 10^6 \times 0.15 \ln(8/7.5)$$
$$= (1.69 \times 10^5) \text{ W or } \underline{\underline{169 \text{ kW}}}$$

PROBLEM 4.5

A vacuum distillation plant operating at 7 kN/m^2 pressure at the top has a boil-up rate of 0.125 kg/s of xylene. Calculate the pressure drop along a 150 mm bore vapour pipe used to connect the column to the condenser. The pipe length may be taken as equivalent to 6 m, $e/d = 0.002$ and $\mu = 0.01$ mN s/m^2.

Solution

From vapour pressure data, the vapour temperature $= 338$ K and the molecular weight of xylene $= 106$ kg/kmol.

In equation 4.55:

$$(G/A)^2 \ln(P_1/P_2) + (P_2^2 - P_1^2)/2P_1v_1 + 4(R/\rho u^2)(l/d)(G/A)^2 = 0$$

Cross-sectional area of pipe, $A = (\pi/4)(0.15)^2 = 1.76 \times 10^{-2}$ m^2

$$G/A = (0.125/1.76 \times 10^{-2}) = 7.07 \text{ kg/m}^2\text{s}$$

The Reynolds number, is $\rho u d/\mu = d(G/A)/\mu$
$$= (0.15 \times 7.07/(0.01 \times 10^{-3}) = 1.06 \times 10^5$$

From Fig. 3.7, with $e/d = 0.002$ and $Re = 1.06 \times 10^5$, $(R/\rho u^2) = 0.003$.

Specific volume, $v_1 = (22.4/106)(338/273)(101.3/7.0) = 3.79$ m^3/kg.

Substituting in equation 4.55:

$$(7.0)^2 \ln(7/P_2) + (P_2^2 - 7^2) \times 10^6/2 \times 7 \times 10^3 \times 3.79 + 4 \times 0.003(6/0.15)(7.07)^2 = 0$$

where P_2 is the pressure at the condenser (kN/m^2).
Solving by trial and error:

$$P_2 = 6.91 \text{ kN/m}^2$$

$$\therefore \qquad (P_1 - P_2) = (7.0 - 6.91) = 0.09 \text{ kN/m}^2 \text{ or } \underline{\underline{90 \text{ N/m}^2}}$$

PROBLEM 4.6

Nitrogen at 12 MN/m² pressure is fed through a 25 mm diameter mild steel pipe to a synthetic ammonia plant at the rate of 0.4 kg/s. What will be the drop in pressure over a 30 m length of pipe assuming isothermal expansion of the gas at 300 K? What is the average quantity of heat per unit area of pipe surface that must pass through the walls in order to maintain isothermal conditions? What would be the pressure drop in the pipe if it were perfectly lagged? $\mu = 0.02$ mNs/m².

Solution

At high pressure, the kinetic energy term in equation 4.55 may be neglected to give:

$$(P_2^2 - P_1^2)/2P_1v_1 + 4(R/\rho u^2)(l/d)(G/A)^2 = 0$$

Specific volume at entry of pipe, $v_1 = (22.4/28)(300/273)(0.1013/12)$
$$= 0.00742 \text{ m}^3/\text{kg}$$

Cross-sectional area of pipe, $A = (\pi/4)(0.025)^2 = 0.00049 \text{ m}^2$

$$\therefore G/A = (0.4/0.00049) = 816 \text{ kg/m}^2\text{s}.$$

Reynolds number, $d(G/A)/\mu = 0.025 \times 816/(0.02 \times 10^{-3}) = 1.02 \times 10^6$

If $e/d = 0.002$ and $Re = 1.02 \times 10^6$, $R/\rho u^2 = 0.0028$ from Fig. 3.7.

Substituting: $(12^2 - P_2^2)10^{12}/(2 \times 12 \times 10^6 \times 0.00742) = 4(0.0028)(30/0.025)(816)^2$

and: $P_2 = 11.93 \text{ MN/m}^2$

and: pressure drop $= (12.0 - 11.93) = 0.07 \text{ MN/m}^2 \equiv \underline{\underline{70 \text{ kN/m}^2}}$

The heat required to maintain isothermal flow is given in Section 4.5.2 as $G\Delta u^2/2$. The velocity at the high pressure end of the pipe = volumetric flow/area

$$= (G/A)v_1 = (816 \times 0.0072) = 6.06 \text{ m/s}$$

and the velocity in the plant is taken as zero.

Thus: $\qquad G\Delta u^2/2 = 0.4 \times (6.06)^2/2 = 7.34 \text{ W}$

Outside area of pipe $= (30 \times \pi \times 0.025) = 2.36$ m^2.

Heat required $= (7.34/2.36) = \underline{\underline{3.12 \text{ W/m}^2}}$

This low value of the heat required stems from the fact that the change in kinetic energy is small and conditions are almost adiabatic. If the pipe were perfectly lagged, the flow would be adiabatic and the pressure drop would then be calculated from equations 4.77 and 4.72. The specific volume at the low pressure end v_2 to be calculated from:

$$8(R/\rho u^2)(l/d) = \left[\frac{\gamma - 1}{2\gamma} + \frac{P_1}{v_1}\left(\frac{A}{G}\right)^2\right]\left[1 - \left(\frac{v_1}{v_2}\right)^2\right] - \frac{\gamma + 1}{\gamma}\ln\left(\frac{v_2}{v_1}\right)$$

(equation 4.77)

For nitrogen, $\gamma = 1.4$ and hence:

$$8(0.0028)(30/0.025) = \left[\frac{1.4 - 1}{2 \times 1.4} + \frac{12 \times 10^6}{0.00742}\left(\frac{1}{816}\right)^2\right]\left[1 - \left(\frac{0.00742}{v_2}\right)^2\right]$$

$$- \frac{1.4 + 1}{1.4}\ln\left(\frac{v_2}{0.00742}\right)$$

Solving by trial and error, $v_2 = 0.00746$ m^3/kg.
Thus:

$$\frac{1}{2}\left(\frac{G}{A}\right)^2 v_1^2 + \left(\frac{\gamma}{\gamma - 1}\right)P_1 v_1 = \frac{1}{2}\left(\frac{G}{A}\right)^2 v_2^2 + \left(\frac{\gamma}{\gamma - 1}\right)P_2 v_2 \qquad \text{(equation 4.72)}$$

Substitution gives:

$$(816)^2(0.00742)^2/2 + [1.4/(1.4 - 1)]12 \times 10^6 \times 0.00742$$

$$= (816)^2(0.00746)^2/2 + [1.4/(1.4 - 1)]P_2 \times 10^6 \times 0.00746$$

and: $P_1 = 11.94$ MN/m^2

The pressure drop for adiabatic flow $= (12.0 - 11.94) = 0.06$ MN/m^2 or $\underline{\underline{60 \text{ kN/m}^2}}$

PROBLEM 4.7

Air, at a pressure of 10 MN/m^2 and a temperature of 290 K, flows from a reservoir through a mild steel pipe of 10 mm diameter and 30 m long into a second reservoir at a pressure P_2. Plot the mass rate of flow of the air as a function of the pressure P_2. Neglect any effects attributable to differences in level and assume an adiabatic expansion of the air. $\mu = 0.018$ mN s/m^2, $\gamma = 1.36$.

Solution

G/A is required as a function of P_2. v_2 cannot be found directly since the downstream temperature T_2 is unknown and varies as a function of the flowrate. For adiabatic flow,

v_2 may be calculated from equation 4.77 using specified values of G/A and substituted in equation 4.72 to obtain the value of P_2. In this way the required data may be calculated.

$$8(R/\rho u^2)(l/d) = \left[\frac{\gamma - 1}{2\gamma} + \frac{P_1}{v_1}\left(\frac{A}{G}\right)^2\right]\left[1 - \left(\frac{v_1}{v_2}\right)^2\right] - \frac{\gamma + 1}{\gamma}\ln\left(\frac{v_2}{v_1}\right)$$

(equation 4.77)

$$0.5(G/A)^2 v_1^2 + [\gamma/(\gamma - 1)]P_1 v_1 = 0.5(G/A)^2 v_2^2 + [\gamma/(\gamma - 1)]P_2 v_2$$

(equation 4.72)

or:

$$\frac{0.5(G/A)^2(v_1^2 - v_2^2) + [\gamma/(\gamma - 1)]P_1 v_1}{[\gamma/(\gamma - 1)]v_2} = P_2$$

When $P_2 = P_1 = 10$ MN/m^2, $G/A = 0$.

If G/A is 2000 kg/m^{2}s, then:

$$Re = (0.01 \times 2000/0.018 \times 10^{-3}) = 1.11 \times 10^6$$

When $e/d = 0.0002$, $R/\rho u^2 = 0.0028$ from Fig. 3.7 and:

$$v_1 = (22.4/29)(290/273)(0.1013/10) = 0.0083 \text{ m}^3/\text{kg}$$

Substituting in equation 4.77:

$$8(0.0028)(30/0.01) = \left[\frac{0.36}{2 \times 1.36} + \frac{10 \times 10^6}{0.0083}\left(\frac{1}{2000}\right)^2\right]$$

$$\times \left[1 - \left(\frac{0.0083}{v_2}\right)^2\right] - \frac{2.36}{1.36}\ln\left(\frac{v_2}{0.0083}\right)$$

and: $v_2 = 0.00942$ m^3/kg.

Substituting for v_2 in equation 4.72 gives:

$$P_2 = [0.5(2000)^2(0.0083^2 - 0.00942^2) + (1.36/0.36)10 \times 10^{6 \times 0.0083}]/(1.36/0.36)$$

$$\times 0.00942$$

and: $P_2 = 8.75$ MN/m^2.

In a similar way the following table may be produced.

G/A(kg/m^{2}s)	v_2(m^3/kg)	P_2(MN/m^2)
0	0.0083	10.0
2000	0.00942	8.75
3000	0.012	6.76
3500	0.0165	5.01
4000	0.025	3.37
4238	0.039	2.04

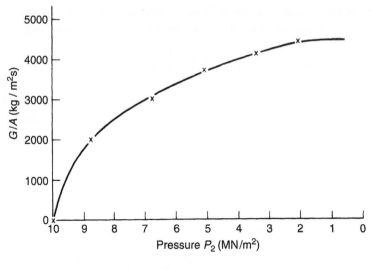

Figure 4a.

These data are plotted in Fig. 4a. It is shown in Section 4.5.4, Volume 1, that the maximum velocity which can occur in a pipe under adiabatic flow conditions is the sonic velocity which is equal to $\sqrt{\gamma P_2 v_2}$.

From the above table $\sqrt{\gamma P_2 v_2}$ at maximum flow is:

$$\sqrt{1.36 \times 2.04 \times 10^6 \times 0.039} = 329 \text{ m/s}$$

The temperature at this condition is given by $P_2 v_2 = \mathbf{R}T/M$, and:

$$T_2 = (29 \times 0.039 \times 2.04 \times 10^6/8314) = 227 \text{ K}$$

The velocity of sound in air at 227 K $= 334$ m/s, which serves as a check on the calculated data.

PROBLEM 4.8

Over a 30 m length of 150 mm vacuum line carrying air at 293 K, the pressure falls from 1 kN/m^2 to 0.1 kN/m^2. If the relative roughness e/d is 0.002, what is approximate flowrate?

Solution

The specific volume of air at 293 K and 1 kN/m^2 is:

$$v_1 = (22.4/29)(293/273)(101.3/1.0) = 83.98 \text{ m}^3/\text{kg}$$

It is necessary to assume a Reynolds number to determine $R/\rho u^2$ and then calculate a value of G/A which should correspond to the original assumed value. Assume a Reynolds number of 1×10^5.

When $e/d = 0.002$ and $Re = 10^5$, $R/\rho u^2 = 0.003$ from Fig. 3.7.

$$(G/A)^2 \ln(P_1/P_2) + (P_2^2 - P_1^2)/2P_1v_1 + 4(R/\rho u^2)(l/d)(G/A)^2 = 0 \quad \text{(equation 4.55)}$$

Substituting:

$$(G/A)^2 \ln(1.0/0.1) + (0.1^2 - 1^2) \times 10^6/(2 \times 1 \times 10^3 \times 83.98)$$

$$+ 4(0.003)(30/0.15)(G/A)^2 = 0$$

and: $(G/A) = 1.37$ kg/m^{2}s.

The viscosity of air is 0.018 mN s/m^2.

$$\therefore Re = (0.15 \times 1.37)/(0.018 \times 10^{-3}) = 1.14 \times 10^4$$

Thus the chosen value of Re is too high. When $Re = 1 \times 10^4$, $R/\rho u^2 = 0.0041$ and $G/A = 1.26$ kg/m^{2}s.

Re now equals 1.04×10^4 which agrees well with the assumed value.

Thus:
$$G = 1.26 \times (\pi/4) \times (0.15)^2 = \underline{\underline{0.022 \text{ kg/s}}}$$

PROBLEM 4.9

A vacuum system is required to handle 10 g/s of vapour (molecular weight 56 kg/kmol) so as to maintain a pressure of 1.5 kN/m^2 in a vessel situated 30 m from the vacuum pump. If the pump is able to maintain a pressure of 0.15 kN/m^2 at its suction point, what diameter of pipe is required? The temperature is 290 K, and isothermal conditions may be assumed in the pipe, whose surface can be taken as smooth. The ideal gas law is followed. Gas viscosity $= 0.01$ mN s/m^2.

Solution

Use is made of equation 4.55 to solve this problem. It is necessary to assume a value of the pipe diameter d in order to calculate values of G/A, the Reynolds number and $R/\rho u^2$.

If $d = 0.10$ m, $A = (\pi/4)(0.10)^2 = 0.00785$ m^2

$\therefore \qquad G/A = (10 \times 10^{-3}/0.00785) = 1.274$ kg/m^{2}s

and $\qquad Re = d(G/A)/\mu = 0.10 \times 1.274/(0.01 \times 10^{-3}) = 1.274 \times 10^4$

For a smooth pipe, $R/\rho u^2 = 0.0035$, from Fig. 3.7.

Specific volume at inlet, $v_1 = (22.4/56)(290/273)(101.3/1.5) = 28.7$ m^3/kg

$$(G/A)^2 \ln(P_1/P_2) + (P_2^2 - P_1^2)/2P_1v_1 + 4(R/\rho u^2)(l/d)(G/A)^2 = 0 \text{ (equation 4.55)}$$

Substituting gives:

$$(1.274)^2 \ln(1.5/0.15) + (0.15^2 - 1.5^2) \times 10^6/(2 \times 1.5 \times 10^3 \times 28.7)$$

$$+ (0.0035)(30/0.10)(1.274)^2 = -16.3$$

and the chosen value of d is too large.

A further assumed value of $d = 0.05$ m gives a value of the right hand side of equation 4.55 of 25.9 and the procedure is repeated until this value is zero.

This occurs when $d = 0.08$ m or <u>80 mm.</u>

PROBLEM 4.10

In a vacuum system, air is flowing isothermally at 290 K through a 150 mm diameter pipeline 30 m long. If the relative roughness of the pipewall e/d is 0.002 and the downstream pressure is 130 N/m^2, what will the upstream pressure be if the flowrate of air is 0.025 kg/s? Assume that the ideal gas law applies and that the viscosity of air is constant at 0.018 mN s/m^2.

What error would be introduced if the change in kinetic energy of the gas as a result of expansion were neglected?

Solution

As the upstream and mean specific volumes v_1 and v_m are required in equations 4.55 and 4.56 respectively, use is made of equation 4.57:

$$(G/A)^2 \ln(P_1/P_2) + (P_2^2 - P_1^2)/(2\mathbf{R}T/M) + 4(R/\rho u^2)(l/d)(G/A)^2 = 0$$

$\mathbf{R} = 8.314$ kJ/kmol K and hence:

$$2\mathbf{R}T/M = (2 \times 8.314 \times 10^3 \times 290)/29 = 1.66 \times 10^5 \text{ J/kg}$$

The second term has units of $(N/m^2)^2/(J/kg) = kg^2/s^2 \ m^4$ which is consistent with the other terms.

$$A = (\pi/4)(0.15)^2 = 0.0176 \text{ m}^2$$

∴ $G/A = (0.025/0.0176) = 1.414$

and $Re = d(G/A)/\mu = (0.15 \times 1.414)/(0.018 \times 10^{-3}) = 1.18 \times 10^4$

For smooth pipes and $Re = 1.18 \times 10^4$, $R/\rho u^2 = 0.0040$ from Fig. 3.7. Substituting in equation 4.57 gives:

$$(1.414)^2 \ln(P_1/130) + (130^2 - P_1^2)/1.66 \times 10^5 + 4 \times 0.0040(30/0.15)(1.414)^2 = 0$$

Solving by trial and error, the upstream pressure, $P_1 = $ <u>1.36 kN/m^2</u>

If the kinetic energy term is neglected, equation 4.57 becomes:

$$(P_2^2 - P_1^2)/(2\mathbf{R}T/M) + 4(R/\rho u^2)(l/d)(G/A)^2 = 0 \text{ and } P_1 = \underline{1.04 \text{ kN/m}^2}$$

Thus a considerable error would be introduced by this simplifying assumption.

PROBLEM 4.11

Air is flowing at the rate of 30 kg/m²s through a smooth pipe of 50 mm diameter and 300 m long. If the upstream pressure is 800 kN/m², what will the downstream pressure be if the flow is isothermal at 273 K? Take the viscosity of air as 0.015 mN s/m² and assume that volume occupies 22.4 m³. What is the significance of the change in kinetic energy of the fluid?

Solution

$$(G/A)^2 \ln(P_1/P_2) + (P_2^2 - P_1^2)/2P_1v_1 + 4(R/\rho u^2)(l/d)(G/A)^2 = 0 \quad \text{(equation 4.55)}$$

The specific volume at the upstream condition is:

$$v_1 = (22.4/29)(273/273)(101.3/800) = 0.098 \text{ m}^3/\text{kg}$$

$$G/A = 30 \text{ kg/m}^2\text{s}$$

$$\therefore \qquad Re = (0.05 \times 30)/(0.015 \times 10^{-3}) = 1.0 \times 10^5$$

For a smooth pipe, $R/\rho u^2 = 0.0032$ from Fig. 3.7.
   Substituting gives:

$$(30)^2 \ln(800/P_2) + (P_2^2 - 800^2) \times 10^6/(2 \times 800 \times 10^3 \times 0.098)$$
$$+ 4(0.0032)(300/0.05)(30)^2 = 0$$

and the downstream pressure, $P_2 = \underline{\underline{793 \text{ kN/m}^2}}$

   The kinetic energy term $= (G/A)^2 \ln(800/793) = 7.91 \text{ kg}^2/\text{m}^4\text{s}^2$
   This is insignificant in comparison with $69,120 \text{ kg}^2/\text{m}^4\text{s}^2$ which is the value of the other terms in equation 4.55.

PROBLEM 4.12

If temperature does not change with height, estimate the boiling point of water at a height of 3000 m above sea-level. The barometer reading at sea-level is 98.4 kN/m² and the temperature is 288.7 K. The vapour pressure of water at 288.7 K is 1.77 kN/m². The effective molecular weight of air is 29 kg/kmol.

Solution

The air pressure at 3000 m is P_2 and the pressure at sea level, $P_1 = 98.4 \text{ kN/m}^2$.

$$\int v \, dP + \int g \, dz = 0$$
$$v = v_1(P/P_1)$$

and:
$$P_1 v_1 \int \frac{dP}{P} + \int g\,dz = 0$$

and:
$$P_1 v_1 \ln(P_2/P_1) + g(z_2 - z_1) = 0$$

$$v_1 = (22.4/29)(288.7/273)(101.3/98.4) = 0.841 \text{ m}^3/\text{kg}.$$

$\therefore$
$$98{,}400 \times 0.841 \ln(P_2/98.4) + 9.81(3000 - 0) = 0$$

and:
$$P_2 = 68.95 \text{ kN/m}^2$$

The relationship between vapour pressure and temperature may be expressed as:

$$\log P = a + bT$$

When,
$$T = 288.7, P = 1.77 \text{ kN/m}^2$$

and when,
$$T = 373, P = 101.3 \text{ kN/m}^2$$

$\therefore$
$$\log P = -5.773 + 0.0209\, T$$

When $P_2 = 68.95$, $T = \underline{\underline{364 \text{ K}}}$.

PROBLEM 4.13

A 150 mm gas main is used for transferring gas (molecular weight 13 kg/kmol and kinematic viscosity 0.25 cm²/s) at 295 K from a plant to a storage station 100 m away, at a rate of 1 m³/s. Calculate the pressure drop if the pipe can be considered to be smooth.

If the maximum permissible pressure drop is 10 kN/m², is it possible to increase the flowrate by 25%?

Solution

If the flow of 1 m³/s is at STP, the specific volume of the gas is:

$$(22.4/13) = 1.723 \text{ m}^3/\text{kg}.$$

The mass flowrate, $G = (1.0/1.723) = 0.58$ kg/s.
  Cross-sectional area, $A = (\pi/4)(0.15)^2 = 0.0176 \text{ m}^2$

$\therefore$
$$G/A = 32.82 \text{ kg/m}^2\text{s}$$

$$\mu/\rho = 0.25 \text{ cm}^2/\text{s} = 0.25 \times 10^{-4} \text{ m}^2/\text{s}$$

and
$$\mu = (0.25 \times 10^{-4})(1/1.723) = 1.45 \times 10^{-5} \text{ N s/m}^2$$

$\therefore$
$$Re = (0.15 \times 32.82/1.45 \times 10^{-5}) = 3.4 \times 10^5$$

For smooth pipes, $R/\rho u^2 = 0.0017$, from Fig. 3.7.
  The pressure drop due to friction is:

$$4(R/\rho u^2)(l/d)(G/A)^2 = 4(0.0017)(100/0.15)(32.82)^2 = 4883 \text{ kg}^2/\text{m}^4 \text{ s}^2$$

and: $-\Delta P = (4883/1.723) = 2834$ N/m² or $\underline{\underline{2.83 \text{ kN/m}^2}}$.

If the flow is increased by 25%, $G = (1.25 \times 0.58) = 0.725$ kg/s

$$G/A = 41.19 \text{ kg/m}^2 \text{ s}$$

and: $$Re = (0.15 \times 41.9)/(1.45 \times 10^5) = 4.3 \times 10^5$$

and, from Fig. 3.7, $R/\rho u^2 = 0.00165$

The pressure drop $= 4(0.00165)(100/0.15)(41.19)^2 1.723 = \underline{\underline{4.33 \text{ kN/m}^2}}$ (which is less than 10 kN/m^2)

It is therefore possible to increase the flowrate by 25%.

SECTION 5

Flow of Multiphase Mixtures

PROBLEM 5.1

It is required to transport sand of particle size 1.25 mm and density 2600 kg/m^3 at the rate of 1 kg/s through a horizontal pipe 200 m long. Estimate the air flowrate required, the pipe diameter, and the pressure drop in the pipe-line.

Solution

For conventional pneumatic transport in pipelines, a solids-gas mass ratio of about 5 is employed. Mass flow of air $= (1/5) = 0.20$ kg/s

and, taking the density of air as 1.0 kg/m^3, volumetric flowrate of air $= (1.0 \times 0.20)$

$$= 0.20 \text{ m}^3/\text{s}$$

In order to avoid excessive pressure drops, an air velocity of 30 m/s seems reasonable. Ignoring the volume occupied by the sand (which is about 0.2% of that occupied by the air), the cross-sectional area of pipe required $= (0.20/30) = 0.0067$ m^2, equivalent to a pipe diameter of $\sqrt{(4 \times 0.0067/\pi)} = 0.092$ m or 92 mm.

Thus a pipe diameter of <u>101.6 mm</u> (100 mm) would be specified.

From Table 5.3 for sand of particle size 1.25 mm and density 2600 kg/m^3, the free-falling velocity is:

$$u_0 = 4.7 \text{ m/s}$$

In equation 5.37, $(u_G - u_s) = 4.7/[0.468 + 7.25\sqrt{(4.7/2600)}] = 6.05$ m/s
The cross-sectional area of a 101.6 mm i.d. pipe $= (\pi \times 0.1016^2/4) = 0.0081$ m^2.

$\therefore$ air velocity, $u_G = (0.20/0.0081) = 24.7$ m/s

and: $u_s = (24.7 - 6.05) = 18.65$ m/s

Taking the viscosity and density of air as 1.7×10^{-5} N s/m^2 and 1.0 kg/m^3 respectively, the Reynolds number for the air flow alone is:

$$Re = (0.102 \times 24.7 \times 1.0)/(1.7 \times 10^{-5}) = 148,000$$

and from Fig. 3.7, the friction factor $\phi = 0.002$.

$$-\Delta P_{\text{air}} = 4\phi(l/d)\rho u^2 \qquad\qquad \text{(equation 3.18)}$$
$$= (4 \times 0.002(200/0.102) \times 1.0 \times 24.7^2) = 9570 \text{ N/m}^2 \text{ or } 9.57 \text{ kN/m}^2$$

assuming isothermal conditions and incompressible flow.

$$(-\Delta P_x/ - \Delta P_{\text{air}})(u_s^2/F) = (2805/u_0) \qquad\qquad \text{(equation 5.38)}$$
$$\therefore \qquad -\Delta P_x = (2805(-\Delta P_{\text{air}})F)/(u_0 u_s^2) = (2805 \times 9.57 \times 1.0)/(4.7 \times 18.65^2)$$
$$= \underline{\underline{16.4 \text{ kN/m}^2}}$$

PROBLEM 5.2

Sand of a mean diameter 0.2 mm is to be conveyed in water flowing at 0.5 kg/s in a 25 mm ID horizontal pipe 100 m long. What is the maximum amount of sand which may be transported in this way if the head developed by the pump is limited to 300 kN/m^2? Assume fully suspended heterogeneous flow.

Solution

See Volume 1, Example 5.2.

PROBLEM 5.3

Explain the various mechanisms by which particles may be maintained in suspension during hydraulic transport in a horizontal pipeline and indicate when each is likely to be important.

A highly concentrated suspension of flocculated kaolin in water behaves as a pseudo-homogeneous fluid with shear-thinning characteristics which can be represented approximately by the Ostwald–de Waele power-law, with an index of 0.15. It is found that, if air is injected into the suspension when in laminar flow, the pressure gradient may be reduced, even though the flowrate of suspension is kept constant. Explain how this is possible in "slug" flow, and estimate the possible reduction in pressure gradient for equal volumetric, flowrates of suspension and air.

Solution

If u is the superficial velocity of slurry, then:

For slurry alone:

The pressure drop in a pipe of length l is: $Ku^n l$.

If the air: slurry volumetric ratio is R, there is no slip between the slurry and the air and the system consists of alternate slugs of air and slurry, then:

The linear velocity of slurry is $(R + 1)u$

Fraction of pipe occupied by slurry slugs is $\dfrac{1}{R+1}$

Assuming that the pressure drop is the sum of the pressure drops along the slugs, then:

the new pressure drop is: $K\{(R+1)u\}^n \left(\dfrac{l}{R+1}\right) = Ku^n l(R+1)^{n-1}$

Then: $\quad r = \dfrac{\text{pressure gradient with air}}{\text{pressure gradient without air}} = \dfrac{Ku^n l(R+1)^{n-1}}{Ku^n l} = (R+1)^{n-1}$

For $\quad n = 0.15$ and: $R = 1$

$\quad r = 2^{-0.85} = \underline{\underline{0.55}}$

SECTION 6

Flow and Pressure Measurement

PROBLEM 6.1

Sulphuric acid of density 1300 kg/m^3 is flowing through a pipe of 50 mm internal diameter. A thin-lipped orifice, 10 mm diameter, is fitted in the pipe and the differential pressure shown by a mercury manometer is 10 cm. Assuming that the leads to the manometer are filled with the acid, calculate (a) the mass of acid flowing per second, and (b) the approximate loss of pressure caused by the orifice. The coefficient of discharge of the orifice may be taken as 0.61, the density of mercury as 13,550 kg/m^3, and the density of water as 1000 kg/m^3.

Solution

See Volume 1, Example 6.2.

PROBLEM 6.2

The rate of discharge of water from a tank is measured by means of a notch, for which the flowrate is directly proportional to the height of liquid above the bottom of the notch. Calculate and plot the profile of the notch if the flowrate is 0.1 m^3/s when the liquid level is 150 mm above the bottom of the notch.

Solution

The velocity of fluid discharged as a height h above the bottom of the notch is:

$$u = \sqrt{(2gh)}$$

The velocity therefore varies from zero at the bottom of the notch to a maximum value at the free surface.

For a horizontal element of fluid of width $2w$ and depth dh at a height h above the bottom of the notch, the discharge rate of fluid is given by:

$$dQ = \sqrt{(2gh)}2wdh$$

If the discharge rate is linearly related to the height of the liquid over the notch, H, w will be a function of h and it may be supposed that:

$$w = kh^n$$

where k is a constant.

77

Substituting for w in the equation for dQ and integrating to give the discharge rate over the notch Q then:

$$Q = 2\sqrt{(2g)}\ k \int_0^H h^n h^{0.5}\ dh$$

$$= 2\sqrt{(2g)}\ k \int_0^H h^{n+0.5}\ dh$$

$$= 2\sqrt{(2g)}\ k[1/(n+1.5)]H^{(n+1.5)}$$

Since it is required that $Q \propto H$:

$$n + 1.5 = 1$$

and:

$$n = -0.5$$

Thus:

$$Q = 2\sqrt{(2g)}\ kH$$

Since $Q = 0.1$ m³/s when $H = 0.15$ m:

$$k = (0.1/0.15)[1/(2\sqrt{(2g)}] = 0.0753 \text{ m}^{1.5}$$

Thus, with w and h in m: $\qquad w = 0.0753h^{-0.5}$

and, with w and h in mm: $\qquad w = 2374h^{-0.5}$

and using this equation, the profile is plotted as shown in Figure 6a.

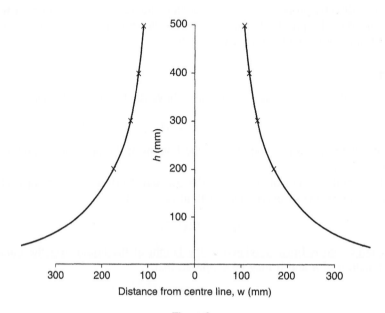

Figure 6a.

PROBLEM 6.3

Water flows at between 3 1 and 4 l/s through a 50 mm pipe and is metered by means of an orifice. Suggest a suitable size of orifice if the pressure difference is to be measured with a simple water manometer. What is the approximate pressure difference recorded at the maximum flowrate?

Solution

Equations 6.19 and 6.21 relate the pressure drop to the mass flowrate. If equation 6.21 is used as a first approximation, $G = C_D A_0 \rho \sqrt{(2gh)}$.

For the maximum flow of 4 l/s, $G = 4$ kg/s. The largest practicable height of a water manometer will be taken as 1 m and equation 6.21 is then used to calculate the orifice area A_0. If the coefficient of discharge C_D is taken as 0.6, then:

$$4.0 = 0.6 A_0 \times 1000 \sqrt{(2 \times 9.81 \times 1.0)}, A_0 = 0.0015 \text{ m}^2 \text{ and } d_0 = 0.0438 \text{ m}$$

The diameter, d_0, is comparable with the pipe diameter and hence the area correction term must be included and:

$$[1 - (A_0/A_1)^2] = [1 - (43.8^2/50^2)^2] = 0.641.$$

Therefore the value of A_0 must be recalculated as:

$$4.0 = 0.6 A_0 \times 1000 \sqrt{(2 \times 9.8 \times 1.0)/[1 - (A_0/A_1)^2]}$$

from which $A_0 = 0.00195$ m^2 and $d = 0.039$ m or <u>39 mm</u>

$$\sqrt{[1 - (A_0/A_1)^2]} = \sqrt{[1 - (39^2/50^2)^2]} = 0.793$$

Substituting in equation 6.19:

$$4.0 = (0.6 \times 0.00195) \times 1000 \sqrt{(2 \times 0.001(-\Delta P)/0.793)}$$

and: $$-\Delta P = 12320 \text{ N/m}^2 \text{ or } \underline{12.3 \text{ kN/m}^2}$$

PROBLEM 6.4

The rate of flow of water in a 150 mm diameter pipe is measured by means of a venturi meter with a 50 mm diameter throat. When the drop in head over the converging section is 100 mm of water, the flowrate is 2.7 kg/s. What is the coefficient for the converging cone of the meter at that flowrate and what is the head lost due to friction? If the total loss of head over the meter is 15 mm water, what is the coefficient for the diverging cone?

Solution

The equation relating the mass flowrate G and the head loss across a venturi meter is given by:

$$G = \frac{C_D A_0}{v}\sqrt{\frac{2v(P_1 - P_2)}{1 - (A_0/A_1)^2}} \qquad \text{(equation 6.19)}$$

$$G = C_D \rho \frac{A_1 A_2}{\sqrt{(A_1^2 - A_1^2)}}\sqrt{(2v(P_1 - P_2))} \qquad \text{(equation 6.32)}$$

$$G = C_D \rho C'\sqrt{(2gh_v)} \qquad \text{(equation 6.33)}$$

where C' is a constant for the meter and h_v is the loss in head over the converging cone expressed as height of fluid.

$$A_1 = (\pi/4)(0.15)^2 = 0.0176 \text{ m}^2$$

$$A_2 = (\pi/4)(0.05)^2 = 0.00196 \text{ m}^2$$

$$C' = (0.0176 \times 0.00196/\sqrt{(0.0176^2 - 0.00196^2)}) = 0.00197 \text{ m}^2$$

$$h_v = 0.1 \text{ m}$$

$$\therefore \qquad 2.7 = (C_D \times 1000 \times 0.00197)\sqrt{(2 \times 9.81 \times 0.10)} \text{ and } \underline{\underline{C_D = 0.978}}$$

In equation 6.33, if there were no losses, the coefficient of discharge of the meter would be unity, and for a flowrate G the loss in head would be $(h_v - h_f)$ where h_f is the head loss due to friction.

Thus: $$G = \rho C'\sqrt{[2g(h_v - h_f)]}$$

Dividing this equation by equation 6.33 and squaring gives:

$$1 - (h_f/h_v) = C_D^2 \text{ and } h_f = h_v(1 - C_D^2)$$

$$\therefore \qquad h_f = 100(1 - 0.978^2) = \underline{\underline{4.35 \text{ mm}}}$$

If the head recovered over the diverging cone is h_v' and the coefficient of discharge for the converging cone is C_D', then $G = C_D' \rho C'\sqrt{(2gh_v')}$

If the whole of the excess kinetic energy is recovered as pressure energy, the coefficient C_D' will equal unity and G will be obtained with a recovery of head equal to h_v' plus some quantity h_f', $G = \rho C'\sqrt{[2g(h_v' + h_f')}$

Equating these two equations and squaring gives:

$$C_D'^2 = 1 + (h_f'/h_v') \text{ and } h_f' = h_v'(C_D'^2 - 1)$$

Thus the coefficient of the diverging cone is greater than unity and the total loss of head $= h_f + h_f'$.

Head loss over diverging cone $= (15.0 - 4.35) = 10.65$ mm

The coefficient of the diverging cone C_D' is given by:

$$G = C_D' \rho C' \sqrt{(2gh_v')}$$

$$h_v' = (100 - 15) = 85 \text{ mm}$$

and: $2.7 = (C_D' \times 1000 \times 0.00197)\sqrt{(2 \times 9.81 \times 0.085)}$ or $\underline{C_D' = 1.06}$

PROBLEM 6.5

A venturi meter with a 50 mm throat is used to measure a flow of slightly salt water in a pipe of inside diameter 100 mm. The meter is checked by adding 20 cm^3/s of normal sodium chloride solution above the meter and analysing a sample of water downstream from the meter. Before addition of the salt, 1000 cm^3 of water requires 10 cm^3 of 0.1 M silver nitrate solution in a titration. 1000 cm^3 of the downstream sample required 23.5 cm^3 of 0.1 M silver nitrate. If a mercury-under-water manometer connected to the meter gives a reading of 221 mm, what is the discharge coefficient of the meter? Assume that the density of the liquid is not appreciably affected by the salt.

Solution

If the flow of the solution is x m^3/s, then a mass balance in terms of sodium chloride gives:

$$(x \times 0.0585) + (20 \times 10^{-6} \times 58.5) = 0.1375(20 \times 10^{-6} + x)$$

and: $$x = 0.0148 \text{ m}^3/\text{s}$$

or, assuming the density of the solution is 1000 kg/m^3, the mass flowrate is:

$$(0.0148 \times 1000) = 14.8 \text{ kg/s}$$

For the venturi meter, the area of the throat is given by:

$$A_1 = (\pi/4)(50/1000)^2 = 0.00196 \text{ m}^2$$

and the area of the pipe is:

$$A_2 = (\pi/4)(100/1000)^2 = 0.00785 \text{ m}^2$$

From equations 6.32 and 6.33:

$$C' = A_1 A_2 / \sqrt{(A_1^2 - A_2^2)} = 0.00204 \text{ m}^2$$

$$h = 221 \text{ mm Hg}_{\text{under-water}} = 0.221(13500 - 1000)/1000 = 2.78 \text{ m water}$$

and hence: $14.8 = (C_D \times 1000 \times 0.00204)\sqrt{(2 \times 9.81 \times 2.78)}$

and: $C_D = \underline{0.982}$

PROBLEM 6.6

A gas cylinder containing 30 m^3 of air at 6 MN/m^2 pressure discharges to the atmosphere through a valve which may be taken as equivalent to a sharp edged orifice of 6 mm diameter (coefficient of discharge $= 0.6$). Plot the rate of discharge against the pressure in the cylinder. How long will it take for the pressure in the cylinder to fall to (a) 1 MN/m^2, and (b) 150 kN/m^2? Assume an adiabatic expansion of the gas through the valve and that the contents of the cylinder remain at 273 K.

Solution

Area of orifice $= (\pi/4)(0.006)^2 = 2.828 \times 10^{-5}$ m^2.

The critical pressure ratio w_c is:

$$w_c = [2/(k+1)]^{k/(k-1)} \qquad \text{(equation 4.43)}$$

Taking $k = \gamma = 1.4$ for air, $w_c = 0.527$.

Thus sonic velocity will occur until the cylinder pressure falls to a pressure of:

$P_2 = (101.3/0.527) = 192.2$ kN/m^2.

For pressures in excess of 192.2 kN/m^2, the rate of discharge is given by:

$$G = C_D A_0 \sqrt{(kP_1/v_1)(2/(k+1))^{(k+1)/(k-1)}} \qquad \text{(equation 6.29)}$$

If $k = 1.4$, $G = 1.162 \times 10^{-5} \sqrt{(P_1/v_1)}$

If P_a and v_a are atmospheric pressure and the specific volume at atmospheric pressure respectively, $P_a v_a = P_1 v_1$ and $v_1 = P_a v_a/P_1$

$P_a = 101,300$ N/m^2 and $v_a = (22.4/29) = 0.773$ m^3/kg

$\therefore \qquad v_1 = (101,300 \times 0.773/P_1) = (78,246/P_1)$

and: $\qquad G = 1.162 \times 10^{-5} \sqrt{(P_1^2/78,246)} = 4.15 \times 10^{-8} P_1$ kg/s

If P_1 is expressed in MN/m^2, then: $G = \underline{0.0415\ P_1}$ kg/s.

For pressures lower than 192.2 kN/m^2:

$$G^2 = (A_0 C_D/v_2)^2 2 P_1 v_1 (k/k-1)[1 - (P_2/P_1)^{(k-1)/k}] \qquad \text{(equation 6.26)}$$

$$v_2 = v_a = 0.773 \text{ m}^3/\text{kg}$$

$$P_2 = P_a = 101,300 \text{ N/m}^2$$

$$v_1 = P_a v_a/P_1$$

Substituting gives: $\qquad G^2 = \underline{\underline{2.64 \times 10^{-4}[1 - (P_a/P_1)^{0.286}]}}$

Thus a table of G as a function of pressure may be produced as follows:

$P < 192.2$ kN/m^2		$P > 192.2$ kN/m^2	
P (MN/m^2)	G (kg/s)	P (MN/m^2)	G (kg/s)
0.1013	0	0.2	0.0083
0.110	0.0024	0.5	0.0208
0.125	0.0039	1.0	0.0416
0.150	0.0053	2.0	0.0830
0.175	0.0062	6.0	0.249

These data are plotted in Fig. 6b, from which discharge rate is seen to be linear until the cylinder pressure falls to 0.125 MN/m^2.

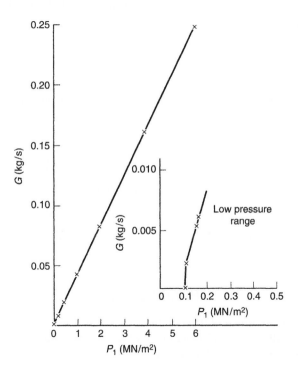

Figure 6b.

If m is the mass of air in the cylinder at any pressure P_1 over the linear part of the curve, $G = \mathrm{d}m/\mathrm{d}t = 0.0415P_1$.

$$\therefore \qquad\qquad\qquad \mathrm{d}t = \mathrm{d}m/0.0415P_1$$

$$m = (29/22.4)(P_1/0.1013) \times 30 = 383.4P_1 \text{ kg}$$

$$\therefore \qquad\qquad \mathrm{d}t = 383.4\mathrm{d}m/0.0415m = 9240\mathrm{d} \text{ m/m}$$

and $\qquad\qquad t = 9240\ln(m_1/m_2)$

At 6 MN/m^2 and 1 MN/m^2, the masses of air in the cylinder are 2308 and 383.4 kg respectively.

$\therefore$ The time for the pressure to fall to 1 MN/m^2 = 9240 ln(2308/3834.4)

$$= 16,600 \text{ s } (4.61 \text{ h})$$

As 0.15 MN/m^2 is still within the linear region, the time for the pressure to fall to this value is 34,100 s .(9.47 h)

PROBLEM 6.7

Air, at 1500 kN/m^2 and 370 K, flows through an orifice of 30 mm^2 to atmospheric pressure. If the coefficient of discharge is 0.65, the critical pressure ratio 0.527, and the ratio of the specific heats is 1.4, calculate the mass flowrate.

Solution

If the critical pressure ratio w_c is 0.527 (from Problem 6.6), sonic velocity will occur until the pressure falls to (101.3/0.527) = 192.2 kN/m^2. For pressures above this value, the mass flowrate is given by:

$$G = C_D A_0 \sqrt{(kP_1/v_1)[2/(k+1)]^{(k+1)/(k-1)}} \qquad \text{(equation 6.29)}$$

If $k = 1.4$, $G = C_D A_0 \sqrt{(1.4P_1/v_1)(2/2.4)^{2.4/0.4}} = C_D A_0 \sqrt{(0.468P_1/v_1)}$

$$P_1 = 1,500,000 \text{ N/m}^2$$

and: $v_1 = (22.4/29)(370/273)(101.3/1500) = 0.0707 \text{ m}^3/\text{kg}$

Substituting gives: $G = (0.65 \times 30 \times 10^{-6})\sqrt{(0.486 \times 1,500,000/0.0707)} = 0.061 \text{ kg/s}$

PROBLEM 6.8

Water flows through an orifice of 25 mm diameter situated in a 75 mm pipe at the rate of 300 cm^3/s. What will be the difference in level on a water manometer connected across the meter? Viscosity of water is 1 mN s/m^2.

Solution

See Volume 1, Example 6.1.

PROBLEM 6.9

Water flowing at 1.5 l/s in a 50 mm diameter pipe is metered by means of a simple orifice of diameter 25 mm. If the coefficient of discharge of the meter is 0.62, what will be the reading on a mercury-under-water manometer connected to the meter? What is the Reynolds number for the flow in the pipe? Density of water = 1000 kg/m^3. Viscosity of water = 1 mN s/m^2.

Solution

Mass flowrate, $G = (1500 \times 10^{-6} \times 1000) = 1.5$ kg/s.

Area of orifice, $A_0 = (\pi/4)(0.025)^2 = 0.00049$ m^2.

Area of pipe, $A_1 = (\pi/4)(0.050)^2 = 0.00196$ m^2.

Reynolds number $= \rho u d/\mu = d(G/A_1)/\mu$
$$= 0.05(1.5/0.00196)/(1 \times 10^{-3}) = \underline{\underline{3.83 \times 10^4}}$$

The orifice meter equations are 6.19 and 6.21; the latter being used when $\sqrt{[1 - (A_0/A_1)^2]}$ approaches unity.

Thus: $$\sqrt{[1 - (A_0/A_1)^2]} = \sqrt{[1 - (25^2/50^2)^2]} = 0.968$$

Using equation 6.21, $G = C_D A_0 \rho \sqrt{(2gh)}$ gives:

$$1.5 = 0.62 \times 0.00049 \times 1000\sqrt{(2 \times 9.81h)}, \text{ and } h = 1.24 \text{ m of water}$$

Using equation 6.19 in terms of h gives:

$$1.5 = (0.62 \times 0.00049 \times 1000/0.968)\sqrt{(2gh)} \text{ and } h = 1.16 \text{ m of water}$$

This latter value of h should be used. The height of a mercury-under-water manometer would then be $1.16/((13.55 - 1.00)/1.00) = 0.092$ m or $\underline{\underline{92 \text{ mm Hg}}}$.

PROBLEM 6.10

What size of orifice would give a pressure difference of 0.3 m water gauge for the flow of a petroleum product of density 900 kg/m^3 at 0.05 m^3/s in a 150 mm diameter pipe?

Solution

As in previous problems, equations 6.19 and 6.21 may be used to calculate the flow through an orifice. In this problem the size of the orifice is to be found so that the simpler equation will be used in the first instance.

$$G = C_D A_0 \rho \sqrt{(2gh)} \qquad\qquad \text{(equation 6.21)}$$

$$G = (0.05 \times 900) = 45.0 \text{ kg/s}$$

$$\rho = 900 \text{ kg/m}^3$$

$$h = 0.3 \text{ m of water or } (0.3/0.9) = 0.333 \text{ m of petroleum product}$$

$$C_D = 0.62 \text{ (assumed)}$$

$$\therefore \qquad 45.0 = (0.62 \times A_0 \times 900)\sqrt{(2 \times 9.81 \times 0.333)}$$

Thus: $\qquad A_0 = 0.3155 \text{ m}^2 \text{ and } d_0 = 0.2 \text{ m}.$

This orifice diameter is larger than the pipe size so that it was clearly wrong to use the simpler equation.

Thus: $\qquad G = C_D A_0 \rho \sqrt{[2gh/(1 - (A_0/A_1)^2)]}$ $\qquad$ (equation 6.19)

$$A_1 = (\pi/4)(0.15)^2 = 0.0177 \text{ m}^2$$

$$\therefore \qquad 45.0 = (0.62 \times A_0 \times 900)\sqrt{[2 \times 9.81 \times 0.33/(1 - (A_0/0.0177)^2)]}$$

Thus: $\qquad A_0 = 0.154 \text{ m}^2 \text{ and } d_0 = \underline{\underline{0.14 \text{ m}}}$

PROBLEM 6.11

The flow of water through a 50 mm pipe is measured by means of an orifice meter with a 40 mm aperture. The pressure drop recorded is 150 mm on a mercury-under-water manometer and the coefficient of discharge of the meter is 0.6. What is the Reynolds number in the pipe and what would the pressure drop over a 30 m length of the pipe be expected to be? Friction factor, $\phi = R/\rho u^2 = 0.0025$. Density of mercury = 13,600 kg/m^3. Viscosity of water = 1 mN s/m^2.

What type of pump would be used, how would it be driven and what material of construction would be suitable?

Solution

Area of pipe, $A_1 = (\pi/4)(0.05)^2 = 0.00197 \text{ m}^2$.

Area of orifice, $A_0 = (\pi/4)(0.04)^2 = 0.00126 \text{ m}^2$.

$h = 150$ mmHg under water $= 0.15 \times (13600 - 1000)/1000 \equiv 1.88$ m of water.

$1 - (A_0/A)^2 = 0.591$, and hence:

$$G = C_D A_0 \rho \sqrt{[2gh/(1 - (A_0/A)^2)]} \qquad \text{(equation 6.19)}$$

$$= (0.6 \times 0.00126 \times 1000)\sqrt{2 \times 9.81 \times 1.88/0.591} = 5.97 \text{ kg/s}$$

Reynolds number, $\rho u d/\mu = d(G/A_1)/\mu = 0.05(6.22/0.00197)/(1 \times 10^{-3}) = \underline{\underline{1.52 \times 10^5}}$

The pressure drop is given by:

$$-\Delta P/v = 4(R/\rho u^2)(l/d)(G/A)^2$$

$$= 4(0.0025)(30/0.05)(5.97/0.00197)^2 = 5.74 \times 10^7 \text{ kg}^2/\text{m}^4\text{s}^2$$

$$-\Delta P = 5.74 \times 10^7 \times (1/1000)$$
$$= (5.74 \times 10^4) \text{ N/m}^2 \text{ or } \underline{57.4 \text{ kN/m}^2}$$

Power required = head loss (m) $\times G \times g$
$$= (5.74 \times 10^4/1000 \times 9.81)(5.97 \times 9.81) = 343 \text{ W}$$

For a pump efficiency of 60%, the actual power requirement = $(343/0.6) = 571$ W.

Water velocity = $5.97/(0.00197 \times 1000) = 3.03$ m/s.

For this low-power requirement at a low head and comparatively low flowrate, a centrifugal pump, electrically driven and made of stainless steel, would be suitable.

PROBLEM 6.12

A rotameter has a tube 0.3 m long which has an internal diameter of 25 mm at the top and 20 mm at the bottom. The diameter of the float is 20 mm, its effective density is 4800 kg/m^3, and its volume 6.6 cm^3. If the coefficient of discharge is 0.72, at what height will the float be when metering water at 100 cm^3/s?

Solution

See Volume 1, Example 6.4.

PROBLEM 6.13

Explain why there is a critical pressure ratio across a nozzle at which, for a given upstream pressure, the flowrate is a maximum. Obtain an expression for the maximum flow for a given upstream pressure for isentropic flow through a horizontal nozzle. Show that for air (ratio of specific heats, $\gamma = 1.4$) the critical pressure ratio is 0.53 and calculate the maximum flow through an orifice of area 30 mm^2 and coefficient of discharge 0.65 when the upstream pressure is 1.5 MN/m^2 and the upstream temperature 293 K.
Kilogram molecular volume = 22.4 m^3.

Solution

The reasons for critical pressure ratios are discussed in Section 4.4.1. The maximum rate of discharge is given by:

$$G_{max} = C_D A_0 \sqrt{(kP_1/v_1)(2/(k+1))^{(k+1)/(k-1)}} \qquad \text{(equation 6.29)}$$

For an isentropic process, $k = \gamma = 1.4$ for air.

The critical pressure ratio, $\qquad w_c = (2/k+1)^{k/(k-1)} \qquad$ (equation 4.43)

Substituting for $k = \gamma = 1.4$, $w_c = (2/2.4)^{1.4/0.4} = \underline{0.523}$

The maximum rate of discharge is given by equation 6.29.

$$P_1 = 1.5 \times 10^6 \text{ N/m}^2$$

$$A_0 = 30 \times 10^{-6} \text{ m}^2$$

$$k = 1.4 \text{ and } C_D = 0.65$$

At $P_1 = 1.5$ MN/m^2 and $T_1 = 293$ K, the specific volume v_1 is:

$$v_1 = (22.4/29)(293/273)(0.1013/1.5) = 0.056 \text{ m}^3/\text{kg}$$

Substituting, $G_{max} = 0.65 \times 30 \times 10^{-6} \sqrt{(1.4 \times 1.5 \times 10^6/0.056)(2/2.4)^{2.4/0.4}}$
$$= 0.069 \text{ kg/s}$$

PROBLEM 6.14

A gas cylinder containing air discharges to atmosphere through a valve whose charac-teristics may be considered similar to those of a sharp-edged orifice. If the pressure in the cylinder is initially 350 kN/m^2, by how much will the pressure have fallen when the flowrate has decreased to one-quarter of its initial value? The flow through the valve may be taken as isentropic and the expansion in the cylinder as isothermal. The ratio of the specific heats at constant pressure and constant volume is 1.4.

Solution

From equation 4.43:

the critical pressure ratio, $w_c = [2/(k+1)]^{k/(k-1)} = (2/2.4)^{1.4/0.4} = 0.528$

If the cylinder is discharging to atmospheric pressure, sonic velocity will occur until the cylinder pressure has fallen to $(101.3/0.528) = 192$ kN/m^2

The maximum discharge when the cylinder pressure exceeds 192 kN/m^2 is given by:

$$G_{max} = C_D A_0 \sqrt{\frac{kP_1}{v_1} \left(\frac{2}{(k+1)}\right)^{(k+1)/(k-1)}} \qquad \text{(equation 6.29)}$$

If P_a and v_a are the pressure and specific volume at atmospheric pressure, then:

$$1/v_1 = P_1/P_a v_a$$

and:
$$G_{max} = C_D A_0 \sqrt{\frac{kP_1^2}{P_a v_a} \left(\frac{2}{k+1}\right)^{(k+1)/(k-1)}}$$

$$= C_D A_0 P_1 \sqrt{[(k/P_a v_a)(2/k+1)]^{(k+1)/(k-1)}}$$

If G_{350} and G_{192} are the rates of discharge at 350 and 192 kN/m^2 respectively, then:

$$G_{350}/G_{192} = (350/192) = 1.82$$

or:
$$G_{192} = 0.55 G_{350}$$

For pressures below 192 kN/m^2:

$$G = \frac{C_D A_0}{v_2} \sqrt{2P_1 v_1 \left(\frac{k}{k-1}\right)\left[1 - \left(\frac{P_2}{P_1}\right)^{(k-1)/k}\right]} \qquad \text{(equation 6.26)}$$

Substituting for $1/v_1 = P_1/P_a v_a$ and $v_2 = v_a$ gives:

$$G = \frac{C_D A_0}{v_a} \sqrt{2P_a v_a \left(\frac{k}{k-1}\right)\left[1 - \left(\frac{P_2}{P_1}\right)^{(k-1)/k}\right]}$$

and:
$$G^2 = (C_D A_0/v_a)^2 2P_a v_a [k/(k-1)][1 - (P_2/P_1)^{(k-1)/k}]$$
$$= (C_D A_0/v_a)^2 2P_a v_a \times 3.5[1 - (P_2/P_1)^{0.286}]$$

When $P_1 = 192$ kN/m^2, $G_{192} = 0.55 G_{350}$, P_2, atmospheric pressure, 101.3 kN/m^2 and:

$$(0.55 G_{350})^2 = (C_D A_0/v_a)^2 2P_a v_a \times 3.5[1 - (101.3/192)^{0.286}]$$

When the final pressure P_1 is reached, the flowrate is $0.25 G_{350}$.

$$\therefore \qquad (0.25 G_{350})^2 = (C_D A_0/v_a)^2 2P_a v_a \times 3.5(1 - (101.3/P_1)^{0.286})$$

Dividing these two equations gives:

$$\left(\frac{0.55}{0.25}\right)^2 = \frac{1 - (101.3/192)^{0.286}}{1 - (101.3/P_1)^{0.286}}$$

and:
$$P_1 = \underline{\underline{102.3 \text{ kN/m}^2}}$$

PROBLEM 6.15

Water discharges from the bottom outlet of an open tank 1.5 m by 1 m in cross-section. The outlet is equivalent to an orifice 40 mm diameter with a coefficient of discharge of 0.6. The water level in the tank is regulated by a float valve on the feed supply which shuts off completely when the height of water above the bottom of the tank is 1 m and which gives a flowrate which is directly proportional to the distance of the water surface below this maximum level. When the depth of water in the tank is 0.5 m the inflow and outflow are directly balanced. As a result of a short interruption in the supply, the water level in the tank falls to 0.25 m above the bottom but is then restored again. How long will it take the level to rise to 0.45 m above the bottom?

Solution

The mass flowrate G is related to the head h for the flow through an orifice when the area of the orifice is small in comparison with the area of the pipe by:

$$G = C_D A_0 \rho \sqrt{(2gh)} \qquad \text{(equation 6.21)}$$

If h is the distance of the water level below the maximum depth of 1 m, then the head above the orifice is equal to $(1 - h)$ and:

$$G = C_D A_0 \rho \sqrt{[2g(1 - h)]}$$

When the tank contains 0.5 m of water, the flowrate is given by:

$$G = (0.6 \times (\pi/4)(0.04)^2 \times 1000\sqrt{(2 \times 9.81 \times 0.5)}) = 2.36 \text{ kg/s}$$

The input to the tank is stated to be proportional to h, and when the tank is half full the inflow is equal to the outflow,

or: $2.36 = (K \times 0.5)$ and $K = 4.72$ kg/ms

Thus the inflow $= 4.72h$ kg/s and the outflow $= C_D A_0 \rho \sqrt{2g}\sqrt{(1 - h)}$ kg/s.

The net rate of filling $= 4.72h - C_D A_0 \rho \sqrt{2g}\sqrt{(1 - h)} = 4.72h - 3.34\sqrt{(1 - h)}$

Time to fill the tank = (mass of water/rate of filling)
$$= 1 \times 1.5 \times (0.45 - 0.25) \times 1000/\text{rate} = 300/\text{rate}$$

The time to fill from 0.25 to 0.45 m above the bottom of the tank is then:

$$\text{time} = \int_{0.55}^{0.75} \frac{300 dh}{4.72h - 3.34\sqrt{(1 - h)}}$$

This integral is most easily solved graphically as shown in Fig. 6c, where the area under the curve $= 0.233$ s/m and the time $= (300 \times 0.233) = \underline{70 \text{ s}}$.

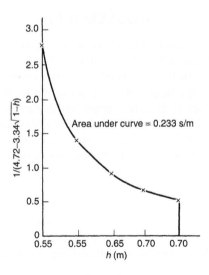

Figure 6c.

PROBLEM 6.16

The flowrate of air at 298 K in a 0.3 m diameter duct is measured with a pitot tube which is used to traverse the cross-section. Readings of the differential pressure recorded on a water manometer are taken with the pitot tube at ten different positions in the cross-section. These positions are so chosen as to be the mid-points of ten concentric annuli each of the same cross-sectional area. The readings are:

Position	1	2	3	4	5
Manometer reading (mm water)	18.5	18.0	17.5	16.8	15.7
Position	6	7	8	9	10
Manometer reading (mm water)	14.7	13.7	12.7	11.4	10.2

The flow is also metered using a 150 mm orifice plate across which the pressure differential is 50 mm on a mercury-under-water manometer. What is the coefficient of discharge of the orifice meter?

Solution

Cross-sectional area of duct $= (\pi/4)(0.3)^2 = 0.0707$ m^2.

Area of each concentric annulus $= 0.00707$ m^2.

If the diameters of the annuli are designated d_1, d_2 etc., then:

$$0.00707 = (\pi/4)(0.3^2 - d_1^2)$$
$$0.00707 = (\pi/4)(d^2 - d_2^2)$$
$$0.00707 = (\pi/4)(d_2^2 - d_3^2) \text{ and so on,}$$

and the mid-points of each annulus may be calculated across the duct.

For a pitot tube, the velocity may be calculated from the head h as $u = \sqrt{(2gh)}$

For position 1, $h = 18.5$ mm of water.

The density of the air $= (29/22.4)(273/298) = 1.186$ kg/m^3.

$$h = (18.5 \times 10^{-3} \times 1000/1.186) = 15.6 \text{ m of air}$$

and: $\quad u = \sqrt{(2 \times 9.81 \times 15.6)} = 17.49$ m/s

In the same way, the velocity distribution across the tube may be found as shown in the following table.

Mass flowrate, $G = (1.107 \times 1.186) = 1.313$ kg/s

For the orifice, $[1 - (A_0/A_1)^2] = [1 - (0.15/0.3)^2] = 0.938$

$\quad h = 50$ mm Hg-under-water

$\quad\quad = (0.05 \times (13.55 - 1) \times 1000/1.186) = 529 \text{ m of air}$

and: $\quad 1.313 = C_D(\pi/4)(0.15)^2 \times 1.186\sqrt{(2 \times 9.81 \times 529/0.938)}$ and $C_D = \underline{0.61}$

Position	Distance from axis of duct (mm)	Manometer reading		Air velocity (u m/s)	Velocity × area of annulus (m³/s)
		Water (mm)	Air (m)		
1	24	18.5	15.6	17.5	0.124
2	57	18.0	15.17	17.3	0.122
3	75	17.5	14.75	17.0	0.120
4	89	16.8	14.16	16.7	0.118
5	101	15.7	13.23	16.1	0.114
6	111	14.7	12.39	15.6	0.110
7	121	13.7	11.55	15.1	0.107
8	130	12.7	10.71	14.5	0.103
9	139	11.4	9.61	13.7	0.097
10	147	10.2	8.60	13.0	0.092
				Total =	1.107

The velocity profile across the duct is plotted in Fig. 6d.

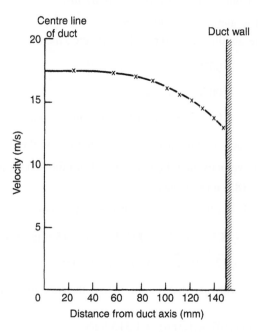

Figure 6d.

PROBLEM 6.17

Explain the principle of operation of the pitot tube and indicate how it can be used in order to measure the total flowrate of fluid in a duct. If a pitot tube is inserted in a circular cross-section pipe in which a fluid is in streamline flow, calculate at what point in the cross-section it should be situated so as to give a direct reading representative of the mean velocity of flow of the fluid.

Solution

The principle of operation of a pitot tube is discussed in Section 6.3.1. It should be emphasised that the pitot tube measures the point velocity of a flowing fluid and not the average velocity so that in order to find the average velocity, a traverse across the duct is necessary. Treatment of typical results is illustrated in Problem 6.16. The point velocity is given by $u = \sqrt{(2gh)}$ where h is the difference of head expressed in terms of the flowing fluid.

For streamline flow, the velocity distribution is discussed in Section 3.3.4 and:

$$u_s/u_{CL} = 1 - (s^2/r^2) \qquad \text{(equation 3.32)}$$

where u_s and u_{CL} are the point velocities at a distance s from the wall and at the axis respectively and r is the radius of the pipe. The average velocity is:

$$u_{av} = u_{max}/2 \qquad \text{(equation 3.36)}$$

When $u_s = u_{av} = u_{max}/2$,

$$u_s/u_{max} = (u_{max}/2)/u_{max} = 1 - (s^2/r^2)$$

and: $\qquad 0.5 = s^2/r^2$ from which $s = \underline{\underline{0.707\ r}}$

PROBLEM 6.18

The flowrate of a fluid in a pipe is measured using a pitot tube, which gives a pressure differential equivalent to 40 mm of water when situated at the centre line of the pipe and 22.5 mm of water when midway between the axis and the wall. Show that these readings are consistent with streamline flow in the pipe.

Solution

For streamline flow in a pipe, a force balance gives:

$$-\Delta P\pi r^2 = -\mu\frac{du}{dr}2\pi rl$$

and: $\qquad -u = \dfrac{-\Delta P}{2\mu l}r$ and $-u = \dfrac{-\Delta P}{2\mu l}\dfrac{r^2}{2} + \text{constant.}$

When $r = a$ (at the wall), $u = 0$, the constant $= -\Delta Pa^2/4\mu l$

and: $\qquad u = -\dfrac{\Delta P}{4\mu l}(a^2 - r^2)$

The maximum velocity, $u_{\max} = \dfrac{-\Delta P a^2}{4 \mu l}$

and:
$$\frac{u}{u_{\max}} = 1 - \left(\frac{r}{a}\right)^2$$

When $r = a/2$, $u/u_{\max} = 0.75$.

The pitot tube is discussed in Section 6.3.1 and:

$$u = k\sqrt{h} \qquad \text{(from equation 6.10)}$$

At the centre-line, $u = u_{\max}$ and $h = 40$ mm.

∴
$$u_{\max} = K\sqrt{40} = 6.32\ K$$

At a point midway between the axis and the wall, $u = u_{1/2}$ and $h = 22.5$ mm.

∴
$$u_{1/2} = K\sqrt{22.5} = 4.74\ K$$

$$u_{1/2}/u_{\max} = (4.74\ K/632\ K) = 0.75$$

and hence the flow is <u>streamline</u>.

PROBLEM 6.19

Derive a relationship between the pressure difference recorded by a pitot tube and the velocity of flow of an incompressible fluid. A pitot tube is to be situated in a large circular duct in which fluid is in turbulent flow so that it gives a direct reading of the mean velocity in the duct. At what radius in the duct should it be located, if the radius of the duct is r?

The point velocity in the duct can be assumed to be proportional to the one-seventh power of the distance from the wall.

Solution

An energy balance for an incompressible fluid in turbulent flow is given by:

$$\Delta u^2/2 + g\Delta z + v\Delta P + F = 0 \qquad \text{(equation 2.55)}$$

Ignoring functional losses and assuming the pitot tube to be horizontal,

$$(u_2^2 - u_1^2)/2 = -v(P_2 - P_1)$$

If the fluid is brought to rest of plane 2, then:

$$-u_1^2/2 = -v(P_2 - P_1)$$

and:
$$u_1 = \sqrt{2v(P_2 - P_1)} = \underline{\sqrt{2gh}} \qquad \text{(equation 6.10)}$$

If the duct radius is r, the velocity u_y at a distance y from the wall (and s from the centreline) is given by the one-seventh power law as:

$$u_y = u_s \left(\frac{y}{r}\right)^{1/7} \qquad \text{(equation 3.59)}$$

where u_s is the velocity at the centreline.

The flow, dQ, through an annulus of thickness dy_1 distance y from the axis is:

$$dQ = 2\pi s\, dy\, u_s \left(\frac{y}{r}\right)^{1/7}$$

Multiplying and dividing through by r^2 gives:

$$dQ = 2\pi r^2 u_s \frac{s}{r} \left(\frac{y}{r}\right)^{1/7} d\left(\frac{y}{r}\right)$$

or, since $s = (r - y)$:

$$= 2\pi r^2 u_s \left(1 - \frac{y}{r}\right) \left(\frac{y}{r}\right)^{1/7} d\left(\frac{y}{r}\right)$$

The total flow is:

$$Q = 2\pi r^2 u_s \int_0^1 \left[\left(\frac{y}{r}\right)^{1/7} - \left(\frac{y}{r}\right)^{8/7}\right] d\left(\frac{y}{r}\right)$$

$$= 2\pi r^2 u_s \left[\frac{7}{8}\left(\frac{y}{r}\right)^{8/7} - \frac{7}{15}\left(\frac{y}{r}\right)^{15/7}\right]_0^1 = 0.817\pi r^2 u_s$$

The average velocity, $u_{av} = Q/\pi r^2 = 0.817 u_s$

Thus:
$$u_y = u_{av}, 0.817 u_s = u_s(y/r)^{1/7}$$

$\therefore$
$$(y/r) = 0.243 \text{ and } s/r = \underline{\underline{0.757}}$$

PROBLEM 6.20

A gas of molecular weight 44 kg/kmol, temperature 373 K and pressure 202.6 kN/m^2 is flowing in a duct. A pitot tube is located at the centre of the duct and is connected to a differential manometer containing water. If the differential reading is 38.1 mm water, what is the velocity at the centre of the duct?

The volume occupied by 1 kmol at 273 K and 101.3 kN/m^2 is 22.4 m^3.

Solution

As shown in section 6.2.5, for a pitot tube:

$$u_1^2/2 + P_1 v = u_2^2/2 + P_2 v$$

$u_2 = 0$, and hence, $u_1 = \sqrt{2(P_2 - P_1)v}$

Difference in head $= 38.1$ mm of water

$$\therefore P_2 - P_1 = ((38.1/1000) \times 1000 \times 9.81) = 373.8 \text{ N/m}^2$$

The specific volume, $v = (22.4/44)(373/273)(101.3/202.6) = 0.348$ m^3/kg

$$\therefore \quad u_1 = \sqrt{2 \times 373.8 \times 0.348} = \underline{16.1 \text{ m/s}}$$

PROBLEM 6.21

Glycerol, of density 1260 kg/m^3 and viscosity 50 mNs/m^2, is flowing through a 50 mm pipe and the flowrate is measured using an orifice meter with a 38 mm orifice. The pressure differential is 150 mm as indicated on a manometer filled with a liquid of the same density as the glycerol. There is reason to suppose that the orifice meter may have become partially blocked and that the meter is giving an erroneous reading. A check is therefore made by inserting a pitot tube at the centre of the pipe. It gives a reading of 100 mm on a water manometer. What does this suggest?

Solution

From the reading taken from the pitot tube, the velocity in the pipe, and hence the mass flowrate, can be calculated. From the orifice meter, the mass flowrate can also be calculated and compared with the accurate value.

$$\text{For the pitot tube, } u = \sqrt{2gh} \qquad \text{(equation 6.10)}$$

where $u = u_{max}$ at the pipe axis, and the head loss h is in m of the liquid flowing.

Now: $h = (100/1000) \times (1000/1260) = 0.0794$ m of glycerol

$$\therefore \quad u_{max} = \sqrt{2 \times 9.81 \times 0.0794} = 1.25 \text{ m/s}$$

Reynolds number $= (1260 \times 1.25 \times 0.05/0.05) = 1575$

$$\therefore \quad u_{av} = 0.5u_{max} = 0.63 \text{ m/s} \qquad \text{(equation 3.36)}$$

Mass flowrate $= (0.63 \times 1260 \times (\pi/4) \times 0.05^2) = 1.56$ kg/s

For the orifice meter:

$$\text{mass flowrate, } G = C_D \frac{A_0}{v} \sqrt{\frac{2v(P_1 - P_2)}{1 - (A_0/A_1)^2}} \qquad \text{(equation 6.19)}$$

$$= C_D A_0 \rho \sqrt{\frac{2gh}{1 - (d_0/d)^4}}$$

$$= (C_D \times (\pi/4) \times 0.038^2 \times 1260) \sqrt{\frac{2 \times 9.81 \times (150/1000)}{1 - (0.038/0.05)^4}} = 2.99 C_D$$

$\therefore C_D = (1.56/2.99) = 0.53$ which confirms that the <u>meter is faulty</u>.

PROBLEM 6.22

The flowrate of air in a 305 mm diameter duct is measured with a pitot tube which is used to traverse the cross-section. Readings of the differential pressure recorded on a water manometer are taken with the pitot tube at ten different positions in the cross-section. These positions are so chosen as to be the mid-points of ten concentric annuli each of the same cross-sectional area. The readings are as follows:

Position	1	2	3	4	5
Manometer reading (mm water)	18.5	18.0	17.5	16.8	15.8
Position	6	7	8	9	10
Manometer reading (mm water)	14.7	13.7	12.7	11.4	10.2

The flow is also metered using a 50 mm orifice plate across which the pressure differential is 150 mm on a mercury-under-water manometer. What is the coefficient of discharge of the orifice meter?

Solution

For a pitot tube, the velocity at any point in the duct is:

$$u = \sqrt{2gh} \qquad \text{(equation 6.10)}$$

where h is the manometer reading in m of the fluid which flows in the duct.

$$\therefore \qquad h = (\text{reading in mm water}/1000) \times (\rho_w/\rho_{\text{air}}) \text{ m}$$

The total volumetric air flowrate is given by:

$Q = (\text{area of duct} \times \text{average velocity})$

$= \text{area} \times (1/10) \sum \sqrt{2gh}$

$= (\pi/4) \times 0.305^2 \times 0.1 \times \sqrt{(\rho_w/\rho_a)/1000} \times \sum \sqrt{\text{manometer reading}}$

$= 0.00102\sqrt{(\rho_w/\rho_a)}(\sqrt{18.5} + \sqrt{18.0} + \sqrt{17.5} + \cdots \sqrt{10.2}) = 0.0394\sqrt{(\rho_w/\rho_a)}$

For the orifice meter, the volumetric flowrate is given by:

$$Q = C_D \frac{A_0}{\sqrt{1 - (A_0/A_1)^2}} \sqrt{2gh}$$

$A_0 = (\pi/4) \times 0.15^2 = 0.0177 \text{ m}^2$, $A_1 = (\pi/4) \times 0.305^2 = 0.0731 \text{ m}^2$, $h = 50$ mm Hg under water

$$= (0.05(13.6 - 1.0)/1.0) = 0.63 \text{ m of water} = 0.63(\rho_w/\rho_a) \text{ m of air.}$$

$$\therefore \qquad Q = C_D \times \frac{0.0177}{\sqrt{1 - (0.0177/0.0731)^2}} \sqrt{2 \times 9.81 \times 0.63(\rho_w/\rho_a)}$$

$$= 0.066\sqrt{(\rho_w/\rho_a)} \times C_D$$

$$\therefore \qquad 0.0394\sqrt{(\rho_w/\rho_a)} = 0.066\sqrt{(\rho_w/\rho_a)} \times C_D \text{ and } C_D = \underline{\underline{0.60}}$$

PROBLEM 6.23

The flow of liquid in a 25 mm diameter pipe is metered with an orifice meter in which the orifice has a diameter of 19 mm. The aperture becomes partially blocked with dirt from the liquid. What fraction of the area can become blocked before the error in flowrate at a given pressure differential exceeds 15 per cent? Assume that the coefficient of discharge of the meter remains constant when calculated on the basis of the actual free area of the orifice.

Solution

If two sections in the pipe are chosen, 1 being upstream and 2 at the orifice, then from an energy balance:

$$u_1^2/2 + P_1 v = u_2^2/2 + P_2 v \qquad \text{(from equation 2.55)}$$

and G, the mass flowrate $= u_2 A_2/v = u_1 A_1/v$

$$\therefore \qquad (u_2^2/2)(A_2/A_1)^2 + P_1 v = u_2^2/2 + P_2 v$$

or:
$$u_2^2 = \frac{2(P_1 - P_2)v}{1 - (A_2/A_1)^2}$$

The volumetric flowrate, $Q = C_D A_2 u_2$

$$\therefore \qquad Q^2 = C_D^2 A_2^2 \times \frac{2(P_1 - P_2)v}{1 - (A_2/A_1)^2}$$

$$= 2C_D^2(P_1 - P_2)v A_1^2 A_2^2/(A_1^2 - A_2^2)$$

or:
$$Q = K \frac{A_1 A_2}{\sqrt{A_1^2 - A_2^2}} \tag{1}$$

If the area of the orifice is reduced by partial blocking, the new orifice area $= r A_2$ where f is the fraction available for flow. The new flowrate $= 0.85\,Q$ when the error is 15 per cent and:

$$0.85Q = \frac{K A_1 f A_2}{\sqrt{A_1^2 - f^2 A_2^2}} \tag{2}$$

$$A_1 = (\pi/4) \times 25^2 = 491 \text{ mm}^2$$

$$A_2 = (\pi/4) \times 19^2 = 284 \text{ mm}^2$$

$\therefore$ Dividing equation (2) by equation (1) and substituting gives:

$$0.85 = \frac{f\sqrt{(491^2 - 284^2)}}{\sqrt{(491^2 - r^2 284^2)}}$$

from which $f = 0.89$ or 11 per cent of the area is blocked.

PROBLEM 6.24

Water is flowing through a 100 mm diameter pipe and its flowrate is metered by means of a 50 mm diameter orifice across which the pressure drop is 13.8 kN/m^2. A second stream, flowing through a 75 mm diameter pipe, is also metered using a 50 mm diameter orifice across which the pressure differential is 150 mm measured on a mercury-under-water manometer. The two streams join and flow through a 150 mm diameter pipe. What would you expect the reading to be on a mercury-under-water manometer connected across a 75 mm diameter orifice plate inserted in this pipe? The coefficients of discharge for all the orifice meters are equal. Density of mercury = 13600 kg/m^3.

Solution

As in Problem 6.23:

$$u_2^2 = \frac{2(P_1 - P_2)v}{1 - (A_1/A_2)^2} \quad \text{and} \quad Q = C_D A_2 \frac{\sqrt{2(P_1 - P_2)v}}{1 - (A_2/A_1)^2}$$

For pipe 1, $A_2 = (\pi/4) \times 0.05^2 = 0.00196$ m^2, $A_1 = (\pi/4) \times 0.10^2 = 0.00785$ m^2,

$(P_1 - P_2) = 13,800$ N/m^2

$$v = (1/1000) = 0.001 \text{ m}^3/\text{kg}$$

$\therefore \qquad Q_1 = C_D \times 0.00196 \sqrt{\dfrac{2 \times 13,800 \times 0.001}{1 - (0.00196/0.00789)^2}} = 0.011 C_D$

For pipe 2, $\qquad\qquad \sqrt{2(P_1 - P_2)v} = \sqrt{2gh} \qquad\qquad$ (equation 6.10)

$$A_2 = 0.00196 \text{ m}^2, A_1 = (\pi/4) \times 0.075^2 = 0.0044 \text{ m}^2$$

Head loss, $h = 150$ mm Hg-under-water or $(150/1000) \times [(13,600 - 1000)/1000]$ = 1.89 m water.

$\therefore \qquad Q_2 = C_D \times 0.00196 \sqrt{\dfrac{2 \times 9.81 \times 1.89}{1 - (0.00196/0.0044)^2}} = 0.0133 C_D$

Total flow in pipe 3, $Q_3 = (Q_1 + Q_2) = (0.011 C_D + 0.0133 C_D) = 0.0243 C_D$

For pipe 3, $A_2 = (\pi/4) \times 0.075^2 = 0.0044$ m^2, $A_1 = (\pi/4) \times 0.15^2 = 0.0176$ m^2

and: $\qquad Q_3 = C_D \times 0.0044 \sqrt{\dfrac{2 \times 9.81 \times h}{(1 - (0.0044/0.0176)^2)}} = 0.020 C_D \sqrt{h}$

$\therefore \qquad\qquad 0.0243 C_D = 0.020 C_D \sqrt{h}$

and: $\qquad\qquad h = 1.476$ m of water

or $\qquad (1.476 \times 1000)/((13600 - 1000)/1000) = \underline{\underline{117 \text{ mm of Hg-under-water.}}}$

PROBLEM 6.25

Water is flowing through a 150 mm diameter pipe and its flowrate is measured by means of a 50 mm diameter orifice, across which the pressure differential is 2.27×10^4 N/m^2. The coefficient of discharge of the orifice meter is independently checked by means of a pitot tube which, when situated at the axis of the pipe, gave a reading of 100 mm on a mercury-under-water manometer. On the assumption that the flow in the pipe is turbulent and that the velocity distribution over the cross-section is given by the Prandtl one-seventh power law, calculate the coefficient of discharge of the orifice meter.

Solution

For the pitot tube:

$$u = \sqrt{2gh} \qquad \text{(equation 6.10)}$$

where h is the manometer reading in m of the same fluid which flows in the pipe.

$\therefore \qquad h = (100/1000) \times ((13.6 - 1.0)/1.0) = 1.26$ m of water

The velocity at the pipe axis, $u = \sqrt{(2 \times 9.81 \times 1.26)} = 4.97$ m/s

For turbulent flow, the Prandtl one-seventh power law can be used to give:

$$u_{av} = 0.82 \times u_{axis} \qquad \text{(equation 3.60)}$$

$\therefore \qquad u_{av} = 0.82 \times 4.97 = 4.08$ m/s

For the orifice meter, the average velocity is:

$$u = \sqrt{\frac{2(P_1 - P_2)v}{1 - (A_2/A_1)^2}}$$

$A_2 = (\pi/4) \times 0.05^2 = 0.00196$ m^2, $A_1 = (\pi/4) \times 0.15^2 = 0.0177$ m^2,

$$v = 0.001 \text{ m}^3/\text{kg}$$

$\therefore \qquad u_{av} = 6.78$ m/s

The coefficient of discharge $= (u_{av}$ from pitot$)/(u_{av}$ from orifice meter$)$.

$$= (4.08/6.78) = \underline{\underline{0.60}}$$

PROBLEM 6.26

Air at 323 K and 152 kN/m^2 flows through a duct of circular cross-section, diameter 0.5 m. In order to measure the flowrate of air, the velocity profile across a diameter of the duct is measured using a pitot-static tube connected to a water manometer inclined at an angle of $\cos^{-1} 0.1$ to the vertical. The following results are obtained:

Distance from duct centre-line (m)	Manometer Reading h_m (mm)
0	104
0.05	100
0.10	96
0.15	86
0.175	79
0.20	68
0.225	50

Calculate the mass flowrate of air through the duct, the average velocity, the ratio of the average to the maximum velocity and the Reynolds number. Comment on these results.

Discuss the application of this method of measuring gas flowrates, with particular emphasis on the best distribution of experimental points across the duct and on the accuracy of the results.

Take the viscosity of air as 1.9×10^{-2} mN s/m^2 and the molecular weight of air as 29 kg/kmol.

Solution

If h_m is the manometer reading, the vertical manometer height will be $0.1h_m$ (mm of water).

For a pitot tube, the velocity at any point is:

$$u = \sqrt{2gh} \qquad \text{(equation 6.10)}$$

where h is the manometer reading in terms of the fluid flowing in the duct.

Thus: $h = (0.1h_m/1000) \times (\rho_w/\rho_{air})$

$\rho_{air} = (29/22.4)(152/101.3)(273/323) = 1.64$ kg/m^3

$\therefore$ $h = (0.1h_m/1000)(1000/1.64) = 0.061h_m$

and: $u = \sqrt{2 \times 9.81 \times 0.061h_m} = 1.09\sqrt{h_m}$ (m/s)

If the duct is divided into a series of elements with the measured radius at the centre-line of the element, the velocity of the element can be found from the previous equation and the volumetric flowrate calculated. By adopting this procedure across the whole section, the required values may be determined.

For example, at 0.05 m, where $h_m = 10$ mm,
Inner radius of element = 0.025 m
Outer radius of element = 0.075 m

Area of element $= \pi(0.075^2 - 0.025^2) = 0.0157$ m^2

$\therefore$ $u = (1.09\sqrt{h_m}) = 1.09\sqrt{100} = 10.9$ m/s

Volumetric flowrate in the element $= (10.9 \times 0.0157) = 0.171$ m^3/s

The following table is constructed in the same way.

Distance from duct centre line (m)	Outer radius of element (m)	Inner radius of element (m)	Area of element $(m)^2$	h_m (mm)	Velocity u (m/s)	Volumetric flowrate ΔQ m^3/s
0	0.025	0	0.00196	104	11.1	0.0218
0.05	0.075	0.025	0.0157	100	10.9	0.171
0.10	0.125	0.075	0.0314	96	10.7	0.336
0.15	0.1625	0.125	0.0339	86	10.1	0.342
0.175	0.1875	0.1625	0.0275	79	9.7	0.293
0.20	0.2125	0.1875	0.0314	68	8.9	0.279
0.225	0.2375	0.2125	0.0353	50	7.7	0.272
0.25	0.25	0.2375	0.0192	0	0	0
			$\sum = 0.1964 \ m^2$			$\sum = 1.715 \ m^3/s$

Average velocity $= (1.715/0.1964) = \underline{\underline{8.73 \ m/s}}$

Mass flowrate $= (1.715 \times 1.64) = \underline{\underline{2.81 \ kg/s}}$

$u_{av}/u_{max} = (8.73/11.1) = \underline{\underline{0.79}}$

$Re = (8.73 \times 1.64 \times 0.05)/(1.9 \times 10^{-5}) = \underline{\underline{3.77 \times 10^4}}$

The velocity distribution in turbulent flow is discussed in Section 3.3.6 where the Prandtl one-seventh power law is used to give:

$$u_{av} = 0.82u_{max} \qquad \text{(equation 3.63)}$$

This is close to that measured in this duct though strictly it only appears at very high values of Re. Reference to Fig. 3.14 shows that, at $Re = 3.8 \times 10^4$, the velocity ratio is about 0.80 which shows remarkably good agreement.

Liquid Mixing

PROBLEM 7.1

A reaction is to be carried out in an agitated vessel. Pilot plant experiments were performed under fully turbulent conditions in a tank 0.6 m in diameter, fitted with baffles and provided with a flat-bladed turbine. It was found that satisfactory mixing was obtained at a rotor speed of 4 Hz, when the power consumption was 0.15 kW and the Reynolds number was 160,000. What should be the rotor speed in order to retain the same mixing performance if the linear scale of the equipment is increased 6 times? What will be the power consumption and the Reynolds number?

Solution

See Volume 1, Example 7.3.

PROBLEM 7.2

A three-bladed propeller is used to mix a fluid in the laminar region. The stirrer is 0.3 m in diameter and is rotated at 1.5 Hz. Due to corrosion, the propeller has to be replaced by a flat two-bladed paddle, 0.75 m in diameter. If the same motor is used, at what speed should the paddle rotate?

Solution

For mixing in the laminar region, the power requirement is:

$$\mathbf{P} = k'N^2D^3 \qquad \text{(equation 7.17)}$$

where $k' = 1964$ for a propeller and 1748 for a flat paddle.
Thus, for *a propeller* 0.3 m in diameter rotating at 1.5 Hz:

$$\mathbf{P} = (1964 \times 1.5^2 \times 0.3^2) = \underline{119.3 \text{ W}}$$

and for *a paddle*, 0.75 m in diameter using the same motor:

$$119.3 = (1748N^2 \times 0.75^3) \text{ and } N = \underline{0.403 \text{ Hz}} \text{ (24 rpm)}$$

PROBLEM 7.3

Compare the capital and operating costs of a three-bladed propeller with those of a constant speed six-bladed turbine, both constructed from mild steel. The impeller diameters are 0.3 and 0.45 m respectively and both stirrers are driven by a 1 kW motor. What is the recommended speed of rotation in each case? Assume operation for 8000 h/year, power costs of £0.01/kWh and interest and depreciation at 15%/year.

Solution

The capital cost of an impeller, $C = F_M C_B \mathbf{P}^n$

where F_M is a factor for the material of construction which for mild steel $= 1.0$, C_B is a base cost (£), $\mathbf{P}$ is the power (kW) and n is an index $(-)$.

For the propeller: $C_B = 960£(1990)$ and $n = 0.34$

$$\therefore \qquad C = (1.0 \times 960 \times 1^{0.34}) = £960$$

$$\left.\begin{array}{l}\text{Interest and depreciation} = (960 \times 15/100) = £144/\text{year} \\ \text{Operating costs} = (1 \times 0.01 \times 8000) = £80/\text{year}\end{array}\right\} \text{ a total of } \underline{\underline{£224/\text{year}}}$$

For the turbine: $C_B = 3160£(1990)$ and $n = 0.10$

$$\therefore \qquad C = (1.0 \times 3160 \times 1^{0.10}) = £3160$$

$$\left.\begin{array}{l}\text{Interest and depreciation} = (3160 \times 15/100) = £474/\text{year} \\ \text{Operating costs} = (1 \times 0.01 \times 8000) = £80/\text{year}\end{array}\right\} \text{ a total of } \underline{\underline{£554/\text{year.}}}$$

In equation 7.13, $k' = 165$ for a propeller and 3245 for a turbine.

For the propeller: $\mathbf{P} = 165N^3 D^5$

or: $1000 = 165N^3 0.3^5$ and $N = \underline{\underline{13.5 \text{ Hz}}}$ (810 rpm)

For the turbine: $\mathbf{P} = 3245N^2 D^5$

or: $1000 = (3245N^3 0.45^5)$ and $N = \underline{\underline{2.54 \text{ Hz}}}$ (152 rpm)

PROBLEM 7.4

In a leaching operation, the rate at which solute goes into solution is given by:

$$dM/dt = k(c_s - c) \text{ kg/s}$$

where M kg is the amount of solute dissolving in t s, k (m³/s) is a constant and c_s and c are the saturation and bulk concentrations of the solute respectively in kg/m³. In a pilot test on a vessel 1 m³ in volume, 75% saturation was attained in 10 s. If 300 kg of a solid containing 28% by mass of a water soluble solid is agitated with 100 m^3 of water, how long will it take for all the solute to dissolve assuming conditions are the same as in the pilot unit? Water is saturated with the solute at a concentration of 2.5 kg/m^3.

Solution

The mass of solute M, dissolving in t s is:

$$dM/dt = k(c_s - c) \text{ kg/s} \tag{i}$$

For a batch of solution, $V \text{ m}^3$ in volume: $dM = V dc$

and substituting for dM in (i): $V dc/dt = k(c_s - c)/V$

Integrating: $$\ln((c_s - c_0)/(c_s - c)) = kt/V \tag{ii}$$

where c_0 is the concentration of the solute when $t = 0$.
For pure water, $c_0 = 0 \text{ kg/m}^3$ when $t = 0$ and hence equation (ii) becomes:

$$c = c_s(1 - e^{-kt/V}) \text{ kg/m}^3 \tag{iii}$$

For the pilot test, the batch volume, $V = 1 \text{ m}^3$, and $c_s = 2.5 \text{ kg/m}^3$ at saturation. When $t = 10$ s, 75% saturation is achieved or:

$$c = (2.5 \times 75/100) = 1.875 \text{ kg/m}^3$$

Therefore, in equation (iii):

$$1.875 = 2.5(1 - e^{-10k/1}) \text{ and } k = 0.138$$

For the full-scale unit, the batch volume, $V = 100 \text{ m}^3$. Mass of solute present $= (300 \times 28/100) = 84$ kg and $c = (84/100) = 0.84 \text{ kg/m}^3$.
  Therefore, in equation (iii): $0.84 = 2.5(1 - e^{-0.138t/100})$ and $t = \underline{\underline{297 \text{ s}}}$

PROBLEM 7.5

For producing an oil-water emulsion, two portable three-bladed propeller mixers are available; a 0.5 m diameter impeller rotating at 1 Hz and a 0.35 m impeller rotating at 2 Hz. Assuming turbulent conditions prevail, which unit will have the lower power consumption?

Solution

Under turbulent conditions, the power requirements for mixing are given by:

$$\mathbf{P} = kN^3D^5 \qquad \text{(equation 7.13)}$$

In this case: $\mathbf{P}_1 = (k1^3 \times 0.5^5) = 0.03125k$ and $\mathbf{P}_2 = (k2^3 \times 0.35^5) = 0.0420k$

$\therefore$ $$\mathbf{P}_1/\mathbf{P}_2 = (0.03125k/0.0420k) = 0.743$$

Thus the 0.5 m diameter impeller will have the lower power consumption; some 75% of that of the 0.35 m diameter impeller.

PROBLEM 7.6

A reaction is to be carried out in an agitated vessel. Pilot-plant experiments were performed under fully turbulent conditions in a tank 0.6 m in diameter, fitted with baffles and provided with a flat-bladed turbine. It was found that satisfactory mixing was obtained at a rotor speed of 4 Hz, when the power consumption was 0.15 kW and the Reynolds number 160,000. What should be the rotor speed in order to retain the same mixing performance if the linear scale of the equipment is increased 6 times? What will be the power consumption and the Reynolds number?

Solution

See Volume 1, Example 7.3.

PROBLEM 7.7

Tests on a small scale tank 0.3 m diameter (Rushton impeller, diameter 0.1 m) have shown that a blending process between two miscible liquids (aqueous solutions, properties approximately the same as water, i.e. viscosity 1 mN s/m^2, density 1000 kg/m^3) is satisfactorily completed after 1 minute using an impeller speed of 250 rev/min. It is decided to scale up the process to a tank of 2.5 m diameter using the criterion of constant tip-speed.

(a) What speed should be chosen for the larger impeller?
(b) What power will be required?
(c) What will be the blend time in the large tank?

Solution

a) In the small scale tank, the 0.1 m diameter impeller is rotated at 250 rev/min or:

$$(250/60) = 4.17 \text{ Hz.}$$

The tip speed is then: $\pi DN = (\pi \times 0.1 \times 4.17) = 1.31$ m/s
If this is the same in the large scale tank, where $D = (2.5/3) = 0.83$ m, then:

$$1.31 = (\pi \times 0.83 \times N)$$

from which the speed of rotation to the larger impeller,

$$N = \underline{\underline{0.346 \text{ Hz}}} \text{ or } \underline{\underline{20.8 \text{ rev/min}}}$$

b) In the large scale tank: $N = 0.346$ Hz, $D = 0.83$ m, $\rho = 1000$ kg/m^3 and $\mu = 1 \times 10^{-3}$ Ns/m^2.

Thus, $Re = D^2 N \rho / \mu = (0.83^2 \times 0.346 \times 1000)/(1 \times 10^{-3}) = 238,360.$

From Fig. 7.6, for a propeller mixer, the Power number, $N_p = 0.6$.

Thus: $0.6 = \mathbf{P}/\rho N^3 D^5$

and:
$$\mathbf{P} = 0.6\rho N^3 D^5 = (0.6 \times 1000 \times 0.346^3 \times 0.83^5)$$
$$= \underline{\underline{9.8 \text{ W}}}$$

c) In the smaller tank: $Re = D^2 N\rho/\mu = (0.1^2 \times 4.17 \times 1000)/(1 \times 10^{-3}) = 41700$
In Equation 7.22, the dimensionless mixing time is:

$$\theta_m = Nt_m = kRe$$

or for $t_m = 60s$: $(4.17 \times 10) = k \times 41700$

and: $$k = 0.0060$$

Thus in the larger tank: $Nt_m = 0.0060 \ Re$

or: $$0.346 \ t_m = (0.0060 \times 238,360)$$

and: $$t_m = 4140 \text{ s or } \underline{\underline{1.15 \text{ min}}}$$

PROBLEM 7.8

An agitated tank with a standard Rushton impeller is required to disperse gas in a solution of properties similar to those of water. The tank will be 3 m diameter (1 m diameter impeller). A power level of 0.8 kW/m³ is chosen. Assuming fully turbulent conditions and that the presence of the gas does not significantly affect the relation between the Power and Reynolds numbers:

(a) What power will be required by the impeller?
(b) At what speed should the impeller be driven?
(c) If a small pilot scale tank 0.3 m diameter is to be constructed to test the process, at what speed should the impeller be driven?

Solution

(a) Assuming that the depth of liquid = tank diameter, then:
$$\text{volume of liquid} = (\pi D^2/4) \ H$$
$$= (\pi \times 3^2 \times 3)/4 = 21.2 \text{ m}^3$$

With a power input of 0.8 kW/m³, the power required be the impeller is:
$$\mathbf{P} = (0.8 \times 21.2) = \underline{\underline{17.0 \text{ kW}}}$$

(b) For fully turbulent conditions and $\mu = 1$ mN s/m², and the power number, from Fig. 7.6 is approximately 0.7. On this basis:
$$\mathbf{P}/\rho N^3 D^5 = 0.7$$

or: $$(17.0 \times 10^3)/(1000N^3 \times 1^5) = 0.7$$

from which: $$N = \underline{\underline{2.90 \text{ Hz}}} \text{ or } \underline{\underline{173 \text{ rev/min}}}$$

(c) For the large tank, from Equation 7.13:

$$P = kN^3D^5$$

or: $(17.0 \times 10^3) = k \times 2.90^3 \times 1^5$

from which: $k = 697$

Thus, for the smaller tank, assuming power/unit volume is constant:

$$\text{volume of fluid} = (\pi/4)0.3^2 \times 0.3 = 0.021 \text{ m}^3$$

and: power supplied, $P = (0.021 \times 0.8 \times 10^3)$

$$= 17 \text{ W}$$

Thus, for the smaller tank:

$$17 = 697 \, N^3 \times 0.1^5$$

and: $N = \underline{13.5 \text{ Hz}}$ or $\underline{807 \text{ rev/min}}$.

SECTION 8

Pumping of Fluids

PROBLEM 8.1

A three-stage compressor is required to compress air from 140 kN/m^2 and 283 K to 4000 kN/m^2. Calculate the ideal intermediate pressures, the work required per kilogram of gas, and the isothermal efficiency of the process. It may be assumed that the compression is adiabatic and interstage cooling is provided to cool the air to the initial temperature. Show qualitatively, by means of temperature–entropy diagrams, the effect of unequal work distribution and imperfect intercooling, on the performance of the compressor.

Solution

It is shown in Section 8.3.4 that the work done is a minimum when the intermediate pressures P_{i1} and P_{i2} are related to the initial and final pressures P_1 and P_2 by:

$$P_{i1}/P_1 = P_{i2}/P_{i1} = P_2/P_{i2} \quad \text{(equation 8.45)}$$

$P_1 = 140$ kN/m^2 and $P_2 = 4000$ kN/m^2.

$$\therefore \qquad P_2/P_1 = 28.57$$

$$\therefore \qquad P_{i2}/P_{i1} = P_2/P_{i2} = \sqrt[3]{28.57} = 3.057,$$

$$P_{i1} = \underline{\underline{428 \text{ kN/m}^2}},$$

and: $\qquad P_{i2} = \underline{\underline{1308 \text{ kN/m}^2}}$

The specific volume of the air at the inlet is:

$$v_1 = (22.4/29)(283/273)(101.3/140) = 0.579 \text{ m}^3/\text{kg}$$

Hence, for 1 kg of air, the minimum work of compression in a compressor of n stages is:

$$W = nP_1v_1\left(\frac{\gamma}{\gamma - 1}\right)\left[\left(\frac{P_2}{P_1}\right)^{(\gamma-1)/n\gamma} - 1\right] \quad \text{(equation 8.46)}$$

Thus: $W = (3 \times 140{,}000 \times 0.579)(1.4/0.4)[(28.57)^{0.4/3 \times 1.4} - 1] = \underline{\underline{319{,}170 \text{ J/kg}}}$

The isothermal work of compression is:

$$W_{\text{iso}} = P_1V_1 \ln(P_2/P_1) \quad \text{(equation 8.36)}$$

$$= (140{,}000 \times 0.579 \ln 28.57) = 271{,}740 \text{ J/kg}$$

The isothermal efficiency $= (100 \times 271{,}740)/319{,}170 = \underline{85.1\%}$

Compression cycles are shown in Figs 8a and 8b. The former indicates the effect of various values of n in PV^n = constant and it is seen that the work done is the area under the temperature–entropy curve. Figure 8b illustrates the three-stage compressor of this problem. The final temperature T_2, found from $T_2/T_1 = (P_2/P_1)^{(\gamma-1)/\gamma}$, is 390 K. The dotted lines illustrate the effect of imperfect interstage cooling.

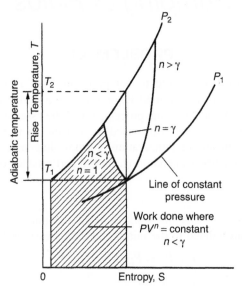

Figure 8a.

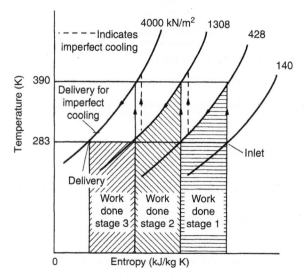

Figure 8b.

PROBLEM 8.2

A twin-cylinder, single-acting compressor, working at 5 Hz, delivers air at 515 kN/m² at the rate of 0.2 m³/s. If the diameter of the cylinder is 20 cm, the cylinder clearance ratio 5%, and the temperature of the inlet air 283 K, calculate the length of stroke of the piston and the delivery temperature.

Solution

For adiabatic conditions, $PV^\gamma = \text{constant}$

and: $P_2/P_1 = (T_2/T_1)^{\gamma/(\gamma-1)}$ or $T_2 = T_1(P_2/P_1)^{(\gamma-1)/\gamma}$

Thus the delivery temperature $= 283(515/101.3)^{0.4/1.4} = \underline{500\ K}$

The volume handled per cylinder $= (0.2/2) = 0.1\ \text{m}^3/\text{s}$.

Volume per stroke $= (0.1/5) = 0.02\ \text{m}^3/\text{s}$ at 515 kN/m².

Volume at the inlet condition $= (0.02 \times 283)/500 = 0.0126\ \text{m}^3/\text{s}$.
From equation 8.42, $0.0126 = V_s[1 + c - c(P_2/P_1)^{1/\gamma}]$ where c is the clearance and V_s the swept volume.

Thus: $0.0126 = V_s[1 \times 0.05 - 0.05(515/101.3)^{1/1.4}]$ and $V_s = 0.0142\ \text{m}^3$

∴ $(\pi/4)(0.2)^2 \times \text{stroke} = 0.0142$ and the stroke $= \underline{0.45\ m}$

PROBLEM 8.3

A single-stage double-acting compressor running at 3 Hz is used to compress air from 110 kN/m² and 282 K to 1150 kN/m². If the internal diameter of the cylinder is 20 cm, the length of stroke 25 cm, and the piston clearance 5%, calculate:

(a) the maximum capacity of the machine, referred to air at the initial temperature and pressure, and
(b) the theoretical power requirements under isentropic conditions.

Solution

The volume per stroke $= (2 \times (\pi/4)(0.2)^2 \times 0.25) = 0.0157\ \text{m}^3$

The compression ratio $= (1150/110) = 10.45$.

The swept volume V_s is given by:

$0.0157 = V_s[1 + 0.05 - 0.05(10.45)^{1/1.4}]$ and $V_s = 0.0217\ \text{m}^3$ (equation 8.42)

The work of compression/cycle is:

$$W = P_1(V_1 - V_4)(\gamma/\gamma - 1)[(P_2/P_1)^{(\gamma-1)/\gamma} - 1]$$ (equation 8.41)

and substituting for $(V_1 - V_4)$, gives:

$$W = P_1 V_s [1 + c - c(P_2/P_1)^{1/\gamma}][\gamma/\gamma - 1][(P_2/P_1)^{(\gamma-1)/\gamma} - 1] \quad \text{(equation 8.43)}$$

$$= (110,000 \times 0.0157)(1.4/0.4)[(10.45)^{0.286} - 1] = 5781 \text{ J}$$

The theoretical power requirement $= (3 \times 5781) = 17,340$ W or $\underline{17.3 \text{ kW}}$

The capacity $= (3 \times 0.0157) = \underline{0.047 \text{ m}^3/\text{s}}$

PROBLEM 8.4

Methane is to be compressed from atmospheric pressure to 30 MN/m^2 in four stages. Calculate the ideal intermediate pressures and the work required per kilogram of gas. Assume compression to be isentropic and the gas to behave as an ideal gas. Indicate on a temperature–entropy diagram the effect of imperfect intercooling on the work done at each stage.

Solution

The ideal intermediate pressures are obtained when the compression ratios in each stage are equal. If the initial, intermediate, and final pressures from this compressor are P_1, P_2, P_3, P_4, and P_5, then:

$$P_2/P_1 = P_3/P_2 = P_4/P_3 = P_5/P_4 = P_5/P_1$$

as in problem 8.1.

$$P_5/P_1 = (30,000/101.3) = 296.2$$

and: $(P_5/P_1)^{0.25} = 4.148$

Hence: $P_2 = 4.148 P_1 = (4.148 \times 101.3) = \underline{420 \text{ kN/m}^2}$

$$P_3 = 4.148 P_2 = \underline{1.74 \text{ MN/m}^2}$$

$$P_4 = 4.148 P_3 = \underline{7.23 \text{ MN/m}^2}$$

The work required per kilogram of gas is:

$$W = n P_1 V_1 \frac{\gamma}{\gamma - 1} \left[\left(\frac{P_5}{P_1} \right)^{(\gamma-1)/n\gamma} - 1 \right] \quad \text{(equation 8.46)}$$

For methane, the molecular mass $= 16$ kg/kmol and the specific volume at STP $= (22.4/16) = 1.40$ m^3/kg.

If $\gamma = 1.4$, the work per kilogram is:

$$W = (4 \times 101,300 \times 1.40)(1.4/0.4)[(296.2)^{0.4/(4 \times 1.4)} - 1]$$

$$= 710,940 \text{ J/kg or } \underline{711 \text{ kJ/kg}}$$

The effect of imperfect cooling is shown in Figs 8a and 8b.

PROBLEM 8.5

An air-lift raises 0.01 m^3/s of water from a well 100 m deep through a 100 mm diameter pipe. The level of the water is 40 m below the surface. The air consumed is 0.1 m^3/s of free air compressed to 800 kN/m^2. Calculate the efficiency of the pump and the mean velocity of the mixture in the pipe.

Solution

See Volume 1, Example 8.6

PROBLEM 8.6

In a single-stage compressor: Suction pressure $= 101.3$ kN/m^2. Suction temperature $= 283$ K. Final pressure $= 380$ kN/m^2. If each new charge is heated 18 deg K by contact with the clearance gases, calculate the maximum temperature attained in the cylinder.

Solution

The compression ratio $= (380/101.3) = 3.75$.

On the first stroke, the air enters at 283 K and is compressed adiabatically to 380 kN/m^2.

Thus:
$$T_2/T_1 = (P_2/P_1)^{(\gamma-1)/\gamma}$$
$$= 3.75^{0.286} = 1.459$$

Hence, the exit temperature is: $T_2 = (1.459 \times 283) = 413$ K

The clearance volume gases which remain in the cylinder are able to raise the temperature of the next cylinder full of air by 18 deg K leaving the cylinder and its contents at $(283 + 18) = 301$ K. After compression, the exit temperature is:

$$T = (301 \times 3.75^{0.286}) = 439.2 \text{ K}$$

On each subsequent stroke, the inlet temperature is always 301 K and hence the maximum temperature attained is 439.2 K.

PROBLEM 8.7

A single-acting reciprocating pump has a cylinder diameter of 115 mm and a stroke of 230 mm. The suction line is 6 m long and 50 mm diameter and the level of the water in the suction tank is 3 m below the cylinder of the pump. What is the maximum speed at which the pump can run without an air vessel if separation is not to occur in the suction line? The piston undergoes approximately simple harmonic motion. Atmospheric pressure is equivalent to a head of 10.4 m of water and separation occurs at pressure corresponding to a head of 1.22 m of water.

Solution

The tendency for separation to occur will be greatest at the inlet to the cylinder and at the beginning of the suction stroke.

If the maximum speed of the pump is N Hz, the angular velocity of the driving mechanism is $2\pi N$ radian/s.

The acceleration of the piston $= (0.5 \times 0.23)(\pi N)^2 \cos(2\pi N)$ m/s^2.

The maximum acceleration, when $t = 0$, is $4.54 \, N^2$ m/s^2.

Maximum acceleration of the liquid in the suction pipe is:

$$(0.115/0.05)^2 (4.54 N^2) = 24.02 \, N^2 \text{ m/s}$$

Accelerating force on the liquid $= (24.02 N^2 (\pi/2)(0.05)^2 \times 6 \times 1000)$.
Pressure drop in suction line due to acceleration

$$= (24.02 N^2 \times 6 \times 1000) = 1.44 \times 10^5 \, N^2 \text{ N/m}^2$$

$$= (1.44 \times 10^5 \, N^2 / 1000 \times 9.81) = 14.69 \, N^2 \text{ m of water}$$

Pressure head at the cylinder when separation is about to occur:

$$1.22 = (10.4 - 3.0 - 14.69 \, N^2) \text{ m of water and} : N = \underline{\underline{0.65 \text{ Hz}}}$$

PROBLEM 8.8

An air-lift pump is used for raising 0.8 l/s of a liquid of density 1200 kg/m^3 to a height of 20 m. Air is available at 450 kN/m^2. If the efficiency of the pump is 30%, calculate the power requirement, assuming isentropic compression of the air ($\gamma = 1.4$).

Solution

Volume flow of liquid $= 800$ cm^3/s or 800×10^{-6} m^3/s

Mass of flowrate of liquid $= (800 \times 10^{-6} \times 1200 = 0.96$ kg/s

Work done per second $= (0.96 \times 20 \times 9.81) = 188.4$ W

Actual work of expansion of air $= (188.4/0.3) = 627.8$ W.

The mass of air required per unit time is:

$$W = P_a v_a m \ln(P/P_a) = P_a V_a \ln(P/P_a) \qquad \text{(equation 8.49)}$$

where V_a is the volume of air at STP,

Thus: $627.8 = 101{,}300 V_a \ln(450/101.3)$ and $V_a = 0.0042$ m^3

The work done in the isentropic compression of this air is:

$$P_1 V_1 [\gamma/(\gamma - 1)][(P_2/P_1)^{(\gamma-1)/\gamma} - 1] \qquad \text{(equation 8.37)}$$

$$= (101{,}300 \times 0.0042)(1.4/0.4)[(450/101.3)^{0.286} - 1] = 792 \text{ J}$$

Power required $= 792$ J/s $= 792$ W or $\underline{\underline{0.79 \text{ kW}}}$.

PROBLEM 8.9

A single-acting air compressor supplies 0.1 m³/s of air (at STP) compressed to 380 kN/m²
from 101.3 kN/m² pressure. If the suction temperature is 288.5 K, the stroke is 250 mm,
and the speed is 4 Hz, find the cylinder diameter. Assume the cylinder clearance is 4%
and compression and re-expansion are isentropic ($\gamma = 1.4$). What is the theoretical power
required for the compression?

Solution

See Volume 1, Example 8.3.

PROBLEM 8.10

Air at 290 K is compressed from 101.3 to 2000 kN/m² pressure in a two-stage compressor
operating with a mechanical efficiency of 85%. The relation between pressure and volume
during the compression stroke and expansion of the clearance gas is $PV^{1.25}$ = constant.
The compression ratio in each of the two cylinders is the same and the interstage cooler
may be taken as perfectly efficient. If the clearances in the two cylinders are 4% and 5%
respectively, calculate:

- (a) the work of compression per unit mass of gas compressed;
- (b) the isothermal efficiency;
- (c) the isentropic efficiency ($\gamma = 1.4$);
- (d) the ratio of the swept volumes in the two cylinders.

Solution

See Volume 1, Example 8.4.

PROBLEM 8.11

Explain briefly the significance of the "specific speed" of a centrifugal or axial-flow pump.
A pump is designed to be driven at 10 Hz and to operate at a maximum efficiency when
delivering 0.4 m³/s of water against a head of 20 m. Calculate the specific speed. What
type of pump does this value suggest? A pump built for these operating conditions has
a measured overall efficiency of 70%. The same pump is now required to deliver water
at 30 m head. At what speed should the pump be driven if it is to operate at maximum
efficiency? What will be the new rate of delivery and the power required?

Solution

Specific speed is discussed in Section 8.2.3 of Volume 1, where it is shown to be
$N_s = NQ^{1/2}/(gh)^{3/4}$. This expression is dimensionless providing that the pump speed,
throughput, and head are expressed in consistent units.

In this problem, $N = 10$ Hz, $Q = 0.4$ m^3/s, and $h = 20$ m.

Thus:
$$N_s = (10 \times (0.4)^{0.5}/(9.81 \times 20)^{0.75}) = \underline{0.121}$$

Reference should be made to specialist texts on pumps where classifications of pump types as a function of specific speed are presented. A centrifugal pump is suggested here.

$$Q \propto N \text{ and } Q_1/Q_2 = N_1/N_2 \qquad \text{(equation 8.15)}$$

and:
$$h \propto N^2 \text{ and } h_1/h_2 = (N_1/N_2)^2 \qquad \text{(equation 8.16)}$$

Thus:
$$(20/30) = (10/N_2)^2 \text{ from which } N_2 = \underline{12.24 \text{ Hz}}$$

and:
$$0.4/Q_2 = (10/12.24) \text{ from which } Q_2 = \underline{0.49 \text{ m}^3/\text{s}}$$

Power required $= (1/n)(\text{mass flow} \times \text{head} \times g)$
$$= (1/0.7)(0.49 \times 1000 \times 30 \times 9.81) = \underline{206 \text{ W}}$$

PROBLEM 8.12

A centrifugal pump is to be used to extract water from a condenser in which the vacuum is 640 mm of mercury. At the rated discharge, the net positive suction head must be at least 3 m above the cavitation vapour pressure of 710 mm mercury vacuum. If losses in the suction pipe account for a head of 1.5 m, what must be the least height of the liquid level in the condenser above the pump inlet?

Solution

The system is illustrated in Fig. 8c. From an energy balance, the head at the suction point of the pump is:
$$h_i = (P_0/\rho h) + x - (u_i^2/2g) - h_f$$

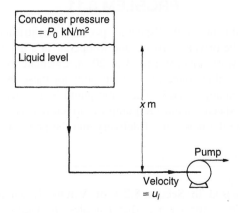

Figure 8c.

The losses in the suction pipe $= 1.5$ m, and $(u_i^2/2g) = h_f = 1.5$

The net positive suction head (NPSH) is discussed in Section 8.2.3 where it is shown that:

$$\text{NPSH} = h_i - (P_v/\rho g)$$

where P_v is the vapour pressure of the liquid being pumped. The minimum height x is then obtained from:

$$3 = (P_0/\rho g) + x - 1.5 - (P_v/\rho g)$$

$$P_0 = (760 - 640) = 120 \text{ mm Hg} = 16{,}000 \text{ N/m}^2$$

$$P_v = (760 - 710) = 50 \text{ mm Hg} = 6670 \text{ N/m}^2$$

$$\rho = 1000 \text{ kg/m}^3, g = 9.81 \text{ m/s}^2$$

$\therefore$
$$x = 3 + 1.5 - (16{,}000 + 6670)/(1000 \times 9.81) = \underline{\underline{3.55 \text{ m}}}$$

PROBLEM 8.13

What is meant by the Net Positive Suction Head (NPSH) required by a pump? Explain why it exists and how it can be made as low as possible. What happens if the necessary NPSH is not provided?

A centrifugal pump is to be used to circulate liquid of density 800 kg/m^3 and viscosity 0.5 mN s/m^2 from the reboiler of a distillation column through a vaporiser at the rate of 400 cm^3/s, and to introduce the superheated liquid above the vapour space in the reboiler which contains liquid to a depth of 0.7 m. Suggest a suitable layout if a smooth-bore 25 mm pipe is to be used. The pressure of the vapour in the reboiler is 1 kN/m^2 and the NPSH required by the pump is 2 m of liquid.

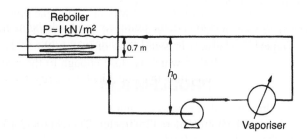

Figure 8d.

Solution

See Volume 1, Example 8.2

PROBLEM 8.14

1250 cm^3/s of water is to be pumped through a steel pipe, 25 mm diameter and 30 m long, to a tank 12 m higher than its reservoir. Calculate the approximate power required. What type of pump would you install for the purpose and what power motor (in kW) would you provide? Viscosity of water = 1.30 mN s/m^2. Density of water = 1000 kg/m^3.

Solution

For a 25 mm bore pipe, cross-sectional area = $(\pi/4)(0.025)^2 = 0.00049$ m^2.

Viscosity, $u = (1250 \times 10^{-6})/0.00049 = 2.54$ m/s.

and $Re = \rho u d/\mu = (1000 \times 2.54 \times 0.025)/(1.3 \times 10^{-3}) = 48,900$

From Table 3.1 the roughness e of a steel pipe will be taken as 0.045 mm. Hence $e/d = (0.046/25) = 0.0018$.

When $e/d = 0.0018$ and $Re = 4.89 \times 10^4$, from Fig. 3.7 $R/\rho u^2 = 0.0032$

The pressure drop is then calculated from the energy balance equation and equation 3.19. For turbulent flow of an incompressible fluid:

$$\Delta u^2/2 + g\Delta z + v(P_2 - P_1) + 4(R/\rho u^2)(l/d)u^2 = 0$$

The pressure drop is:

$$(P_1 - P_2) = \rho[\Delta u^2/2 + g\Delta z + 4(R/\rho u^2)(l/d)u^2] = \rho\{[0.5 + 4(R/\rho u^2)(l/d)]u^2 + g\Delta z\}$$

since the velocity in the tank is equal to zero.
Substituting:

$$(P_1 - P_2) = 1000\{[0.5 + 4(0.0032(30/0.025)]2.54^2 + (9.81 \times 12)\}$$

$$= 219,500 \text{ N/m}^2 \text{ or } \underline{219.5 \text{ kN/m}^2}$$

$$\text{Power} = G[(u^2/2) + g\Delta z + F] = (\text{kg/s})(\text{m}^2/\text{s}^2) = (\text{m}^2/\text{s})(\text{N/m}^2) \qquad \text{(equation 8.60)}$$

$$= (1.25 \times 10^{-3})(2.195 \times 10^5) = 275 \text{ W}$$

If a pump efficiency of 60% is assumed, the pump motor should be rated at $(275/0.6) = \underline{458 \text{ W}}$. A single stage centrifugal pump would be suitable for this duty.

PROBLEM 8.15

Calculate the pressure drop in, and the power required to operate, a condenser consisting of 400 tubes, 4.5 m long and 10 mm internal diameter. The coefficient of contraction at the entrance of the tubes is 0.6, and 0.04 m^3/s of water is to be pumped through the condenser.

Solution

Flow of water through each tube = $(0.04/400) = 0.0001$ m^3/s.

Cross-sectional area of each tube = $(\pi/4)(0.01)^2 = 0.0000785$ m^2.

Water velocity $= (0.0001/0.0000785) = 1.273$ m/s.

Entry pressure drop is:

$$\Delta P_f = -\frac{\rho u^2}{2}\left(\frac{1}{C_c} - 1\right)^2 \qquad \text{(from equation 3.78)}$$

$$= (1000 \times 1.273^2/2)[(1/0.6) - 1]^2 = 360 \text{ N/m}^2$$

$Re = \rho u d/\mu = (1000 \times 1.273 \times 0.01)/(1.0 \times 10^{-3}) = 1.273 \times 10^4$

If e is taken as 0.046 mm from Table 3.1, then $e/d = 0.0046$ and from Fig. 3.7:

$$R/\rho u^2 = 0.0043.$$

The pressure drop due to friction is:

$$\Delta P_f = 4(R/\rho u^2)(l/d)(\rho u^2) \qquad \text{(equation 3.18)}$$

$$= (4 \times 0.0043(4.5/0.01)(1000 \times 1.273^2)) = 12,540 \text{ N/m}^2$$

Total pressure drop across one tube $= (12,540 + 360) = 12,900$ N/m^2 or $\underline{12.9 \text{ kN/m}^2}$

If tubes are connected in parallel, power required $= \left(\begin{array}{c}\text{pressure} \\ \text{drop}\end{array}\right) \times \left(\begin{array}{c}\text{volumetric} \\ \text{flowrate}\end{array}\right)$

$$= (12,900 \times 0.04)$$

$$= \underline{\underline{516 \text{ W}}}$$

PROBLEM 8.16

75% sulphuric acid, of density 1650 kg/m^3 and viscosity 8.6 mN s/m^2, is to be pumped for 0.8 km along a 50 mm internal diameter pipe at the rate of 3.0 kg/s, and then raised vertically 15 m by the pump. If the pump is electrically driven and has an efficiency of 50%, what power will be required? What type of pump would you use and of what material would you construct the pump and pipe?

Solution

Cross-sectional area of pipe $= (\pi/4)(0.05)^2 = 0.00196$ m^2.

Velocity, $u = 3.0/(1650 \times 0.00196) = 0.93$ m/s.

$Re = \rho u d/\mu = (1650 \times 0.93 \times 0.05)/8.6 \times 10^{-3} = 8900$

If e is taken as 0.046 mm from Table 3.1, $e/d = 0.00092$.

From Fig. 3.7, $R/\rho u^2 = 0.0040$.

Head loss due to friction is:

$$h_f = -\Delta P_f/\rho g = 4(R/\rho u^2)(l/d)(u^2/g) \qquad \text{(equation 3.20)}$$

$$= (4 \times 0.004)(800/0.05)(0.93^2/9.81) = 22.6 \text{ m}$$

Total head $= (22.6 + 15) = 37.6$ m.

$$\text{Power} = (\text{mass flowrate} \times \text{head} \times g) \qquad \text{(equation 8.61)}$$

$$= (3.0 \times 37.6 \times 9.81) = 1105 \text{ W}$$

If the pump is 50% efficient, power required $= (1105/0.5) = 2210$ W or 2.2 kW.

For this duty a PTFE lined pump and lead piping would be suitable.

PROBLEM 8.17

60% sulphuric acid is to be pumped at the rate of 4000 cm^3/s through a lead pipe 25 mm diameter and raised to a height of 25 m. The pipe is 30 m long and includes two right-angled bends. Calculate the theoretical power required. The density of the acid is 1531 kg/m^3 and its kinematic viscosity is 4.25×10^{-5} m^2/s. The density of water may be taken as 1000 kg/m^3.

Solution

Cross-sectional area of pipe $= (\pi/4)(0.025)^2 = 0.00049$ m^2

Velocity, $u = (4000 \times 10^{-6}/0.00049) = 8.15$ m/s.

$Re = \rho u d/\mu = u d/(\mu/\rho) = (8.15 \times 0.025)/(4.25 \times 10^{-5}) = 4794$

If e is taken as 0.05 mm from Table 3.1, $e/d = 0.002$ and from Fig. 3.7, $R/\rho u^2 = 0.0047$.

Head loss due to friction is given by:

$$h_f = 4(R/\rho u^2)(l/d)(u^2/g) \qquad \text{(equation 3.20)}$$

$$= (4 \times 0.0047)(30/0.025)(8.15^2/9.81) = 152.8 \text{ m and } \Delta z = 25.0 \text{ m}$$

From Table 3.2, 0.8 velocity heads $(u^2/2g)$ are lost through each 90° bend so that the loss through two bends is 1.6 velocity heads or $(1.6 \times 8.15^2)/(2 \times 9.81) = 5.4$ m.

Total head loss $= (152.8 + 25 + 5.4) = 183.2$ m.

Mass flowrate $= (4000 \times 10^{-6} \times 1.531 \times 1000) = 6.12$ kg/s.

From equation 8.61 the theoretical power requirement $= (6.12 \times 183.2 \times 9.81) = 11,000$ W or 11.0 kW.

PROBLEM 8.18

1.3 kg/s of 98% sulphuric acid is to be pumped through a 25 mm diameter pipe, 30 m long, to a tank 12 m higher than its reservoir. Calculate the power required and indicate the type of pump and material of construction of the line that you would choose. Viscosity of acid $= 0.025$ N s/m^2. Density $= 1840$ kg/m^3.

Solution

Cross-sectional area of pipe $= (\pi/4)(0.0025)^2 = 0.00049$ m^2.

Volumetric flowrate $= 1.3/(1.84 \times 1000) = 0.00071$ m^3/s.

Velocity in the pipe, $u = (0.00071/0.00049) = 1.45$ m/s

$$Re = \rho u d/\mu = ((1.84 \times 1000) \times 1.45 \times 0.025)/0.025 = 2670$$

This value of the Reynolds number lies within the critical zone. If the flow were laminar, the value of $R/\rho u^2$ from Fig. 3.7 would be 0.003. If the flow were turbulent, the value of $R/\rho u^2$ would be considerably higher, and this higher value should be used in subsequent calculation to provide a margin of safety. If the roughness is taken as 0.05 mm, $e/d = (0.05/25) = 0.002$ and, from Fig 3.7, $R/\rho u^2 = 0.0057$.

The head loss due to friction,

$$h_f = 4(R/\rho u^2)(l/d)(u^2/g) \qquad \text{(equation 3.20)}$$

$$= (4 \times 0.0057(30/0.025)(1.45^2/9.81) = 5.87 \text{ m}$$

$\Delta z = 12$ m so that the total head $= 17.87$ m.

The theoretical power requirement, from equation 8.61, is:

$$\text{power} = (17.87 \times 1.3 \times 9.81) = 227 \text{ W}$$

If the pump is 50% efficient, actual power $= (227/0.5) = \underline{454 \text{ W}}$

A PTFE lined centrifugal pump and lead or high silicon iron pipe would be suitable for this duty.

PROBLEM 8.19

A petroleum fraction is pumped 2 km from a distillation plant to storage tanks through a mild steel pipeline, 150 mm in diameter, at the rate of 0.04 m^3/s. What is the pressure drop along the pipe and the power supplied to the pumping unit if it has an efficiency of 50%? The pump impeller is eroded and the pressure at its delivery falls to one half. By how much is the flowrate reduced? Density of the liquid $= 705$ kg/m^3. Viscosity of the liquid $= 0.5$ mN s/m^2. Roughness of pipe surface $= 0.004$ mm.

Solution

Cross-sectional area of pipe $= (\pi/4)0.15^2 = 0.0177$ m^2.

Velocity in the pipe $= (0.04/0.0177) = 2.26$ m/s.

Reynolds number $= (0.705 \times 1000 \times 2.26 \times 0.15)/(0.5 \times 10^{-3}) = 4.78 \times 10^5$

$e = 0.004$ mm, $e/d = (0.004/150) = 0.000027$ and from Fig. 3.7, $R/\rho u^2 = 0.00165$

The pressure drop is: $-\Delta P_f = 4(R/\rho u^2)(l/d)(\rho u^2)$ \qquad (equation 3.18)

$$-\Delta P_f = (4 \times 0.00165)(2000/0.15)(705 \times 2.26^2) = 316,900 \text{ N/m}^2 \text{ or } \underline{\underline{320 \text{ kN/m}^2}}$$

If the pump efficiency is 50%,

$$\text{power} = (\text{head} \times \text{mass flowrate} \times g)/0.5$$

$$= \text{pressure drop (N/m}^2) \times \text{volumetric flowrate (m}^3\text{s)}/0.5$$

$$= (316,900 \times 0.04)/0.5 = 25,350 \text{ W or } \underline{25.4 \text{ W}}$$

If, due to impeller erosion, the delivery pressure is halved, the new flowrate may be found from:

$$(R/\rho u^2)\,Re^2 = -\Delta P_f d^3 \rho/4l\mu^2 \qquad \text{(equation 3.23)}$$

The new pressure drop $= (316,900/2) = 158,450 \text{ N/m}^2$ and:

$$(R/\rho u^2)\,Re^2 = (158,450 \times 0.15^3 \times 705)/(4 \times 2000 \times 0.5^2 \times 10^{-6}) = 1.885 \times 10^8$$

From Fig. 3.8, when $(R/\rho u^2)\,Re^2 = 1.9 \times 10^8$ and $e/d = 0.000027$, $Re = 3.0 \times 10^5$

and: $(3.0 \times 10^5) = (705 \times 0.15 \times u)/(0.5 \times 10^{-3})$ and: $u = 1.418$ m/s

The volumetric flowrate is now: $(1.418 \times 0.0177) = \underline{\underline{0.025 \text{ m}^3/\text{s}}}$

PROBLEM 8.20

Calculate the power required to pump oil of density 850 kg/m³ and viscosity 3 mN s/m² at 4000 cm³/s through a 50 mm pipeline 100 m long, the outlet of which is 15 m higher than the inlet. The efficiency of the pump is 50%. What effect does the nature of the surface of the pipe have on the resistance?

Solution

Cross-sectional area of pipe $= (\pi/4)0.05^2 = 0.00196 \text{ m}^2$

Velocity of oil in the pipe $= (4000 \times 10^{-6})/0.00196 = 2.04$ m/s.

$Re = \rho u d/\mu = (0.85 \times 1000 \times 2.04 \times 0.05)/(3 \times 10^{-3}) = 2.89 \times 10^4$
If the pipe roughness e is taken to be 0.05 mm, $e/d = 0.001$, and from Fig. 3.7, $R/\rho u^2 = 0.0031$.
Head loss due to friction is:

$$h_f = 4(R/\rho u^2)(l/d)(u^2/g) \qquad \text{(equation 3.20)}$$

$$= (4 \times 0.0031)(100/0.05)(2.04^2/9.81) = 10.5 \text{ m}$$

The total head $= (10.5 + 15) = 25.5$ m

The mass flowrate $= (4000 \times 10^{-6} \times 850) = 3.4$ kg/s

Power required $= (25.5 \times 3.4 \times 9.81/0.5) = 1700$ W or $\underline{1.7 \text{ kW}}$

The roughness of the pipe affects the ratio e/d. The rougher the pipe surface, the higher will be e/d and there will be an increase in $R/\rho u^2$. This will increase the head loss due to friction and will ultimately increase the power required.

PROBLEM 8.21

600 litres/s of water at 320 K is pumped in a 40 mm i.d. pipe through a length of 150 m in a horizontal direction and up through a vertical height of 10 m. In the pipe there is a control valve which may be taken as equivalent to 200 pipe diameters and other pipe fittings equivalent to 60 pipe diameters. Also in the line there is a heat exchanger across which there is a loss in head of 1.5 m of water. If the main pipe has a roughness of 0.0002 m, what power must be delivered to the pump if the unit is 60% efficient?

Solution

Mass flowrate of water $= (600 \times 10^{-6} \times 1000) = 0.6$ kg/s.

Cross-sectional area of pipe $= (\pi/4)0.04^2 = 0.00126$ m^2.

Velocity of water in the pipe $= (600 \times 10^{-6}/0.00126) = 0.476$ m/s.

$Re = \rho u d/\mu = (1000 \times 0.476 \times 0.04)/(1 \times 10^{-3}) = 1.9 \times 10^4$.

If $e = 0.0002$ m, $e/d = 0.005$, and from Fig. 3.7, $R/\rho u^2 = 0.0042$.

The valve and fittings are equivalent to 260 pipe diameters which is equal to $(260 \times 0.04) = 10.4$ m of pipe.

The equivalent length of pipe is therefore $(150 + 10.4) = 160.4$ m. The head loss due to friction is:

$$h_f = 4(R/\rho u^2)(l/d)(u^2/g) \qquad \text{(equation 3.20)}$$

$$= (4 \times 0.0042)(160.4/0.04)(0.476^2/9.81) = 1.56 \text{ m}$$

$\therefore \qquad$ total head $= (1.56 + 1.5 + 10) = 13.06$ m.

and: $\qquad$ power required $= (13.06 \times 0.6 \times 9.81)/0.6 = \underline{\underline{128 \text{ W}}}$

PROBLEM 8.22

A pump developing a pressure of 800 kN/m^2 is used to pump water through a 150 mm pipe 300 m long to a reservoir 60 m higher. With the valves fully open, the flowrate obtained is 0.05 m^3/s. As a result of corrosion and scaling the effective absolute roughness of the pipe surface increases by a factor of 10. By what percentage is the flowrate reduced? Viscosity of water $= 1$ mN s/m^2.

Solution

800 kN/m^2 is equivalent to a head of $80,000/(1000 \times 9.81) = 81.55$ m of water. If the pump is required to raise the water through a height of 60 m, then neglecting kinetic energy losses, the head loss due to friction in the pipe $= (81.55 - 60) = 21.55$ m.

The flowrate under these conditions is 0.05 m^3/s.

The cross-sectional area of the pipe $= (\pi/4)0.15^2 = 0.0177$ m^2.

Velocity of the water $= (0.05/0.0177) = 2.82$ m/s.

Head loss due to friction is:

$$h_f = 8(R/\rho u^2)(l/d)(u^2/2g) \qquad \text{(equation 3.20)}$$

Thus: $21.55 = 8(R/\rho u^2)(300/0.15)(2.82^2/(2 \times 9.81))$

and: $R/\rho u^2 = 0.0033$

$Re = \rho u d/\mu = (1000 \times 2.82 \times 0.15)/10^{-3} = 4.23 \times 10^5$

From Fig. 3.7, e/d is 0.003.

If, as a result of scaling and fouling, the roughness increases by a factor of 10, the new value of $e/d = 0.03$. Fig. 3.7 can no longer be used since the new velocity, and hence the Reynolds number, is unknown. Use is made of equation 3.23 and Fig. 3.8 to find the new velocity.

The maximum head loss due to friction is still equal to 21.55 m as the pump head is unchanged.

Thus: 21.55 m $= (21.55 \times 1000 \times 9.81) = 211,410$ N/m^2

$$(R/\rho u^2)Re^2 = -\Delta P_f d^3 \rho/4l\mu^2 \qquad \text{(equation 3.23)}$$

$$= (211,410 \times 0.15^3 \times 1000/4 \times 300 \times 10^{-6}) = 6.0 \times 10^8$$

From Fig. 3.8, $Re = 2.95 \times 10^5$ when $e/d = 0.03$.

Hence the new velocity $= (2.95 \times 10^5 \times 10^{-3})/(1000 \times 0.15) = 1.97$ m/s

Reduction in flow $= (100(2.82 - 1.97)/2.82) = \underline{\underline{30.1 \text{ per cent.}}}$

Heat Transfer

PROBLEM 9.1

Calculate the time taken for the distant face of a brick wall, of thermal diffusivity, $D_H = 0.0042$ cm^2/s and thickness $l = 0.45$ m, initially at 290 K, to rise to 470 K if the near face is suddenly raised to a temperature of $\theta' = 870$ K and maintained at that temperature. Assume that all the heat flow is perpendicular to the faces of the wall and that the distant face is perfectly insulated.

Solution

The temperature at any distance x from the near face at time t is given by:

$$\theta = \sum_{N=0}^{N=\infty} (-1)^N \theta' \{\text{erfc}[(2lN + x)/(2\sqrt{D_H t})] + \text{erfc}[2(N+1)l - x/(2\sqrt{D_H t})]\}$$

(equation 9.37)

and the temperature at the distant face is:

$$\theta = \sum_{N=0}^{N=\infty} (-1)^N \theta' \{2\,\text{erfc}[(2N+1)l]/(2\sqrt{D_H t})\}$$

Choosing the temperature scale such that the initial temperature is everywhere zero,

$$\theta/2\theta' = (470 - 290)/2(870 - 290) = 0.155$$

$$D_H = 0.0042 \text{ cm}^2/\text{s} \quad \text{or} \quad 4.2 \times 10^{-7} \text{ m}^2/\text{s}, \sqrt{D_H} = 6.481 \times 10^4 \quad \text{and} \quad l = 0.45 \text{ m}$$

Thus: $0.155 = \displaystyle\sum_{N=0}^{N=\infty} (-1)\,\text{erfc}\,347(2N+1)/t^{0.5}$

$$= \text{erfc}(347t^{-0.5}) - \text{erfc}(1042t^{-0.5}) + \text{erfc}(1736t^{-0.5})$$

Considering the first term only, $347t^{-0.5} = 1.0$ and $t = 1.204 \times 10^5$ s

The second and higher terms are negligible compared with the first term at this value of t and hence: $\underline{t = 0.120 \text{ Ms}}$ (33.5 h)

PROBLEM 9.2

Calculate the time for the distant face to reach 470 K under the same conditions as Problem 9.1, except that the distant face is not perfectly lagged but a very large thickness of material of the same thermal properties as the brickwork is stacked against it.

Solution

This problem involves the conduction of heat in an infinite medium where it is required to determine the time at which a point 0.45 m from the heated face reaches 470 K.

The boundary conditions are therefore:

$$\theta = 0, \quad t = 0; \quad \theta = \theta', t > 0 \quad \text{for all values of } x$$

$$\theta = (870 - 290) = 580 \deg \text{ K}, \quad x = 0, t > 0$$

$$\theta = 0, \quad x = \infty, \quad t > 0$$

$$\theta = 0, \quad x = 0, \quad t = 0$$

$$\frac{\partial \theta}{\partial t} = D_H \left(\frac{\partial^2 \theta}{\partial x^2} + \frac{\partial^2 \theta}{\partial y^2} + \frac{\partial^2 \theta}{\partial z^2} \right)$$

$$= D_H \frac{\partial^2 \theta}{\partial x^2} \quad \text{(for unidirectional heat transfer)} \quad \text{(equation 9.29)}$$

The Laplace transform of: $\qquad \theta = \bar{\theta} = \int_0^\infty \theta e^{-pt} dt \qquad$ (i)

and hence: $\qquad \dfrac{d^2 \bar{\theta}}{dx^2} = \dfrac{p}{D_H} \bar{\theta} - \dfrac{\bar{\theta}_{t=0}}{D_H} \qquad$ (ii)

Integrating equation (ii): $\bar{\theta} = B_1 e^{x\sqrt{(p/D_H)}} + B_2 e^{-x\sqrt{(p/D_H)}} + \theta_{t=0}/p \qquad$ (iii)

and: $\qquad \dfrac{d\bar{\theta}}{dx} = B_1 \sqrt{(p/D_H)} e^{x\sqrt{(p/D_H)}} - B_2 \sqrt{(p/D_H)} e^{-x\sqrt{(p/D_H)}} \qquad$ (iv)

In this case, $\qquad \bar{\theta}_{\substack{t>0 \\ x=0}} = \int_0^\infty \theta' e^{-pt} dt = \theta'/p$

and: $\qquad \left(\dfrac{\overline{\partial \theta}}{\partial t} \right)_{\substack{t>0 \\ x=0}} = \int_0^\infty \left(\dfrac{\partial \theta}{\partial t} \right) e^{-pt} dt = 0$

Substituting the boundary conditions in equations (iii) and (iv):

$$\bar{\theta}_{\substack{t>0 \\ x=0}} = \theta'_{\substack{t>0 \\ x=0}}/p = B_1 + B_2 + \theta_{t=0}/p \quad \text{or} \quad B_1 + B_2 = \theta'_{\substack{t>0 \\ x=0}}/p$$

and: $\qquad \left(\dfrac{\overline{\partial \theta}}{\partial t} \right)_{\substack{t>0 \\ x=0}} = 0 = B_1 \sqrt{(p/D_H)} e^\infty - B_2 \sqrt{(p/D_H)} e^{-\infty}$

$\therefore \qquad B_1 \sqrt{(p/D_H)} = 0 \quad \text{and} \quad B_1 = 0, \quad B_2 = \theta'_{\substack{t>0 \\ x=0}}/p$

From (iii), $\qquad \bar{\theta} = B_2 e^{-x\sqrt{(p/D_H)}} = \theta' p^{-1} e^{-k\sqrt{p}}$ where $k = x/\sqrt{D_H}$

The Laplace transform of $p^{-1} e^{-k\sqrt{p}} = \operatorname{erfc} k/2\sqrt{t}$ (from Volume 1, Appendix).

and: $$\theta = \theta'_{\substack{t>0 \\ x=0}} \operatorname{erfc}\left[\frac{x}{2\sqrt{D_H t}}\right] \qquad \text{(v)}$$

When $x = 0.45$ m, $\theta = (470 - 290) = 180 \deg$ K, and hence in (v), with $D_H = 4.2 \times 10^{-7}$ m^2/s,

$$(180/580) = \operatorname{erfc}\{[0.45/(6.481 \times 10^{-4})][1/(2\sqrt{t})]\} = 0.31$$

$\therefore \qquad (0.45/6.481 \times 10^{-4})/2\sqrt{t} = 0.73$

and: $\qquad t = 2.26 \times 10^5$ s $\quad$ or $\quad \underline{0.226 \text{ Ms } (62.8 \text{ h})}$

As an alternative method of solution, Schmidt's method is used with the construction shown in Fig. 9a. In this case $\Delta x = 0.1$ m and it is seen that at $x = 0.45$ m, the temperature is 470 K after a time 20Δ t.

In equation 9.43: $\Delta t = (0.1)^2/(2 \times 4.2 \times 10^{-7}) = 1.191 \times 10^4$ s
and hence the required time, $t = (20 \times 1.191 \times 10^4) = 2.38 \times 10^5$ s $= \underline{0.238 \text{ Ms } (66.1 \text{ h})}$

The difference here is due to inaccuracies resulting from the coarse increments of x.

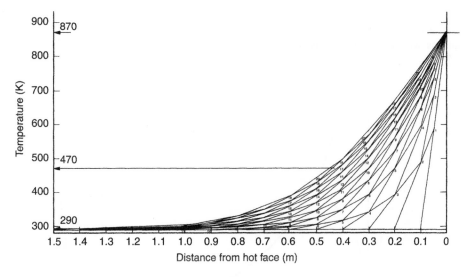

Figure 9a.

PROBLEM 9.3

Benzene vapour, at atmospheric pressure, condenses on a plane surface 2 m long and 1 m wide maintained at 300 K and inclined at an angle of 45° to the horizontal. Plot the thickness of the condensate film and the point heat transfer coefficient against distance from the top of the surface.

Solution

At 101.3 kN/m^2, benzene condenses at $T_s = 353$ K. With a wall temperature of $T_w = 300$ K, the film properties at a mean temperature of 327 K are:

$$\mu = 4.3 \times 10^{-4} \text{ N s/m}^2, \rho = 860 \text{ kg/m}^3, k = 0.151 \text{ W/m K and}$$
$$\lambda = 423 \text{ kJ/kg} = 4.23 \times 10^5 \text{ J/kg}$$

Thus:

$$s = \{[4\mu k(T_s - T_w)x]/(g \sin \phi \lambda \rho^2)\}^{0.25} \qquad \text{(equation 9.168)}$$
$$= \{[4 \times 4.3 \times 10^{-4} \times 0.151(353 - 300)x]/(9.81 \sin 45° \times 4.23 \times 10^5 \times 860^2)\}^{0.25}$$
$$= 2.82 \times 10^{-4}x^{0.25} \text{ m}$$

Similarly:

$$h = \{(\rho^2 g \sin \phi \lambda k^3)/[4\mu(T_s - T_w)x]\}^{0.25} \qquad \text{(equation 9.169)}$$
$$= \{(860^2 \times 9.81 \sin 45° \times 4.23 \times 10^5 \times 0.151^3)/[4 \times 4.3 \times 10^{-4}(353 - 300)x]\}^{0.25}$$
$$= 535x^{-0.25} \text{ W/m}^2\text{K}$$

Values of x between 0 and 2.0 m in increments of 0.20 m are now substituted in these equations with the following results, which are plotted in Fig. 9b.

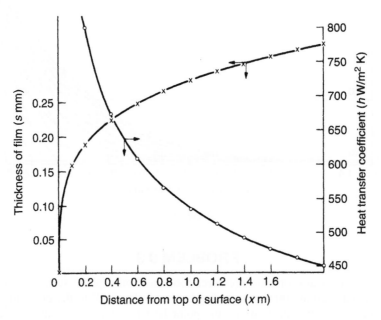

Figure 9b.

x (m)	$x^{0.25}$	$x^{-0.25}$	s (m)	h (W/m² K)
0	0	∞	0	∞
0.1	0.562	1.778	1.58×10^{-4}	951
0.2	0.669	1.495	1.89×10^{-4}	800
0.4	0.795	1.258	2.24×10^{-4}	673
0.6	0.880	1.136	2.48×10^{-4}	608
0.8	0.946	1.057	2.67×10^{-4}	566
1.0	1.000	1.000	2.82×10^{-4}	535
1.2	1.047	0.956	2.95×10^{-4}	512
1.4	1.088	0.919	3.07×10^{-4}	492
1.6	1.125	0.889	3.17×10^{-4}	476
1.8	1.158	0.863	3.27×10^{-4}	462
2.0	1.189	0.841	3.35×10^{-4}	450

PROBLEM 9.4

It is desired to warm 0.9 kg/s of air from 283 to 366 K by passing it through the pipes
of a bank consisting of 20 rows with 20 pipes in each row. The arrangement is in-line
with centre to centre spacing, in both directions, equal to twice the pipe diameter. Flue
gas, entering at 700 K and leaving at 366 K, with a free flow mass velocity of 10 kg/m²s,
is passed across the outside of the pipes. Neglecting gas radiation, how long should the
pipes be?

For simplicity, outer and inner pipe diameters may be taken as 12 mm. Values of k
and μ, which may be used for both air and flue gases, are given below. The specific heat
capacity of air and flue gases is 1.0 kJ/kg K.

Temperature (K)	Thermal conductivity k(W/m K)	Viscosity μ(mN s/m²)
250	0.022	0.0165
500	0.044	0.0276
800	0.055	0.0367

Solution

Heat load, $Q = 0.9 \times 1.0(366 - 283) = 74.7$ kW

Temperature driving force, $\theta_1 = (700 - 366) = 334$ deg K,

$$\theta_2 = (366 - 283) = 83 \text{ deg K}$$

and in equation 9.9, $\theta_m = (334 - 83)/\ln(334/83) = 180$ deg K

Film coefficients

Inside:

$$h_i d_i / k = 0.023 (dG'/\mu)^{0.8} (C_p \mu/k)^{0.4} \qquad \text{(equation 9.61)}$$

$$d_i = 12 \text{ mm or } 1.2 \times 10^{-2} \text{ m.}$$

The mean air temperature $= 0.5(366 + 283) = 325$ K and $k = 0.029$ W/m K.

Cross-sectional area of one tube $= (\pi/4)(1.2 \times 10^{-2})^2 = 1.131 \times 10^{-4}$ m^2

Area for flow $= (20 \times 20)1.131 \times 10^{-4} = 4.52 \times 10^{-2}$ m^2.

Thus, mass velocity $G = 0.9/(4.52 \times 10^{-2}) = 19.9$ kg/m^{2}s.

At 325 K, $\mu = 0.0198$ mN s/m^2 or 1.98×10^{-5} N s/m^2

$$C_p = (1.0 \times 10^3)\text{J/kg K}$$

Thus: $(h_i \times 1.2 \times 10^{-2})/(2.9 \times 10^{-2}) = 0.023(1.2 \times 10^{-2} \times 19.9/1.98 \times 10^{-5})^{0.8}$

$$\times (1.0 \times 10^3 \times 1.98 \times 10^{-5}/0.029)^{0.4}$$

$0.4138 h_i = 0.023(1.206 \times 10^4)^{0.8}(0.683)^{0.4}$ and $h_i = 87.85$ W/m^{2}K

Outside:

$$h_o d_o / k = 0.33 C_h (d_o G'/\mu)_{\text{max}}^{0.6} (C_p \mu/k)^{0.3} \qquad \text{(equation 9.90)}$$

$$d_o = 12.0 \text{ mm} \quad \text{or} \quad 1.2 \times 10^{-2} \text{ m}$$

$$G' = 10 \text{ kg/m}^2 \text{ s for free flow}$$

$$G'_{\text{max}} = YG'/(Y - d_o)$$

where Y, the distance between tube centres $= 2d_o = 2.4 \times 10^{-2}$ m.

$\therefore \qquad G'_{\text{max}} = (2.4 \times 10^{-2} \times 10.0)/(2.4 \times 10^{-2} - 1.2 \times 10^{-2}) = 20$ kg/m^2 s

At a mean flue gas temperature of $0.5(700 + 366) = 533$ K,

$\mu = 0.0286$ mN s/m^2 or 2.86×10^{-5} N s/m^2, $k = 0.045$ W/m K   and $C_p = (1.0 \times 10^3)$ J/kg K

$\therefore \qquad Re_{\text{max}} = (1.2 \times 10^{-2} \times 20.0)/(2.86 \times 10^{-5}) = 8.39 \times 10^3$

From Table 9.3, when $Re_{\text{max}} = 8.39 \times 10^3$, $X = 2d_o$, and $Y = 2d_o$, $C_h = 0.95$.

Thus: $(h_o \times 1.2 \times 10^{-2})/(4.5 \times 10^{-2}) = (0.33 \times 0.95)(8.39 \times 10^3)^{0.6}$

$$\times (1.0 \times 10^3 \times 2.86 \times 10^{-5})/0.045)^{0.3}$$

or: $0.267 h_o = 0.314(8.39 \times 10^3)^{0.6}(0.836)^{0.3}$ and $h_o = 232$ W/m^2 K

Overall:

Ignoring wall and scale resistances, then:

$$1/U = 1/h_o + 1/h_i = (0.0114 + 0.0043) = 0.0157$$

and: $U = 63.7$ W/m^2 K

Area required

In equation 9.1, $A = Q/U\theta_m = (74.7 \times 10^3)/(63.7 \times 180) = 6.52 \text{ m}^2$.

Area/unit length of tube $= (\pi/4)(12 \times 10^{-2}) = 9.43 \times 10^{-3} \text{ m}^2/\text{m}$ and hence: total length of tubing required $= 6.52/(9.43 \times 10^{-3}) = 6.92 \times 10^2$ m.

The length of each tube is therefore $= (6.92 \times 10^2)/(20 \times 20) = \underline{\underline{1.73 \text{ m}}}$

PROBLEM 9.5

A cooling coil, consisting of a single length of tubing through which water is circulated, is provided in a reaction vessel, the contents of which are kept uniformly at 360 K by means of a stirrer. The inlet and outlet temperatures of the cooling water are 280 K and 320 K respectively. What would be the outlet water temperature if the length of the cooling coil were increased by 5 times? Assume the overall heat transfer coefficient to be constant over the length of the tube and independent of the water temperature.

Solution

$$Q = UA\Delta T_m \qquad\qquad \text{(equation 9.1)}$$

where ΔT_m is the logarithmic mean temperature difference. For the initial conditions:

$$Q_1 = (m_1 \times 4.18)(320 - 280) = U_1 A_1[(360 - 280) - (360 - 320)]/$$
$$[\ln(360 - 280)/(360 - 320)]$$

or: $\qquad\qquad 167.2m_1 = U_1 A_1(80 - 40)/\ln 80/40 = 57.7 U_1 A_1$

and: $\qquad\qquad (m_1/U_1 A_1) = 0.345$

In the second case, $m_2 = m_1$, $U_2 = U_1$, and $A_2 = 5A_1$.

∴ $\qquad Q_2 = (m_1 \times 4.18)(T - 280) = 5U_1 A_1[(360 - 280) - (360 - T)]/$
$$\ln(360 - 280)/(360 - T)$$

or: $\qquad 4.18(m_1/U_1 A_1)(T - 280)/5 = (80 - 360 + T)/[\ln[80/360 - T)]]$

Substituting for $(m_1/U_1 A_1)$,

$$0.289(T - 280) = (T - 280)/[\ln 80/(360 - T)]$$

or: $\qquad\qquad \ln[80/(360 - T)] = 3.467 \quad \text{and} \quad \underline{\underline{T = 357.5 \text{ K}}}$

PROBLEM 9.6

In an oil cooler, 216 kg/h of hot oil enters a thin metal pipe of diameter 25 mm. An equal mass of cooling water flows through the annular space between the pipe and a larger concentric pipe; the oil and water moving in opposite directions. The oil enters at 420 K and is to be cooled to 320 K. If the water enters at 290 K, what length of pipe will be required? Take coefficients of 1.6 kW/m^2 K on the oil side and 3.6 kW/m^2 K on the water side and 2.0 kJ/kg K for the specific heat of the oil.

Solution

Heat load

Mass flow of oil $= 6.0 \times 10^{-2}$ kg/s.

and hence, $Q = (6.0 \times 10^{-2} \times 2.0)(420 - 320) = 12$ kW

Thus the water outlet temperature is given by:

$$12 = (6.0 \times 10^{-2} \times 4.18)(T - 290) \text{ or } T = 338 \text{ K}$$

Logarithmic mean temperature driving force

In equation 9.9:

$$\theta_1 = (420 - 338) = 82 \deg \text{ K}, \quad \theta_2 = (320 - 290) = 30 \deg \text{ K}$$

and: $\theta_m = (82 - 30)/\ln(82/30) = 51.7 \deg \text{ K}$

Overal coefficient

The pipe wall is thin and hence its thermal resistance may be neglected.
   Thus in equation 9.8:

$$1/U = 1/h_o + 1/h_i = (1/1.6 + 1/3.6) \quad \text{and} \quad U = 1.108 \text{ kW/m}^2 \text{ K}$$

Area

In equation 9.1, $A = Q/U\theta_m = 12/(1.108 \times 51.7) = 0.210$ m^2

   Tube diameter $= 25 \times 10^{-3}$ m (assuming a mean value)

area/unit length $= (\pi \times 25 \times 10^{-3} \times 1.0) = 7.85 \times 10^{-2}$ m^2/m

and the tube length required $= 0.210/(7.85 \times 10^{-2}) = \underline{2.67 \text{ m}}$

PROBLEM 9.7

The walls of a furnace are built of a 150 mm thickness of a refractory of thermal conductivity 1.5 W/m K. The surface temperatures of the inner and outer faces of the refractory are 1400 K and 540 K respectively. If a layer of insulating material 25 mm thick of thermal conductivity 0.3 W/m K is added, what temperatures will its surfaces attain assuming the inner surface of the furnace to remain at 1400 K? The coefficient of heat transfer from the outer surface of the insulation to the surroundings, which are at 290 K, may be taken as 4.2, 5.0, 6.1, and 7.1 W/m^2 K for surface temperatures of 370, 420, 470, and 520 K respectively. What will be the reduction in heat loss?

Solution

Heat flow through the refractory, $Q = kA(T_1 - T_2)/x$ (equation 9.12)

Thus for unit area, $Q = (1.5 \times 1.0)(1400 - T_2)/(150 \times 10^{-3})$

$$= (14{,}000 - 10T_2) \text{ W/m}^2 \qquad \text{(i)}$$

where T_2 is the temperature at the refractory–insulation interface.
Similarly, the heat flow through the insulation is:

$$Q = (0.3 \times 1.0)(T_2 - T_3)/(25 \times 10^{-3}) = (12T_2 - 12T_3) \text{ W/m}^2 \qquad \text{(ii)}$$

The flow of heat from the insulation surface at T_3 K to the surroundings at 290 K, is:

$$Q = hA(T_3 - 290) \text{ or } (T_3 - 290) \; h\text{W/m}^2 \qquad \text{(iii)}$$

where h is the coefficient of heat transfer from the outer surface.
The solution is now made by trial and error. A value of T_3 is selected and h obtained by interpolation of the given data. This is substituted in equation (iii) to give Q. T_2 is then obtained from equation (ii) and a second value of Q is then obtained from equation (i). The correct value of T_3 is then given when these two values of Q coincide. The working is as follows and the results are plotted in Fig. 9c.

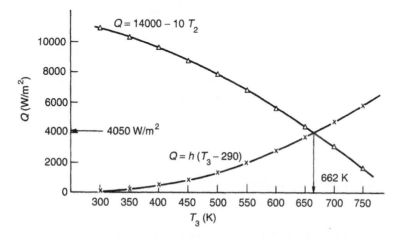

Figure 9c.

T_3 (K)	h (W/m^2 K)	$Q = h(T_3 - 290)$ (W/m^2)	$T_2 = T_3 + Q/12$ (K)	$Q = (14,000 - 10T_2)$ (W/m^2)
300	3.2	32	302.7	10,973
350	3.9	234	369.5	10,305
400	4.7	517	443.1	9569
450	5.6	896	524.7	8753
500	6.5	1355	612.9	7871
550	7.8	2028	719.0	6810
600	9.1	2821	835.1	5649
650	10.4	3744	962.0	4380
700	11.5	4715	1092.9	3071
750	12.7	5842	1236.8	1632

A balance is obtained when $T_3 = 662$ K, at which $Q = 4050$ W/m^2.

In equation (i): $4050 = (14,000 - 10T_2)$ or $T_2 = 995$ K

Thus the temperatures at the inner and outer surfaces of the insulation are

<u>995 K and 662 K respectively</u>

With no insulation, $Q = (1.5 \times 1.0)(1400 - 540)/(150 \times 10^{-3}) = 8600$ W/m^2
and hence the reduction in heat loss is $(8600 - 4050) = \underline{4550 \text{ W/m}^2}$

or: $(4540 \times 100)/8600 = \underline{52.9\%}$

PROBLEM 9.8

A pipe of outer diameter 50 mm, maintained at 1100 K, is covered with 50 mm of insulation of thermal conductivity 0.17 W/m K. Would it be feasible to use a magnesia insulation, which will not stand temperatures above 615 K and has a thermal conductivity of 0.09 W/m K, for an additional layer thick enough to reduce the outer surface temperature to 370 K in surroundings at 280 K? Take the surface coefficient of heat transfer by radiation and convection as 10 W/m^2 K.

Solution

For convection to the surroundings

$$Q = hA_3(T_3 - T_4) \text{ W/m}$$

where A_3 is area for heat transfer per unit length of pipe, (m^2/m).

The radius of the pipe, $r_1 = (50/2) = 25$ mm or 0.025 m.

The radius of the insulation, $r_2 = (25 + 50) = 75$ mm or 0.075 m.

The radius of the magnesia, $r_3 = (75 + x) = (0.075 + 0.001x)$ m where x mm is the thickness of the magnesia.

Hence the area at the surface of the magnesia, $A_3 = 2\pi(0.075 + 0.001x)$ m²/m

and $Q = 10[2\pi(0.075 + 0.001x)](370 + 280) = (424.1 + 5.66x)$ W/m   (i)

For conduction through the insulation

$$Q = k(2\pi r_m l)(T_1 - T_2)/(r_2 - r_1) \qquad \text{(equation 9.22)}$$

where $r_m = (r_2 - r_1)/\ln r_2/r_1$.

$\therefore \qquad Q = 0.17[2\pi \times 1.0(r_2 - r_1)](1100 - T_2)/[(r_2 - r_1)\ln(0.075/0.025)]$

$\qquad = 0.972(1100 - T_2)$ W/m   (ii)

For conduction through the magnesia

In equation 9.22:

$$Q = 0.09[2\pi \times 1.0(r_3 - r_2)](T_2 - 370)/[(r_3 - r_2)\ln(0.075 + 0.001x)/0.075]$$

$$= 0.566(T_2 - 370)/\ln(1 + 0.013x) \qquad \text{(iii)}$$

For a value of x, Q is found from (i) and hence T_2 from (ii). These values are substituted in (iii) to give a second value of Q, with the following results:

x (mm)	$Q = (424.1 + 5.66x)$ (W/m)	$T_2 = (1100 - 1.028Q)$ (K)	$Q = 0.566(T_2 - 370)/$ $\ln(1 + 0.013x)$ (W/m)
5.0	452.4	635	2380
7.5	466.6	620	1523
10.0	480.7	606	1092
12.5	494.9	591	832
15.0	509.0	577	657
17.5	523.2	562	531
20.0	537.3	548	435

From a plot of the two values of Q, a balance is attained when $x = 17.5$ mm. With this thickness, $T_2 = 560$ K which is below the maximum permitted and hence the use of the magnesia would be feasible.

PROBLEM 9.9

In order to heat 0.5 kg/s of a heavy oil from 311 K to 327 K, it is passed through tubes of inside diameter 19 mm and length 1.5 m forming a bank, on the outside of which steam is condensing at 373 K. How many tubes will be needed?

In calculating Nu, Pr, and Re, the thermal conductivity of the oil may be taken as 0.14 W/m K and the specific heat as 2.1 kJ/kg K, irrespective of temperature. The viscosity is to be taken at the mean oil temperature. Viscosity of the oil at 319 and 373 K is 154 and 19.2 mN s/m^2 respectively.

Solution

Heat load

$$Q = (0.5 \times 2.1)(327 - 311) = 16.8 \text{ kW}$$

Logarithmic mean driving force

$$\theta_1 = (373 - 311) = 62 \deg \text{ K}, \theta_2 = (373 - 327) = 46 \deg \text{ K}$$

∴ in equation 9.9, $\theta_m = (62 - 46)/\ln(62/46) = 53.6 \deg \text{ K}$

A preliminary estimate of the overall heat transfer coefficient may now be obtained from Table 9.18.

For condensing steam, $h_o = 10{,}000$ W/m^2 K and for oil, $h_i = 250$ W/m^2 K (say). Thus $1/U = 1/h_o + 1/h_i = 0.0041$, $U = 244$ W/m^2 K and from equation 9.1, the preliminary area:

$$A = (16.8 \times 10^3)/(244 \times 53.6) = 1.29 \text{ m}^2$$

The area/unit length of tube is$(\pi \times 19.0 \times 10^{-3} \times 1.0) = (5.97 \times 10^{-2})$m^2/m

and: total length of tubing $= 1.29/(5.97 \times 10^{-2}) = 21.5$ m

Thus: number of tubes $= (21.5/1.5) = 14.3$, say 14 tubes

Film coefficients

The inside coefficient is controlling and hence this must be checked to ascertain if the preliminary estimate is valid.

The Reynolds number, $Re = d_i G'/\mu = (19.0 \times 10^{-3})G'/\mu$

At a mean oil temperature of $0.5(327 + 311) = 319$ K, $\mu = 154 \times 10^{-3}$ N s/m^2.

Area for flow per tube $= (\pi/4)(19.0 \times 10^{-3})^2 = 2.835 \times 10^{-4}$ m^2.

∴   total area for flow $= (14 \times 2.835 \times 10^{-4}) = 3.969 \times 10^{-3}$ m^2

and hence: $G' = 0.5/(3.969 \times 10^{-3}) = 1.260 \times 10^2$ kg/m^{2}s

Thus: $Re = (19.0 \times 10^{-3} \times 1.260 \times 10^2)/(154 \times 10^{-3}) = 15.5$

That is, the flow is streamline and hence:

$$(h_i d_i/k)(\mu_s/\mu)^{0.14} = 2.01(GC_p/kl)^{0.33} \qquad \text{(equation 9.85)}$$

At a mean wall temperature of $0.5(373 + 319) = 346$ K, $\mu_s = 87.0 \times 10^{-3}$ N s/m^2.

The mass flow, $G = 0.5$ kg/s.

$\therefore$
$$((h_i \times 19.0 \times 10^{-3})/0.14)(87.0 \times 10^{-3}/154 \times 10^{-3})^{0.14}$$
$$= 2.01((0.5 \times 2.1 \times 10^3)/(0.14 \times 1.5))^{0.33}$$

or:
$$(0.13h_i \times 0.923) = (2.01 \times 16.6) \text{ and } h_i = 266 \text{ W/m}^2 \text{ K}$$

This is sufficiently close to the assumed value and hence <u><u>14 tubes</u></u> would be specified.

PROBLEM 9.10

A metal pipe of 12 mm outer diameter is maintained at 420 K. Calculate the rate of heat loss per metre run in surroundings uniformly at 290 K, (a) when the pipe is covered with 12 mm thickness of a material of thermal conductivity 0.35 W/mK and surface emissivity 0.95, and (b) when the thickness of the covering material is reduced to 6 mm, but the outer surface is treated so as to reduce its emissivity to 0.10. The coefficients of radiation from a perfectly black surface in surroundings at 290 K are 6.25, 8.18, and 10.68 W/m^2 K at 310 K, 370 K, and 420 K respectively. The coefficients of convection may be taken as $1.22(\theta/d)^{0.25}$ W/m^2 K, where θ(K) is the temperature difference between the surface and the surrounding air and d(m) is the outer diameter.

Solution

Case (a)

Assuming that the heat loss is q W/m and the surface temperature is T K, for conduction through the insulation, from equation 9.12, $q = kA_m(420 - T)/x$
The mean diameter is 18 mm or 0.018 m, and hence:

$$A_m = (\pi \times 0.018) = 0.0566 \text{ m}^2/\text{m} \quad x = 0.012 \text{ m}$$

$\therefore$
$$q = (0.35 \times 0.0566)(420 - T)/0.012 = (693.3 - 1.67T)\text{W/m} \qquad \text{(i)}$$

For convection and radiation from the surface, from equation 9.119:

$$q = (h_r + h_c)A_2(T - 290) \text{ W/m}$$

where h_r is the film coefficient equivalent to the radiation and h_c the coefficient due to convection given by:

$$h_c = 1.22[(T - 290)/d]^{0.25} \text{ where } d = 36 \text{ mm or } 0.036 \text{ m}$$

$\therefore$
$$h_c = 2.80(T - 290)^{0.25} \text{ W/m}^2 \text{ K}$$

If h_b is the coefficient equivalent to radiation from a black body, $h_r = 0.95h_b$ W/m^2 K
The outer diameter is 0.036 m and hence:

$$A_2 = (\pi \times 0.036 \times 1.0) = 0.1131 \text{ m}^2/\text{m}$$

$\therefore$
$$q = [0.95h_b + 2.80(T - 290)^{0.25}]0.1131(T - 290)$$
$$= 0.1074h_b(T - 290) + 0.317(T - 290)^{1.25} \text{ W/m} \qquad \text{(ii)}$$

Values of T are now assumed and together with values of h_b from the given data substituted into (i) and (ii) until equal values of q are obtained as follows:

T (K)	$q = (693.3 - 1.67T)$ (W/m)	h_b (W/m^2 K)	$0.1074h_b(T - 290)$ (W/m)	$0.317(T - 290)^{1.25}$ (W/m)	q (W/m)
300	193.3	6.0	6.5	5.7	12.2
320	160.0	6.5	20.9	22.2	43.1
340	126.7	7.1	38.1	42.1	80.2
360	93.3	7.8	58.7	64.2	122.9
380	60.0	8.55	82.7	87.9	170.6
400	26.7	9.55	112.8	113.0	225.8

A balance is obtained when $\underline{T = 350 \text{ K}}$ and $\underline{q = 106 \text{ W/m}}$.

Case (b)

For conduction through the insulation, $x = 0.006$ m and the mean diameter is 15 mm or 0.015 m.

$$\therefore \qquad A_m = (\pi \times 0.015 \times 1.0) = 0.0471 \text{ m}^2/\text{m}$$

$$\therefore \qquad q = (0.35 \times 0.0471)(420 - T)/0.006 = (1154 - 2.75T) \text{ W/m} \qquad \text{(i)}$$

The outer diameter is now 0.024 m and $A_2 = (\pi \times 0.024 \times 1.0) = 0.0754 \text{ m}^2/\text{m}$
The coefficient due to convection is:

$$h_c = 1.22[(T - 290)/0.024]^{0.25} = 3.10(T - 290)^{0.25} \text{ W/m}^2 \text{ K}$$

The emissivity is 0.10 and hence $h_r = 0.10h_b$ W/m^2 K

$$\therefore \qquad q = [0.10h_b + 3.10(T - 290)^{0.25}]0.0754(T - 290)$$

$$= 0.00754h_b(T - 290) + 0.234(T - 290)^{1.25} \text{ W/m} \qquad \text{(ii)}$$

Making the calculation as before:

T (K)	$q = (1154 - 2.75T)$ (W/m)	h_{b_2} (W/m^2 K)	$0.0075h_b(T - 290)$ (W/m)	$0.234(T - 290)^{1.25}$ (W/m)	q (W/m)
300	329.0	6.0	0.5	4.2	4.7
320	274.0	6.5	1.5	16.4	17.9
340	219.0	7.1	2.7	31.1	33.8
360	164.0	7.8	4.2	47.4	51.6
380	109.0	8.55	5.8	64.9	70.7
400	54.0	9.55	7.9	83.4	91.3

A balance is obtained when $\underline{T = 390 \text{ K}}$ and $\underline{q = 81 \text{ W/m}}$.

PROBLEM 9.11

A condenser consists of 30 rows of parallel pipes of outer diameter 230 mm and thickness 1.3 mm with 40 pipes, each 2 m long in each row. Water, at an inlet temperature of 283 K, flows through the pipes at 1 m/s and steam at 372 K condenses on the outside of the pipes. There is a layer of scale 0.25 mm thick of thermal conductivity 2.1 W/m K on the inside of the pipes. Taking the coefficients of heat transfer on the water side as 4.0 and on the steam side as 8.5 kW/m^2 K, calculate the water outlet temperature and the total mass flow of steam condensed. The latent heat of steam at 372 K is 2250 kJ/kg. The density of water is 1000 kg/m^3.

Solution

Overall coefficient

$$\frac{1}{U} = \frac{1}{h_i} + \frac{1}{h_o} + \frac{x_w}{k_w} + \frac{x_r}{k_r} \qquad \text{(equation 9.201)}$$

where x_r and k_r are the thickness and thermal conductivity of the scale respectively. Considering these in turn, $h_i = 4000$ W/m^2 K.

The inside diameter, $d_i = 230 - (2 \times 1.3) = 227.4$ mm or 0.2274m.

Therefore basing the coefficient on the outside diameter:

$$h_{io} = (4000 \times 0.2274/0.230) = 3955 \text{W/m}^3 \text{ K}$$

For conduction through the wall, $x_w = 1.3$ mm, and from Table 9.1, $k_w = 45$ W/m K for steel and $(k_w/x_w) = (45/0.0013) = 34615$ W/m^2 K

The mean wall diameter $= (0.230 + 0.2274)/2 = 0.2287$ m and hence the coefficient equivalent to the wall resistance based on the tube o.d. is: $(34615 \times 0.2287/0.230) = 34419$ W/m^2/K

For conduction through the scale, $x_r = (0.25 \times 10^{-3})$ m, $k_r = 2.1$ W/m K and hence:

$$k_r/x_r = (2.1/0.25 \times 10^{-3}) = 8400 \text{ W/m}^2 \text{ K}$$

The mean scale diameter $= (227.4 - 0.25) = 227.15$ mm or 0.2272 m and hence the coefficient equivalent to the scale resistance based on the tube o.d. is:

$$(8400 \times 0.2272/0.230) = 8298 \text{ W/m}^2 \text{ K}$$

$\therefore$ $\qquad 1/U = (1/3955 + 1/8500 + 1/34419 + 1/8298) = 5.201 \times 10^{-4}$

and: $\qquad U = 1923$ W/m^2 K

Temperature driving force

If water leaves the unit at T K:

$$\theta_1 = (372 - 283) = 89 \deg \text{ K}, \theta_2 = (372 - T)$$

and in equation 9.9:

$$\theta_m = [89 - (372 - T)]/\ln[89/(372 - T)] = (T - 283)/\ln[89/(372 - T)]$$

Area

For 230 mm o.d. tubes, outside area per unit length $= (\pi \times 0.230 \times 1.0) = 0.723 \text{ m}^2/\text{m}$.

Total length of tubes $= (30 \times 40 \times 2) = 2400$ m and hence heat transfer area, $A = (2400 \times 0.723) = 1735.2 \text{ m}^2$.

Heat load

The cross-sectional area for flow/tube $= (\pi/4)(0.230)^2 = 0.0416 \text{ m}^2/\text{tube}$.

Assuming a single-pass arrangement, there are 1200 tubes per pass and hence area for flow $= (1200 \times 0.0416) = 49.86 \text{ m}^2$.

For a velocity of 1.0 m/s, the volumetric flow $= (0.1 \times 49.86) \text{ m}^3/\text{s}$ and the mass flow $= (1000 \times 4.986) = 4986$ kg/s.

Thus the heat load, $Q = 4986 \times 4.18(T - 283) = 20,840(T - 283)$ kW or $2.084 \times 10^7(T - 283)$ W.

Substituting for Q, U, A, and θ_m in equation 9.1:

$$(2.084 \times 10^7)(T - 283) = (1923 \times 1735.2)(T - 283)/\ln[89/(372 - T)]$$

or: $\ln[89/(372 - T)] = 0.1601$ and $\underline{T = 296 \text{ K}}$

The total heat load is, therefore, $Q = 20,840(296 - 283) = 2.71 \times 10^5$ kW
and the mass of steam condensed $= (2.71 \times 10^5)/2250 = \underline{120.4 \text{ kg/s}}$.

PROBLEM 9.12

In an oil cooler, water flows at the rate of 360 kg/h per tube through metal tubes of outer diameter 19 mm and thickness 1.3 mm, along the outside of which oil flows in the opposite direction at the rate of 6.675 kg/s per tube. If the tubes are 2 m long and the inlettemperatures of the oil and water are 370 K and 280 K respectively, what will be the outlet oil temperature? The coefficient of heat transfer on the oil side is 1.7 kw/m^2 K and on the water side 2.5 kW/m^2 K and the specific heat of the oil is 1.9 kJ/kg K.

Solution

In the absence of information as to the geometry of the unit, the solution will be worked on the basis of *one tube* — a valid approach as the number of tubes effectively appears on both sides of equation 9.1.

If T_w and T_o are the outlet temperature of the water and the oil respectively, then:

Heat load

$$Q = ((360/3600) \times 4.18)(T_w - 280) = 0.418(T_w - 280) \text{ kW for water}$$

and: $Q = ((75/1000) \times 1.9)(370 - T_o) = 0.143(370 - T_o)$ kW for the oil.

From these two equations, $T_w = (406.5 - 0.342T_o)$ K

Area

For 19.0 mm o.d. tubes, surface area $= (\pi \times 0.019 \times 1.0) = 0.0597 \text{ m}^2/\text{m}$ and for one tube, surface area $= (2.0 \times 0.0597) = 0.1194 \text{ m}^2$

Temperature driving force

$$\theta_1 = (370 - T_w), \quad \theta_2 = (T_o - 280)$$

and in equation 9.9: $\theta_m = [(370 - T_w) - (T_o - 280)]/[\ln(370 - T_w)/(T_o - 280)]$

$$= (650 - T_w - T_o)/\ln(370 - T_w)/(T_o - 280)$$

Substituting for T_w:

$$\theta_m = (243.5 - 0.658T_o)/\ln(0.342T_o - 36.5)/(T_o - 280) \text{ K}$$

Overall coefficient

$$h_i = 2.5 \text{ kW/m}^2 \text{ K}$$

$$d_i = 19.0 - (2 \times 1.3) = 16.4 \text{ mm}$$

Therefore the inside coefficient, based on the outside diameter is:

$$h_{io} = (2.5 \times 16.4/19.0) = 2.16 \text{ kW/m}^2 \text{ K}$$

Neglecting the scale and wall resistances then:

$$1/U = (1/2.16 + 1/1.7) = 1.052 \text{ m}^2 \text{ K/kW}$$

and: $U = 0.951 \text{ kW/m}^2 \text{ K}$

Substituting in equation 9.1 gives:

$$0.143(370 - T_o) = (0.951 \times 0.1194)(243.5 - 0.658T_o)/\ln(0.342T_o - 36.5)/(T_o - 280)$$

$\therefore$ $\ln(0.342T_o - 36.5)/(T_o - 280) = 0.523$ and $\underline{\underline{T_o = 324 \text{ K}}}$

PROBLEM 9.13

Waste gases flowing across the outside of a bank of pipes are being used to warm air which flows through the pipes. The bank consists of 12 rows of pipes with 20 pipes, each

0.7 m long, per row. They are arranged in-line, with centre-to-centre spacing equal, in both directions, to one-and-a-half times the pipe diameter. Both inner and outer diameter may be taken as 12 mm. Air with a mass velocity of 8 kg/m² s enters the pipes at 290 K. The initial gas temperature is 480 K and the total mass flow of the gases crossing the pipes is the same as the total mass flow of the air through them.

Neglecting gas radiation, estimate the outlet temperature of the air. The physical constants for the waste gases, assumed the same as for air, are:

Temperature (K)	Thermal conductivity (W/m K)	Viscosity (mN s/m²)
250	0.022	0.0165
310	0.027	0.0189
370	0.030	0.0214
420	0.033	0.0239
480	0.037	0.0260

Specific heat $= 1.00$ kJ/kg K.

Solution

Heat load

The cross area for flow per pipe $= (\pi/4)(0.012)^2 = 0.000113$ m² and therefore for $(12 \times 20) = 240$ pipes, the total flow area $= (240 \times 0.000113) = 0.027$ m².

Thus: flow of air $= (8.0 \times 0.271) = 0.217$ kg/s

which is also equal to the flow of waste gas.

If the outlet temperatures of the air and waste gas are T_a and T_w K respectively, then:

$$Q = (0.217 \times 1.0)(T_a - 290) \text{ kW or } 217(T_a - 290) \text{ W}$$

and: $Q = (0.217 \times 1.0)(480 - T_w)$ kW

from which: $T_w = (770 - T_a)$ K

Area

Surface area/unit length of pipe $= (\pi \times 0.012 \times 1.0) = 0.0377$ m²/m.

Total length of pipe $= (240 \times 0.7) = 168$ m

and hence the heat transfer area, $A = (168 \times 0.0377) = 6.34$ m².

Temperature driving force

$$\theta_1 = (480 - T_a)$$
$$\theta_2 = (T_w - 1290)$$

or, substituting for T_w: $\theta_2 = (480 - T_a) = \theta_1$

Thus in equation 9.9: $\theta_m = (480 - T_a)$

Overall coefficient

The solution is now one of trial and error in that mean temperatures of both streams must be assumed in order to evaluate the physical properties.

Inside the tubes:

a mean temperature of 320 K, will be assumed at which,

$$k = 0.028 \text{ W/m K}, \mu = 0.0193 \times 10^{-3} \text{ N s/m}^2, \text{ and } C_p = 1.0 \times 10^3 \text{ J/kg K}$$

Therefore:

$$h_i d_i / k = 0.023 (d_i G / \mu)^{0.8} (C_p \mu / k)^{0.4} \qquad \text{(equation 9.61)}$$

$$(h_i \times 0.012/0.028) = 0.023(0.012 \times 8.0/0.0193 \times 10^{-3})^{0.8}$$

$$\times (1 \times 10^3 \times 0.0193 \times 10^{-3}/0.028)^{0.4}$$

$\therefore$ $h_i = 0.0537(4.974 \times 10^3)^{0.8}(0.689)^{0.4} = 41.94 \text{ W/m}^2 \text{ K}$

Outside the tubes:

The cross-sectional area of the tube bundle $= (0.7 \times 20)(1.5 \times 0.012) = 0.252 \text{ m}^2$ and hence the free flow mass velocity, $G' = (0.217/0.252) = 0.861 \text{ kg/m}^2 \text{ s}$.

From Fig. 9.27: $Y = (1.5 \times 0.012) = 0.018 \text{ m}$

and therefore: $G'_{max} = (0.861 \times 0.018)/(0.081 - 0.012) = 2.583 \text{ kg/m}^2 \text{ s}$

At an assumed mean temperature of 450 K, $\mu = 0.0250 \times 10^{-3} \text{ N s/m}^2$ and $k = 0.035 \text{ W/m K}$.

$\therefore$ $Re_{max} = (0.012 \times 2.583)/(0.0250 \times 10^{-3}) = 1.24 \times 10^3$

From Table 9.3: for $X = 1.5 d_o$ and $Y = 1.5 d_o$, $C_h = 0.95$.

In equation 9.90:

$$h_o(0.012/0.035) = (0.33 \times 0.95)(1.24 \times 10^3)^{0.6}(1.0 \times 10^3 \times 0.0250 \times 10^{-3}/0.035)^{0.3}$$

$\therefore$ $h_o = (0.914 \times 71.8 \times 0.714^{0.3}) = 59.3 \text{ W/m}^2 \text{ K}$

Hence, ignoring wall and scale resistances:

$$1/U = (1/41.91 + 1/59.3) = 4.07 \times 10^{-2}$$

and: $U = 24.57 \text{ W/m}^2 \text{ K}$

Thus, in equation 9.1:

$$217(T_a - 290) = 24.57 \times 6.34(480 - T_a)$$

from which: $\underline{\underline{T_a = 369.4 \text{ K}}}$

With this value, the mean air and waste gas temperatures are 330 K and 440 K respectively. These are within 10 deg K of the assumed values in each case. Such a difference would have a negligible effect on the film properties and recalculation is unnecessary.

PROBLEM 9.14

Oil is to be heated from 300 K to 344 K by passing it at 1 m/s through the pipes of a shell-and-tube heat exchanger. Steam at 377 K condenses on the outside of the pipes, which have outer and inner diameters of 48 and 41 mm respectively, though due to fouling, the inside diameter has been reduced to 38 mm, and the resistance to heat transfer of the pipe wall and dirt together, based on this diameter, is 0.0009 m² K/W.

It is known from previous measurements under similar conditions that the oil side coefficients of heat transfer for a velocity of 1 m/s, based on a diameter of 38 mm, vary with the temperature of the oil as follows:

Oil temperature (K)	300	311	322	333	344
Oil side coefficient of heat transfer (W/m² K)	74	80	97	136	244

The specific heat and density of the oil may be assumed constant at 1.9 kJ/kg K and 900 kg/m³ respectively and any resistance to heat transfer on the steam side may be neglected.

Find the length of tube bundle required?

Solution

In the absence of further data, this problem will be worked on the basis of one tube.

Heat load

Cross-sectional area at the inside diameter of the scale $= (\pi/4)(0.038)^2 = 0.00113$ m².

$\therefore$ volumetric flow $= (0.00113 \times 1.0) = 0.00113$ m³/s

and: mass flow $= (0.00113 \times 900) = 1.021$ kg/s

$\therefore$ heat load, $Q = 1.021 \times 1.9(344 - 300) = 85.33$ kW

Temperature driving force

$$\theta_1 = (377 - 300) = 77 \deg \text{ K}, \theta_2 = (377 - 344) = 33 \deg \text{ K}$$

and, in equation 9.9: $\theta_m = (77 - 33)/\ln(77/33) = 52 \deg \text{ K}$

Overall coefficient

Inside:

The mean oil temperature $= 0.5(344 + 300) = 322$ K at which

$$h_i(\text{based on } d_i = 0.038 \text{ m}) = 97 \text{ W/m}^2 \text{ K}.$$

Basing this value on the outside diameter of the pipe:

$$h_{io} = (97 \times 0.038/0.048) = 76.8 \text{ W/m}^2 \text{ K}$$

Outside:
From Table 9.17, h_o for condensing steam will be taken as 10,000 W/m^2 K.
Wall and scale:
The scale resistance based on $d = 0.038$ m is 0.0009 m^2 K/W or:

$$k/x = (1/0.0009) = 1111.1 \text{ W/m}^2 \text{ K}.$$

Basing this on the tube o.d., $k/x = (1111.1 \times 0.038/0.048) = 879.6$ W/m^2 K.

$$1/U = 1/h_{io} + 1/h_o + x/k \qquad \text{(equation 9.201)}$$
$$= 0.0130 + 0.0001 + 0.00114$$

or: $\qquad U = 70.2$ W/m^2 K

Area

$$A = Q/U\theta_m = (85.33 \times 10^3)/(70.2 \times 52) = 23.4 \text{ m}^2$$

Area per unit length of pipe $= (\pi \times 0.048 \times 1.0) = 0.151$ m^2/m
and length of tube bundle $= (23.4/0.151) = \underline{\underline{154.9 \text{ m}}}$

A very large tube length is required because of the very low inside film coefficient and several passes or indeed a multistage unit would be specified. A better approach would be to increase the tube side velocity by decreasing the number of tubes in each pass, though any pressure drop limitations would have to be taken into account. The use of a smaller tube diameter might also be considered.

PROBLEM 9.15

It is proposed to construct a heat exchanger to condense 7.5 kg/s of n-hexane at a pressure of 150 kN/m^2, involving a heat load of 4.5 MW. The hexane is to reach the condenser from the top of a fractionating column at its condensing temperature of 356 K. From experience it is anticipated that the overall heat transfer coefficient will be 450 W/m^2 K. Cooling water is available at 289 K. Outline the proposals that you would make for the type and size of the exchanger, and explain the details of the mechanical construction that you consider require special attention.

Solution

A shell-and-tube unit is suitable with hexane on the shell side. For a heat load of 4.5 MW $= 4.5 \times 10^3$ kW, the outlet water temperature is:

$$4.5 \times 10^3 = m \times 4.18(T - 289)$$

In order to avoid severe scaling in such a case, the maximum allowable water temperature is 320 K and hence 310 K will be chosen as a suitable value for T.

Thus: $4.5 \times 10^3 = 4.18m(310 - 289)$ and $m = 51.3$ kg/s

The next stage is to estimate the heat transfer area required.

$$Q = UA\theta_m \qquad \text{(equation 9.1)}$$

where the heat load $Q = 4.5 \times 10^6$ W

$$U = 450 \text{ W/m}^2 \text{ K}$$

$$\theta_1 = (356 - 289) = 67 \text{ deg K}, \quad \theta_2 = (356 - 310) = 46 \text{ deg K}$$

and from equation 9.9, $\theta_m = (67 - 46)/\ln(67/46) = 55.8$ deg K

No correction factor is necessary as the shell side fluid temperature is constant.

$\therefore$ $A = (4.5 \times 10^6)/(450 \times 55.8) = 179.2$ m^2

A reasonable tube size must now be selected, say 25.4 mm, 14 BWG.

The outside surface area is therefore $(\pi \times 0.0254 \times 1.0) = 0.0798$ m^2/m and hence the total length of tubing required $= (179.2/0.0798) = 2246$ m.

A standard tube length is now selected, say 4.87 m and hence the total number of tubes required $= (2246/4.87) = 460$.

It now remains to decide the number of tubes per pass, and this is obtained from a consideration of the water velocity. For shell and tube units, $u = 1.0 - 1.5$ m/s and a value of 1.25 m/s will be selected.

The water flow, 51.3 kg/s $= (51.3/1000) = 0.0513$ m^3/s.

The tube i.d. is 21.2 mm and hence the cross-sectional area for flow/tube $= (\pi/4)(0.0212)^2 = 0.000353$ m^2.

Area required to give a velocity of 1.25 m/s $= (0.0513/1.25) = 0.0410$ m^2 and hence number of tubes/pass $= (0.0410/0.000353) = 116$ and number of passes $= 460/116 \approx 4$.

As the shell side fluid is clean, triangular pitch might be suitable and 460×25 mm o.d. tubes on 32 mm triangular pitch with 4 tube side passes can be accommodated in a 0.838 m i.d. shell and still allow room for impingement plates.

The proposed unit will therefore consist of:

460, 25.4 mm o.d. tubes $\times$ 14 BWG, 4.87 m long arranged in 4 tube side passes on

32 mm triangular pitch in a 0.838 m i.d. shell.

The general mechanical details of the unit are described in Section 9.9.1 of Volume 1 and points of detail are:

(i) impingement baffles should be fitted under each inlet nozzle;
(ii) segmental baffles are not usually fitted to a condenser of this type;
(iii) the unit should be installed on saddles at say 5° to the horizontal to facilitate drainage of the condensate.

PROBLEM 9.16

A heat exchanger is to be mounted at the top of a fractionating column about 15 m high to condense 4 kg/s of n-pentane at 205 kN/m^2, corresponding to a condensing temperature of 333 K. Give an outline of the calculations you would make to obtain an approximate idea of the size and construction of the exchanger required.

For purposes of standardisation, 19 mm outside diameter tubes of 1.65 mm wall thickness will be used and these may be 2.5, 3.6, or 5 m in length. The film coefficient for condensing pentane on the outside of a horizontal tube bundle may be taken as 1.1 kW/m^2 K. The condensation is effected by pumping water through the tubes, the initial water temperature being 288 K. The latent heat of condensation of pentane is 335 kJ/kg.

For these 19 mm tubes, a water velocity of 1 m/s corresponds to a flowrate of 0.2 kg/s of water.

Solution

The calculations follow the sequence of earlier problems in that heat load, temperature driving force, and overall coefficient are obtained and hence the area evaluated. It then remains to consider the geometry of the unit bearing in mind the need to maintain a reasonable cooling water velocity.

As in the previous example, the n-pentane will be passed through the shell and cooling water through the tubes.

Heat load

$Q = (4.0 \times 335) = 1340$ kW assuming there is no sub-cooling of the condensate.

As in Problem 9.15, the outlet temperature of the cooling water will be taken as 310 K, and for a flow of G kg/s:

$$1340 = G \times 4.18(310 - 288) \text{ or } G = 14.57 \text{ kg/s}$$

Temperature driving force

$$\theta_1 = (333 - 288) = 45 \text{ deg K}, \quad \theta_2 = (333 - 310) = 23 \text{ deg K}$$

and: $\qquad \theta_m = (45 - 23)/\ln(45/23) = 32.8 \text{ deg K}$

Overall coefficient

Inside:

For forced convection to water in tubes:

$$h_i = 4280(0.00488T - 1)u^{0.8}/d_i^{0.2} \text{ W/m}^2 \text{ K} \qquad \text{(equation 9.221)}$$

where T, the mean water temperature $= 0.5(310 + 288) = 299$ K; u, the water velocity will be taken as 1 m/s — a realistic optimum value, bearing in mind the need to limit the pressure drop, and $d_i = (19.0 - (2 \times 1.65)) = 15.7$ mm or 0.0157 m.

$$\therefore \qquad h_i = 4280(0.00488 \times 299 - 1)1.0^{0.8}/0.0157^{0.2}$$

$$= (4280 \times 0.459 \times 1.0)/0.436 = 4506 \text{ W/m}^2 \text{ K}$$

or, based on the outer diameter, $h_{io} = (4.506 \times 0.0157)/0.019 = 3.72$ kW/m^2 K

Wall:

For steel, $k = 45$ W/m K and $x = 0.00163$ m
and hence: $x/k = (0.00163/45) = 0.0000362$ m^2 K/W or 0.0362 m^2 K/kW

Outside:

$$h_o = 1.1 \text{ kW/m}^2 \text{ K}$$

Ignoring scale resistance: $1/U = 1/h_o + x/k + 1/h_{io}$

$$= (0.9091 + 0.0362 + 0.2686)$$

and: $\qquad\qquad U = 0.823$ kW/m^2 K

Area

$$Q = UA\theta_m$$

and hence: $\qquad\qquad A = 1340/(0.823 \times 32.8) = 49.6$ m^2

Outer area of 0.019 m diameter tube $= (\pi \times 0.019 \times 1.0) = 0.0597$ m^2/m and hence total length of tubing required $= (49.6/0.0597) = 830.8$ m.

Thus with 2.5, 3.6, and 5.0 m tubes, the number of tubes will be 332, 231 or 166.

The total cooling water flow $= 14.57$ kg/s

and for $u = 1$ m/s, the flow through 1 tube is 0.20 kg/s

$$\therefore \qquad \text{the number of tubes/pass} = (14.57/0.20) = 73$$

Clearly 3 passes are usually to be avoided, and hence 2 or 4 are suitable, that is 146 or 292 tubes, 5.0 or 2.5 m long.

The former is closer to a standard shell size and 166×19 mm tubes on 25.4 mm square pitch with two tube side passes can be fitted within a 438 mm i.d. shell. In this event, the water velocity would be slightly less than 1 m/s in fact $(1 \times 146/166) = 0.88$ m/s, though this would not affect the overall coefficient to any significant extent.

The proposed unit is therefore $\underline{166 \times 19}$ mm o.d. tubes on 25.4 mm square pitch 5.0 m

long with a 438 mm i.d. shell.

In making such calculations it is good practice to add an overload factor to the heat load, say 10%, to allow for errors in predicting film coefficients, although this is often taken into account in allowing for extra tubes within the shell. In this particular example, the fact that the unit is to be installed 15 m above ground level is of significance in limiting the pressure drop and it may be that in an actual situation space limitations would immediately specify the tube length.

PROBLEM 9.17

An organic liquid is boiling at 340 K on the inside of a metal surface of thermal conductivity 42 W/m K and thickness 3 mm. The outside of the surface is heated by condensing steam. Assuming that the heat transfer coefficient from steam to the outer metal surface is constant at 11 kW/m^2 K, irrespective of the steam temperature, find the value of the steam temperature would give a maximum rate of evaporation.

The coefficients of heat transfer from the inner metal surface to the boiling liquid which depend upon the temperature difference are:

Temperature difference between metal surface and boiling liquid (deg K)	Heat transfer coefficient metal surface to boiling liquid (kW/m^2 K)
22.2	4.43
27.8	5.91
33.3	7.38
36.1	7.30
38.9	6.81
41.7	6.36
44.4	5.73
50.0	4.54

Solution

For a steam temperature T_s K, the heat conducted through the film of condensing steam, $Q = h_c A(T_s - T_1)$, or:

$$Q = (11 \times 1.0)(T_s - T_1) = 11.0(T_s - T_1) \text{ kW/m}^2 \qquad \text{(i)}$$

where T_1 is the temperature at the outer surface of the metal.

For conduction through the metal,

$$Q = kA(T_1 - T_2)/x$$
$$= (42 \times 10^{-3} \times 1.0)(T_1 - T_2)/0.003$$
$$= 14.0(T_1 - T_2) \text{ kW/m}^2 \qquad \text{(ii)}$$

where T_2 is the temperature at the inner surface of the metal.

For conduction through the boiling film:

$$Q = h_b(T_2 - 340) = h_b(T_2 - 340) \text{ kW/m}^2 \qquad \text{(iii)}$$

where h_b kW/m^2 K is the film coefficient to the boiling liquid.

Thus for an assumed value of T_2 the temperature difference $(T_2 - 340)$ is obtained and h_b from the table of data. Q is then obtained from (iii), T_1 from (ii), and hence T_s from (i) as follows:

T_2 (K)	$(T_2 - 340)$ (K)	h_b (kW/m²K)	Q (kW/m²)	T_1 (K)	T_s (K)
362.2	22.2	4.43	98.4	369.2	378.1
367.8	27.8	5.91	164.3	379.5	394.4
373.3	33.3	7.38	245.8	390.9	413.3
376.1	36.1	7.30	263.5	394.9	418.9
378.9	38.9	6.81	264.9	397.8	421.9
381.7	41.7	6.36	265.2	400.7	424.8
384.4	44.4	5.73	254.4	402.6	425.7
390.0	50.0	4.54	227.0	406.2	426.8

It is fairly obvious that the rate of evaporation will be highest when the heat flux is a maximum. On inspection this occurs when $\underline{\underline{T_s = 425 \text{ K}}}$.

PROBLEM 9.18

It is desired to warm an oil of specific heat 2.0 kJ/kg K from 300 to 325 K by passing it through a tubular heat exchanger containing metal tubes of inner diameter 10 mm. Along the outside of the tubes flows water, inlet temperature 372 K, and outlet temperature 361 K.

The overall heat transfer coefficient from water to oil, based on the inside area of the tubes, may be assumed constant at 230 W/m² K, and 0.075 kg/s of oil is to be passed through each tube.

The oil is to make two passes through the heater and the water makes one pass along the outside of the tubes. Calculate the length of the tubes required.

Solution

Heat load

If the total number of tubes is n, there are $n/2$ tubes in one pass on the oil side, that is the oil passes through 2 tubes in traversing the exchanger.

The mass flow of oil is therefore $= (0.075 \times n/2) = 0.0375n$ kg/s and the heat load:

$$Q = 0.0375n \times 2.0(325 - 300) = 1.875n \text{ kW}$$

Temperature driving force

$$\theta_1 = (361 - 300) = 61 \deg \text{ K}, \theta_2 = (372 - 325) = 47 \deg \text{ K}$$

and, in equation 9.9: $\theta_m = (61 - 47)/\ln(61/47) = 53.7 \deg \text{ K}$

In equation 9.213:

$$X = (\theta_2 - \theta_1)/(T_1 - \theta_1) \text{ and } Y = (T_1 - T_2)/(\theta_2 - \theta_1)$$

where T_1 and T_2 are the inlet and outlet temperatures on the shell side and θ_1 and θ_2 are the inlet and outlet temperatures on the tube side.

$\therefore$ $$X = (325 - 300)/(372 - 300) = 0.347$$

and: $$Y = (372 - 361)/(325 - 300) = 0.44$$

For one shell side pass, two tube side passes, Fig. 9.71 applies and $F = 0.98$.

Area

In equation 9.212, $A = Q/UF\theta_m = 1.875n/(0.230 \times 0.98 \times 53.7) = 0.155n$ m^2.

The area per unit length based on 10 mm i.d. $= (\pi \times 0.010 \times 1.0) = 0.0314$ m^2/m and total length of tubing $= 0.155n/0.0314 = 4.94n$ m.

Thus the length of tubes required $= (4.94n/n) = \underline{\underline{4.94 \text{ m}}}$.

PROBLEM 9.19

A condenser consists of a number of metal pipes of outer diameter 25 mm and thickness 2.5 mm. Water, flowing at 0.6 m/s, enters the pipes at 290 K, and it should be discharged at a temperature not exceeding 310 K.

If 1.25 kg/s of a hydrocarbon vapour is to be condensed at 345 K on the outside of the pipes, how long should each pipe be and how many pipes would be needed?

Take the coefficient of heat transfer on the water side as 2.5, and on the vapour side as 0.8 kW/m^2 K and assume that the overall coefficient of heat transfer from vapour to water, based upon these figures, is reduced 20% by the effects of the pipe walls, dirt and scale.

The latent heat of the hydrocarbon vapour at 345 K is 315 kJ/kg.

Solution

Heat load

For condensing the organic at 345 K, $Q = (1.25 \times 315) = 393.8$ kW
If the water outlet temperature is limited to 310 K, then the mass flow of water is given by:

$$393.8 = G \times 4.18(310 - 290) \text{ or } G = 4.71 \text{ kg/s}$$

Temperature driving force

$$\theta_1 = (345 - 290) = 55 \deg \text{ K}, \theta_2 = (345 - 310) = 35 \deg \text{ K}$$

Therefore in equation 9.9, $\theta_m = (55 - 35)/\ln(55/35) = 44.3 \deg$ K.

No correction factor is necessary with isothermal conditions in the shell.

Overall coefficient

Inside: $h_i = 2.5$ kW/m^2 K.

The outside diameter $= 0.025$ m and $d_i = (25 - 2 \times 2.5)/10^3 = 0.020$ m.

Basing the inside coefficient on the outer diameter:

$$h_{io} = (2.5 \times 0.020/0.025) = 2.0 \text{ kW/m}^3 \text{ K}$$

Outside: $h_o = 0.8$ kW/m^2 K

and hence the clean overall coefficient is given by:

$$1/U_c = 1/h_{io} + 1/h_o = 1.75 \text{ m}^2 \text{ K/kW or } U_c = 0.572 \text{ kW/m}^2 \text{ K}$$

Thus allowing for scale and the wall:

$$U_D = 0.572(100 - 20)/100 = 0.457 \text{ kW/m}^2 \text{ K}$$

Area

In equation 9.1: $\qquad A = Q/U\theta_m = 393.8/(0.457 \times 44.3) = 19.45 \text{ m}^2$

$$\text{Outside area} = (\pi \times 0.025 \times 1.0) = 0.0785 \text{ m}^2/\text{m}$$

and hence total length of piping $= (19.45/0.0785) = 247.6$ m.

4.71 kg/s water $\equiv (4.71/1000) = 0.00471$ m^3/s

and hence cross-sectional area/pass to give a velocity of 0.6 m/s

$$= (0.00471/0.6) = 0.00785 \text{ m}^2$$

$$\text{Cross-sectional area of one tube} = (\pi/(0.020)^2 = 0.000314 \text{ m}^2.$$

$$\text{Therefore number of tubes/pass} = (0.00785/0.000314) = 25.$$

Thus:

with 1 tube pass, total tubes $= 25$ and tube length $= (247.6/25) = 9.90$ m

with 2 tube passes, total tubes $= 50$ and tube length $= (247.6/50) = 4.95$ m

with 4 tube passes, total tubes $= 100$ and tube length $= (247.6/100) = 2.48$ m

A tube length of 2.48 m is perhaps the most practical proposition.

PROBLEM 9.20

An organic vapour is being condensed at 350 K on the outside of a bundle of pipes through which water flows at 0.6 m/s; its inlet temperature being 290 K. The outer and inner diameters of the pipes are 19 mm and 15 mm respectively, although a layer of scale, 0.25 mm thick and of thermal conductivity 2.0 W/m K, has formed on the inside of the pipes.

If the coefficients of heat transfer on the vapour and water sides are 1.7 and 3.2 kW/m^2 K respectively and it is required to condense 0.025 kg/s of vapour on each of the pipes, how long should these be, and what will be the outlet temperature of water?

The latent heat of condensation is 330 kJ/kg.

Neglect any resistance to heat transfer in the pipe walls.

Solution

For a total of n pipes, mass flow of vapour condensed $= 25n \times 10^{-3}$ kg/s and hence load, $Q = (0.025n \times 330) = 8.25n$ kW.

For a water outlet temperature of T K and a mass flow of G kg/s:

$$8.25n = G \times 4.18(T - 290) \text{ kW}$$

or

$$G = 1.974n/(T - 290) \text{ kg/s} \qquad \text{(i)}$$

$$\theta_1 = (350 - 290), \; \theta_2 = (350 - T)$$

and hence in equation 9.9:

$$\theta_m = [(350 - 290) - (350 - T)]/\ln[(350 - 290)/(350 - T)]$$

$$= (T - 290)/\ln[60/(350 - T)] \text{ deg K.}$$

Considering the film coefficients: $h_i = 3.2$ kW/m^2 K, $h_o = 1.7$ kW/m^2 K and hence:

$$h_{io} = (3.2 \times 0.015)/0.019 = 2.526 \text{ kW/m}^2 \text{ K.}$$

The scale resistance is:

$$(x/k) = (0.25 \times 10^{-3})/2.0 = 0.000125 \text{ m}^2 \text{ K/W or } 0.125 \text{ m}^2 \text{ K/kW}$$

Therefore the overall coefficient, neglecting the wall resistance is given by:

$$1/U = 1/h_{io} + x/k + 1/h_o$$

$$= (0.5882 + 0.125 + 0.396) = 1.109 \text{ m}^2 \text{ K/kW or } U = 0.902 \text{ kW/m}^2\text{K}$$

Therefore in equation 9.1:

$$A = Q/U\theta_m = 8.25n/\{0.902(T - 290)/\ln[60/(350 - T)]\}\text{m}^2$$

$$= \frac{4.18G(T - 290)\ln[60/(350 - T)]}{0.902(T - 290)} = 4.634G \ln[60/(350 - T)]\text{m}^2 \qquad \text{(ii)}$$

The cross-sectional area for flow $= (\pi/4)(0.015)^2 = 0.000177$ m^2/tube.

$$G \text{ kg/s} \equiv (G/1000) = 0.001G \text{ m}^3/\text{s}$$

and area/pass to give a velocity of 0.6 m/s $= (0.001G/0.6) = 0.00167G$ m^2.

$\therefore$ number of tubes/pass $= (0.00167G/0.000177) = 9.42G \qquad \text{(iii)}$

Area per unit length of tube $= (\pi \times 0.019 \times 1.0) = 0.0597$ m^2/m.

$\therefore$ total length of tubes $= 4.634G \ln[60/350 - T)]/0.0597 = 77.6G \ln[60/350 - T)]$m

length of each tube $= 77.6G \ln[60/350 - T)]/n$ m

and, substituting from (i),

$$\text{tube length} = 77.6 \times 1.974n \ln[60/(350 - T)]/(n(T - 290)]$$

$$= 153.3 \ln[60/(350 - T)]/(T - 290) \text{ m} \qquad \text{(iv)}$$

The procedure is now to select a number of tube passes N and hence m in terms of n from (iii). T is then obtained from (i) and hence the tube length from (iv). The following results are obtained:

No. of tube passes N	Total tubes n	Outlet water temperature T (K)	Tube length (m)
1	9.42G	308.6	3.05
2	18.84G	327.2	3.99
4	37.68G	364.4	—
6	56.52G	401.6	—

Arrangements with 4 and 6 tube side passes require water outlet temperatures in excess of the condensing temperature and are clearly not possible. With 2 tube side passes, $T = 327.2$ K at which severe scaling would result and hence the proposed unit would consist of one tube side pass and a tube length of 3.05 m.

The outlet water temperature would be 308.6 K.

PROBLEM 9.21

A heat exchanger is required to cool continuously 20 kg/s of water from 360 K to 335 K by means of 25 kg/s of cold water, inlet temperature 300 K. Assuming that the water velocities are such as to give an overall coefficient of heat transfer of 2 kW/m² K, assumed constant, calculate the total area of surface required (a) in a counterflow heat exchanger, i.e. one in which the hot and cold fluids flow in opposite directions, and (b) in a multi-pass heat exchanger, with the cold water making two passes through the tubes, and the hot water making one pass along the outside of the tubes. In case (b) assume that the hot-water flows in the same direction as the inlet cold water, and that its temperature over any cross-section is uniform.

Solution

The heat load, $Q = 20 \times 4.18(360 - 335) = 2090$ kW

and the outlet cold water temperature is given by: $2090 = (25 \times 4.18)(T_2 - 300)$ or $T_2 = 320$ K

Case (a)

$$\theta_1 = (360 - 320) = 40 \deg \text{ K}, \quad \theta_2 = (335 - 300) = 35 \deg \text{ K}$$

and in equation 9.9:

$$\theta_m = (40 - 35)/\ln(40/35) = 37.4 \deg \text{ K}$$

As the flow is true counter-flow, no correction factor is necessary and $F = 1.0$. Therefore in equation 9.150:

$$A = Q/UF\theta_m = 2090/(2.0 \times 1.0 \times 37.4) = \underline{27.94 \text{ m}^2}$$

Case (b)

Again, $\theta_m = 37.4$ K. In equation 9.212:

$$X = (\theta_2 - \theta_1)/(T_1 - \theta_1) = (320 - 300)/(360 - 300) = 0.33$$
$$Y = (T_1 - T_2)/(\theta_2 - \theta_1) = (360 - 335)/(320 - 300) = 1.25$$

Hence, from Fig. 9.71, $F = 0.94$
and in equation 9.212:

$$A = 2090/(2.0 \times 0.94 \times 374) = \underline{\underline{29.73 \text{ m}^2}}$$

PROBLEM 9.22

Find the heat loss per unit area of surface through a brick wall 0.5 m thick when the inner surface is at 400 K and the outside at 310 K. The thermal conductivity of the brick may be taken as 0.7 W/m K.

Solution

$$Q = kA(T_1 - T_2)/x \qquad \text{(equation 9.12)}$$
$$= 0.7 \times 1.0(400 - 310)/0.5 = \underline{\underline{126 \text{ W/m}^2}}$$

PROBLEM 9.23

A furnace is constructed with 225 mm of firebrick, 120 mm of insulating brick, and 225 mm of building brick. The inside temperature is 1200 K and the outside temperature 330 K. If the thermal conductivities are 1.4, 0.2, and 0.7 W/m K, find the heat loss per unit area and the temperature at the junction of the firebrick and insulating brick.

Solution

If T_1 K and T_2 K are the temperatures at the firebrick/insulating brick and the insulating brick/building brick junctions respectively, then in equation 9.12, for conduction through the firebrick:

$$Q = 1.4 \times 1.0(1200 - T_1)/0.255 = 6.22(1200 - T_1) \text{ W/m}^2 \qquad \text{(i)}$$

For conduction through the insulating brick:

$$Q = 0.2 \times 1.0(T_1 - T_2)/0.120 = 1.67(T_1 - T_2) \text{ W/m}^2 \qquad \text{(ii)}$$

and for conduction through the building brick:

$$Q = 0.7 \times 1.0(T_2 - 330)/0.225 = 3.11(T_2 - 330) \text{ W/m}^2 \qquad \text{(iii)}$$

The thermal resistances of each material, (x/kA), are: firebrick $= (1/6.22) = 0.161$; insulating brick $= (1/1.67) = 0.60$; building brick $= (1/3.11) = 0.322$ K/Wm^2; and in equation 9.18:

$$(1200 - 330) = (0.161 + 0.60 + 0.322)Q$$

or: $$Q = 803.3 \text{ W/m}^2$$

$\Delta T \text{ firebrick}/\Sigma\Delta T = (x/kA)_{\text{firebrick}}/\Sigma(x/kA)$

$\therefore$ $\quad (1200 - T_1)/(1200 - 330) = 0.161/(0.161 + 0.60 + 0.322) = (0.161/1.083)$

and: $$T_1 = 1071 \text{ K}$$

Similarly for the insulating brick:

$$(1071 - T_2)/(1200 - 330) = (0.60/1.083)$$

and: $$T_2 = 589 \text{ K}$$

PROBLEM 9.24

Calculate the total heat loss by radiation and convection from an unlagged horizontal steam pipe of 50 mm outside diameter at 415 K to air at 290 K.

Solution

Outside area per unit length of pipe $= (\pi \times 0.050 \times 1.0) = 0.157 \text{ m}^2/\text{m}$.

Convection

For natural convection from a horizontal pipe to air, the simplified form of equation 9.102 may be used:

$$h_c = 1.18(\Delta T/d_o)^{0.25}$$

In this case: $\Delta T = (415 - 290) = 125 \text{ deg K}$ and $d_o = 0.050$ m.

$\therefore$ $$h_c = 1.18(125/0.050)^{0.25} = 8.34 \text{ W/m}^2 \text{ K}$$

Thus, heat loss by convection: $q_{c.} = h_c A(T_1 - T_2)$

$$= 8.34 \times 0.157(415 - 290) = 163.7 \text{ W/m}$$

Radiation

An extension of equation 9.118 may be used. Taking the emissivity of the pipe as 0.9:

$$q_r = (5.67 \times 10^{-8} \times 0.9)(415^4 - 290^4) \times 0.157 = 181.0 \text{ W/m}$$

and the total loss is 344.7 W/m length of pipe.

PROBLEM 9.25

Toluene is continuously nitrated to mononitrotoluene in a cast-iron vessel 1 m in diameter fitted with a propeller agitator of 0.3 m diameter driven at 2 Hz. The temperature is maintained at 310 K by circulating cooling water at 0.5 kg/s through a stainless steel coil of 25 mm outside diameter and 22 mm inside diameter wound in the form of a helix of 0.81 m diameter. The conditions are such that the reacting material may be considered to have the same physical properties as 75% sulphuric acid. If the mean water temperature is 290 K, what is the overall heat transfer coefficient?

Solution

The overall coefficient U_o based on the outside area of the coil is given by equation 9.201:

$$1/U_o = 1/h_o + (x_w/k_w)(d_o/d_w) + (1/h_i)(d_o/d_i) + R_o + R_i(d_o/d_i)$$

where d_w is the mean pipe diameter.

Inside

The coefficient on the water side is given by equations 9.202 and 9.203:

$$h_i = (k/d)(1 - 3.5d/d_c)0.023(d_i u\rho/\mu)^{0.8}(C_p\mu/k)^{0.4}$$

where: $\qquad u\rho = 0.5/[(\pi/4) \times 0.022^2] = 1315$ kg/m^2 s

$$d_i = 0.022 \text{ m}, d_c = 0.80 \text{ m}$$

and for water at 290 K: $k = 0.59$ W/m K, $\mu = 0.00108$ Ns/m^2, and $C_p = 4180$ J/kg K.

$\therefore \qquad h_i = (0.59/0.022)(1 + 3.5 \times 0.22/0.80) \times 0.023(0.022 \times 1315/0.00108)^{0.8}$

$\qquad \times (4180 \times 0.00108/0.59)^{0.4}$

$\qquad = 0.680(26,780)^{0.8}(7.65)^{0.4} = 5490$ W/m^2 K

Outside

In equation 9.204:

$$(h_o d_v/k)(\mu_s/\mu)^{0.14} = 0.87(C_p\mu/k)^{0.33}(L^2N\rho/\mu)^{0.62}$$

For 75% sulphuric acid:

$k = 0.40$ W/m K, $\mu_s = 0.0086$ N s/m^2 at 300 K, $\mu = 0.0065$ N s/m^2 at 310 K, $C_p = 1880$ J/kg K and $\rho = 1666$ kg/m^3

$\therefore (h_o \times 1.0/0.40)(0.0086/0.0065)^{0.14} = 0.87(1880 \times 0.0065/0.40)^{0.33}$

$\qquad \times (0.3^2 \times 2.0 \times 1665/0.0065)^{0.62}$

$\therefore \qquad 2.5h_o(1.323)^{0.14} = 0.87(30.55)^{0.33}(46,108)^{0.62}$

and: $\qquad h_o = (0.348 \times 3.09 \times 779)/1.04 = 805.5$ W/m^2 K

Overall

Taking $k_w = 15.9$ W/m K and R_o and R_i as 0.0004 and 0.0002 m^2 K/W respectively, then in equation 9.201:

$$1/U_o = (1/805.5) + (0.0015/15.9)(0.025/0.0235) + (1/5490)(0.025/0.022) + 0.0004$$
$$+ 0.0002(0.025/0.022)$$
$$= (0.00124 + 0.00010 + 0.00021 + 0.00040 + 0.00023) = 0.00218 \text{ } m^2 \text{ K/W}$$

and:
$$U_o = 458.7 \text{ W/m}^2 \text{ K}$$

PROBLEM 9.26

7.5 kg/s of pure iso-butane is to be condensed at a temperature of 331.7 K in a horizontal tubular exchanger using a water inlet temperature of 301 K. It is proposed to use 19 mm outside diameter tubes of 1.6 mm wall arranged on a 25 mm triangular pitch. Under these conditions the resistance of the scale may be taken as 0.0005 m^2 K/W. Determine the number and arrangement of the tubes in the shell.

Solution

The latent heat of vaporisation of isobutane is 286 kJ/kg and hence the heat load:

$$Q = (7.5 \times 286) = 2145 \text{ kW}$$

The cooling water outlet should not exceed 320 K and a value of 315 K will be used. The mass flow of water is then:

$$2145/[4.18(315 - 301)] = 36.7 \text{ kg/s}$$

In order to obtain an approximate size of the unit, a value of 500 W/m^2 K will be assumed for the overall coefficient based on the outside area of the tubes.

$$\theta_1 = (331.7 - 301) = 30.7 \text{ deg K}, \quad \theta_2 = (331.7 - 315) = 16.7 \text{ deg K}$$

and from equation 9.9: $\theta_m = (30.7 - 16.7)/\ln(30.7/16.7) = 23.0$ deg K.

Thus, the approximate area $= (2145 \times 10^3)/(500 \times 23.0) = 186.5 \text{ } m^2$.

The outside area of 0.019 m diameter tubes $= (\pi \times 0.019 \times 1.0) = 0.0597 \text{ } m^2/m$ and hence the total length of tubing $= (186.5/0.0597) = 3125$ m.

Adopting a standard tube length of 4.88 m, number of tubes $= (3125/4.88) = 640$. With the large flow of water involved, a four tube-side pass unit is proposed, and for this arrangement 678 tubes can be accommodated on 25 mm triangular pitch in a 0.78 m i.d. shell. Using this layout, the film coefficients are now estimated and the assumed value of U is checked.

Inside

Water flow through each tube $= 36.7/(678/4) = 0.217$ kg/s.

The tube i.d. $= 19.0 - (2 \times 1.67) = 15.7$ mm

the cross-sectional area for flow $= (\pi/4)(0.0157)^2 = 0.000194$ m^2

and hence the water velocity: $u = 0.217/(1000 \times 0.000194) = 1.12$ m/s.

From equation 9.221 : $h_i = 4280[(0.00488 \times 308) - 1]1.12^{0.8}/0.0157^{0.2}$

$$= (4280 \times 0.503 \times 1.095)/0.436 = 5407 \text{ W/m}^2 \text{ K}$$

or, based on outside diameter: $h_{io} = (5407 \times 0.0157)/0.019$

$$= 4468 \text{ W/m}^2 \text{ K or } 4.47 \text{ kW/m}^2 \text{ K}$$

Outside

The temperature drop across the condensate film, ΔT_f is given by:

(thermal resistance of water film + scale)/(total thermal resistance) $= (\theta_m - \Delta T_f)/\theta_m$

or: $(1/4.47 + 0.5)/(1/0.500) = (23.0 - \Delta T_f)/23.0$

and: $\Delta T_f = 14.7 \deg \text{ K}$

The condensate film is thus at $(331.7 - 14.7) = 317$ K.

The outside film coefficient is given by:

$$h_o = 0.72[(k^3\rho^2 g\lambda)/(jd_o\mu\Delta T_f)]^{0.25} \qquad \text{(equation 9.177)}$$

At 317 K, $k = 0.13$ W/m K, $\rho = 508$ kg/m^3, $\mu = 0.000136$ N s/m^2 and $j = \sqrt{678} = 26.0$.

$\therefore$ $h_o = 0.72[(0.13^3 \times 508^2 \times 9.81 \times 286 \times 10^3)/$

$$(26 \times 19.0 \times 10^{-3} \times 0.000136 \times 14.5)]^{0.25}$$

$$= 814 \text{ W/m}^2 \text{ K or } 0.814 \text{ kW/m}^2 \text{ K}$$

Overall

$$1/U = (1/4.47) + (1/0.814) + 0.50 = 1.952$$

and: $U = 0.512$ kW/m^2 K or 512 W/m^2 K

which is sufficiently near the assumed value. For the proposed unit, the heat load:

$$Q = (0.512 \times 678 \times 4.88 \times 0.0597 \times 23.0) = 2328 \text{ kW}$$

or an overload of: $(2328 - 2145)100/2145 = \underline{\underline{8.5\%}}$

PROBLEM 9.27

37.5 kg/s of crude oil is to be heated from 295 to 330 K by heat transferred from the bottom product from a distillation column. The bottom product, flowing at 29.6 kg/s is to be cooled from 420 to 380 K. There is available a tubular exchanger with an inside shell diameter of 0.60 m, having one pass on the shell side and two passes on the tube side. It has 324 tubes, 19 mm outside diameter with 2.1 mm wall and 3.65 m long, arranged on a 25 mm square pitch and supported by baffles with a 25% cut, spaced at 230 mm intervals. Would this exchanger be suitable?

Solution

Mean temperature of bottom product $= 0.5(420 + 380) = 400$ K.

Mean temperature of crude oil $= 0.5(330 + 295) = 313$ K.

For the crude oil at 313 K: $C_p = 1986$ J/kg K, $\mu = 0.0029$ N s/m^2, $k = 0.136$ W/m K and $\rho = 824$ kg/m^3.

For the bottom product at 400 K: $C_p = 2200$ J/kg K.

Heat loads:

$$\text{tube side: } Q = 37.5 \times 1.986(330 - 295) = 2607 \text{ kW.}$$

$$\text{shell side: } Q = 29.6 \times 2.20(420 - 380) = 2605 \text{ kW.}$$

Outside coefficient

Temperature of wall $= 0.5(400 + 313) = 356.5$ K

and film temperature, $T_f = 0.5(400 + 356.5) = 378$ K.

At 378 K, $\rho = 867$ kg/m^3, $\mu = 0.0052$ N s/m^2, and $k = 0.119$ W/m K

Cross-sectional area for flow $=$ (shell i.d. $\times$ clearance $\times$ baffle spacing)/pitch

$$= (0.60 \times 0.0064 \times 0.23)/0.025 = 0.353 \text{ m}^2$$

(assuming a clearance of 0.0064 m).

$\therefore$ $\qquad\qquad G'_{max} = (29.6/0.0353) = 838.5$ kg/m^2 s

and: $\qquad\qquad Re_{max} = (0.019 \times 838.5)/0.0052 = 306.4$

Therefore in equation 9.90, taking $C_h = 1$:

$$(h_o \times 0.019/0.119) = 0.33 \times 1.0(3064)^{0.6}(2200 \times 0.0052/0.119)^{0.3}$$

or: $\qquad h_o = (2.07 \times 125 \times 3.94) = 1018$ W/m^2 K or 1.02 kW/m^2 K

Inside coefficient

Number tubes per pass $= (324/2) = 162$.

Inside diameter $= [19.0 - (2 \times 2.1)]/1000 = 0.0148$ m

and cross-sectional area for flow $= (\pi/4)(0.0148)^2 = 0.000172 \text{ m}^2$ per tube or:

$$(0.000172 \times 162) = 0.0279 \text{ m}^2 \text{ per pass.}$$

$\therefore$ $\quad\quad\quad G' = (37.5/0.0279) = 1346 \text{ kg/m}^2 \text{ s}$

In equation 9.61:

$$(h_i \times 0.0148/0.136) = 0.023(0.0148 \times 1346/0.0029)^{0.8}(1986 \times 0.0029/0.136)^{0.4}$$

$$h_i = 0.211(6869)^{0.8}(42.4)^{0.4} = 1110 \text{ W/m}^2 \text{ K}$$

or, based on the outside area, $h_{io} = (1110 \times 0.0148)/0.019 = 865 \text{ W/m}^2 \text{ K}$

or: $\quad\quad\quad\quad\quad\quad h_{io} = 0.865 \text{ kW/m}^2 \text{ K.}$

Overall coefficient

Neglecting the wall and scale resistance, the clean overall coefficient is:

$$1/U_c = (1/1.02) + (1/0.865) = 2.136 \text{ m}^2 \text{ K/kW}$$

The area available is $A = (324 \times 3.65 \times \pi \times 0.019) = 70.7 \text{ m}^2$ and hence the minimum value of the design coefficient is:

$$1/U_D = A\theta_m/Q$$

$$\theta_1 = (420 - 330) = 90 \deg \text{ K}, \quad \theta_2 = (380 - 295) = 85 \deg \text{ K}$$

and: $\quad\quad\quad \theta_m = (90 - 85)/\ln(90/85) = 87.5 \deg \text{ K}$

$\therefore$ $\quad\quad\quad 1/U_D = (70.7 \times 87.5)/2607 = 2.37 \text{ m}^2 \text{ K/kW}$

The maximum allowable scale resistance is then:

$$R = (1/U_D) - (1/U_c) = (2.37 - 2.136) = \underline{\underline{0.234 \text{ m}^2 \text{ K/kW}}}$$

This value is very low as seen from Table 9.16, and the exchanger would not give the required temperatures without frequent cleaning.

PROBLEM 9.28

A 150 mm internal diameter steam pipe, carrying steam at 444 K, is lagged with 50 mm of 85% magnesia. What will be the heat loss to the air at 294 K?

Solution

In this case: $d_i = 0.150 \text{ m}$, $d_o = 0.168 \text{ m}$ and $d_w = 0.5(0.150 + 0.168) = 0.159 \text{ m}$. $d_s = (0.168 \times 2 \times 0.050) = 0.268 \text{ m}$ and d_m (the logarithmic mean of d_o and d_s) $= 0.215 \text{ m}$.

The coefficient for condensing steam including any scale will be taken as 8500 W/m^2 K, k_w as 45 W/m K, and k_l as 0.073 W/m K.

The surface temperature of the lagging will be assumed to be 314 K and $(h_r + h_c)$ to be 10 W/m^2 K.

The thermal resistances are therefore:

$$(1/h_i \pi d) = 1/(8500 \times \pi \times 0.150) = 0.00025 \text{ mK/W}$$

$$(x_w/k_w \pi d_w) = 0.009/(45\pi \times 0.159) = 0.00040 \text{ mK/W}$$

$$(x_l/k_l \pi d_m) = 0.050/(0.073\pi \times 0.215) = 1.0130 \text{ mK/W}$$

$$(1/(h_r \times h_c)d_s) = 1/(10 \times 0.268) = 0.1190 \text{ mK/W}$$

Neglecting the first two terms, the total thermal resistance $= 1.132$ mK/W.

From equation 9.261, heat lost per unit length $= (444 - 294)/1.132 = \underline{132.5 \text{ W/m}}$.

The surface temperature of the lagging is given by:

$$\Delta T(\text{lagging})/\Sigma \Delta T = (1.013/1.132) = 0.895$$

and: $$\Delta T(\text{lagging}) = 0.895(444 - 294) = 134 \text{ deg K}$$

Therefore the surface temperature $= (444 - 134) = 310$ K which approximates to the assumed value.

Assuming an emissivity of 0.9:

$$h_r = (5.67 \times 10^{-8} \times 0.9)(310^4 - 294^4)/(310 - 294) = 3.81 \text{ W/m}^2 \text{ K}.$$

For natural convection: $h_c = 1.37(\Delta T/d_s)^{0.25} = 1.37[(310 - 294)/0.268]^{0.25}$
$$= 3.81 \text{ W/m}^2 \text{ K}.$$

$\therefore$ $$(h_r + h_c) = 9.45 \text{ W/m}^2 \text{ K}$$

which again agrees with the assumed value.

In practice forced convection currents are usually present and the heat loss would probably be higher than this value.

For an unlagged pipe and $\Delta T = 150$ K, $(h_r + h_c)$ would be about 20 W/m^2 K and the heat loss, $Q/l = (h_r + h_c)\pi d_0 \Delta T = (20\pi \times 0.168 \times 150) = 1584$ W/m.

Thus the heat loss has been reduced by about 90% by the addition of 50 mm of lagging.

PROBLEM 9.29

A refractory material with an emissivity of 0.40 at 1500 K and 0.43 at 1420 K is at a temperature of 1420 K and is exposed to black furnace walls at a temperature of 1500 K. What is the rate of gain of heat by radiation per unit area?

Solution

In the absence of further data, the system will be considered as two parallel plates.

The radiating source is the furnace walls at $T_1 = 1500$ K and for a black surface, $e_1 = 1.0$.

The heat sink is the refractory at $T_2 = 1420$ K, at which $e_2 = 0.43$.

Putting $A_1 = A_2$ in equation 9.150: $q = (e_1 e_2 \sigma)(T_1^4 - T_2^4)/(e_1 + e_2 - e_1 e_2)$

$$= (1.0 \times 0.43 \times 5.67 \times 10^{-8})(1500^4 - 1420^4)/(1.0 + 0.43 - (0.43 \times 1.0))$$

$$= 2.44 \times 10^{-8}(9.97 \times 10^{11})/1.0 = 2.43 \times 10^4 \text{ W/m}^2 \text{ or } \underline{\underline{24.3 \text{ kW/m}^2}}$$

PROBLEM 9.30

The total emissivity of clean chromium as a function of surface temperature, T K, is given approximately by: $e = 0.38(1 - 263/T)$.

Obtain an expression for the absorptivity of solar radiation as a function of surface temperature, and calculate the values of the absorptivity and emissivity at 300, 400 and 1000 K.

Assume that the sun behaves as a black body at 5500 K.

Solution

It may be assumed that the absorptivity of the chromium at temperature T_1 is the emissivity of the chromium at the geometric mean of T_1 and the assumed temperature of the sun, T_2 where $T_2 = 5500$ K.

Since: $e = 0.38(1 - (263/T))$ (i)

then, taking the geometric mean temperature as $(5500T_1)^{0.5}$:

$$a = 0.38\{1 - [263/(5500T_1)^{0.5}]\} \tag{ii}$$

For the given values of T_1, values of e and a are now calculated from (i) and (ii) respectively to give the following data:

T_1	$(T_1 T_2)^{0.5}$	e	a
300	1285	0.047	0.300
400	1483	0.130	0.312
1000	2345	0.280	0.337

PROBLEM 9.31

Repeat Problem 9.30 for the case of aluminium, assuming the emissivity to be 1.25 times that for chromium.

Solution

In this case:

$$e = (1.25 \times 0.38)(1 - (263/T_1)) = 0.475[(1 - (263/T_1))] \tag{i}$$

and: $a = 0.475 - 1.66T_1^{-0.5}$ (ii)

The following data are obtained by substituting values for T_1 in equations (i) and (ii):

T_1	$(T_1 T_2)^{0.5}$	e	a
300	1285	0.059	0.378
400	1483	0.163	0.391
1000	2345	0.350	0.422

PROBLEM 9.32

Calculate the heat transferred by solar radiation on the flat concrete roof of a building, 8 m by 9 m, if the surface temperature of the roof is 330 K. What would be the effect of covering the roof with a highly reflecting surface, such as polished aluminium, separated from the concrete by an efficient layer of insulation? The emissivity of concrete at 330 K is 0.89, whilst the total absorptivity of solar radiation (sun temperature = 5500 K) at this temperature is 0.60.

Use the data for aluminium from Problem 9.31 which should be solved first.

Solution

The emission from a body with an emissivity, **e**, at a temperature T is given by:

$$I = e\sigma T^4$$

Thus, for the concrete:

$$I = (0.89 \times 5.67 \times 10^{-8} \times 330^4) = 598.5 \text{ W/m}^2$$

Taking $T = 330$ K as the equilibrium temperature, the energy emitted by the concrete must equal the energy absorbed and, since the absorptivity of concrete, **a** = 0.60, the solar flux is then:

$$I_s = (598.5/0.6) = 997.4 \text{ W/m}^2$$

which approximates to the generally accepted figure of about 1 kW/m^2.

With a covering of polished aluminium, then using the data given in Problem 9.31 and an equilibrium surface temperature of T K, the absorptivity is:

$$\mathbf{a} = 0.475 - 1.66 T^{0.5}$$

and, with an area of $(8 \times 9) = 72$ m^2, the energy absorbed is:

$$(1.0 \times 10^3 \times 72)(0.475 - 1.66 T^{0.5}) = 3.42 \times 10^4 - 1.20 \times 10^5 T^{0.5} \text{ W} \qquad \text{(i)}$$

The emissivity is given by:

$$\mathbf{e} = 0.475(1 - 263/T) = 0.475 - 125 T^{-1}$$

and the energy emitted is:

$$(72 \times 5.67 \times 10^{-8} T^4)(0.475 - 125 T^{-1}) = 1.94 \times 10^{-6} T^4 - 5.10 \times 10^{-4} T^3 \text{ W} \qquad \text{(ii)}$$

Equating (i) and (ii):

$$1.94 \times 10^{-6} T^4 - 5.10 \times 10^{-4} T^3 + 1.20 \times 10^5 T^{0.5} = 3.42 \times 10^4$$

or: $$5.67 \times 10^{-11} T^4 - 1.49 \times 10^{-8} T^3 + 3.51 T^{0.5} = 1$$

Solving by trial and error, the equilibrium temperature of the aluminium is:

$$T = \underline{\underline{438 \text{ K}}}.$$

Substituting $T = 438$ K in (i), the energy absorbed and emitted is then 2847 W which represents an increase of some $\underline{\underline{375 \text{ per cent}}}$ compared with the value for the concrete alone.

PROBLEM 9.33

A rectangular iron ingot 15 cm × 15 cm × 30 cm is supported at the centre of a reheating furnace. The furnace has walls of silica-brick at 1400 K, and the initial temperature of the ingot is 290 K. How long will it take to heat the ingot to 600 K?

It may be assumed that the furnace is large compared with the ingot, and that the ingot is always at uniform temperature throughout its volume. Convection effects are negligible.

The total emissivity of the oxidised iron surface is 0.78 and both emissivity and absorptivity may be assumed independent of the surface temperature. (Density of iron = 7.2 Mg/m^3. Specific heat capacity of iron = 0.50 kJ/kg K.)

Solution

As there are no temperature gradients within the ingot, the rate of heating is dependent on the rate of radiative heat transfer to the surface. In addition, since the dimensions of the ingot are much smaller than those of the surrounding surfaces, the ingot may be treated as a black body.

Volume of ingot = $(15 \times 15 \times 30) = 6750$ cm^3 or 0.00675 m^3.

Mass of ingot = $(7.2 \times 10^3 \times 0.00675) = 48.6$ kg.

For an ingot temperature of T K, the increase in enthalpy = $d(mC_pT)/dt$ or $mC_p dT/dt$ where t is the time and C_p the specific heat of the ingot.

The heat received by radiation = $A\sigma \mathbf{a}(T_f^4 - T^4)$ where the area, $A = (4 \times 30 \times 15) + (2 \times 15 \times 15) = 2250$ cm^2 or 0.225 m^2.

The absorptivity $\mathbf{a}$ will be taken as the emissivity = 0.78

and the furnace temperature, $T_f = 1400$ K.

Thus: $$mC_p dT/dt = A\sigma \mathbf{a}(T_f^4 - T^4)$$

or: $$\int_0^t dt = \frac{mC_p}{\mathbf{a}A\sigma} \int_{290}^{600} \frac{dT}{(T_f^4 - T^4)}$$

$$\therefore \qquad t = \left(\frac{(48.6 \times 0.50 \times 10^3)}{(0.78 \times 0.225 \times 5.67 \times 10^{-8})} \right) \left(\frac{1}{(4 \times 1400^3)} \right)$$

$$\times \left(\ln \frac{(1400 + T)}{(1400 - T)} + 2 \tan^{-1} \frac{T}{1400} \right)_{290}^{600} = \underline{\underline{200 \text{ s}}}$$

PROBLEM 9.34

A wall is made of brick, of thermal conductivity 1.0 W/m K, 230 mm thick, lined on the inner face with plaster of thermal conductivity 0.4 W/m K and of thickness 10 mm. If a temperature difference of 30 K is maintained between the two outer faces, what is the heat flow per unit area of wall?

Solution

For an area of 1 m^2,

thermal resistance of the brick:　　$(x_1/k_1A) = 0.230/(1.0 \times 1.0) = 0.230$ K/W

thermal resistance of the plaster:　　$(x_2/k_2A) = 0.010/(0.4 \times 1.0) = 0.0025$ K/W

and in equation 9.18: $30 = (230 + 0.0025)Q$ or $\underline{\underline{Q = 129 \text{ W}}}$

PROBLEM 9.35

A 50 mm diameter pipe of circular cross-section and with walls 3 mm thick is covered with two concentric layers of lagging, the inner layer having a thickness of 25 mm and a thermal conductivity of 0.08 W/m K, and the outer layer a thickness of 40 mm and a thermal conductivity of 0.04 W/m K. What is the rate of heat loss per metre length of pipe if the temperature inside the pipe is 550 K and the outside surface temperature is 330 K?

Solution

From equation 9.22, the thermal resistance of each component is: $(r_2 - r_1)/k(2\pi r_m l)$

　　Thus *for the wall:*　$r_2 = (0.050/2) + 0.003 = 0.028$ m

　　　　　　　　　$r_1 = (0.050/2) = 0.025$ m

and:　　　　　　　$r_m = (0.028 - 0.025)/(\ln 0.028/0.025) = 0.0265$ m.

　　Taking $k = 45$ W/m K and $l = 1.0$ m the thermal resistance is:

$$= (0.028 - 0.025)/(45 \times 2\pi \times 0.0265 \times 1.0) = 0.00040 \text{ K/W}.$$

For the *inner lagging:* $r_2 = (0.028 + 0.025) = 0.053$ m

$$r_1 = 0.028 \text{ m}$$

and: $r_m = (0.053 - 0.028)/(\ln 0.053/0.028) = 0.0392$ m.

Therefore the thermal resistance $= (0.053 - 0.028)/(0.08 \times 2\pi \times 0.0392 \times 1.0)$
$= 1.2688$ K/W

For the *outer lagging:* $r_2 = (0.053 + 0.040) = 0.093$ m

$$r_1 = 0.053 \text{ m}$$

and: $r_m = (0.093 - 0.053)/(\ln 0.093/0.053) = 0.0711$ m

Therefore the thermal resistance $= (0.093 - 0.053)/(0.04 \times 2\pi \times 0.0711 \times 1.0)$
$= 2.2385$ K/W

From equation 9.19: $Q = (550 - 330)/(0.0004 + 1.2688 + 2.2385) = \underline{\underline{62.7 \text{ W/m}}}$

PROBLEM 9.36

The temperature of oil leaving a co-current flow cooler is to be reduced from 370 to
350 K by lengthening the cooler. The oil and water flowrates, the inlet temperatures and
the other dimensions of the cooler will remain constant. The water enters at 285 K and
oil at 420 K. The water leaves the original cooler at 310 K. If the original length is 1 m,
what must be the new length?

Solution

For the *original cooler*, for the oil: $Q = G_o C_{po}(420 - 370)$

and for the water: $Q = G_w C_{pw}(310 - 285)$

$\therefore$ $(G_o C_p/G_w C_p) = (25/50) = 0.5$

where G_o and G_w are the mass flows and C_{po} and C_{pw} the specific heat capacities of the
oil and water respectively.

$\theta_1 = (420 - 285) = 135$ deg K, $\theta_2 = (370 - 310) = 60$ deg K for co-current flow, and
from equation 9.9: $\theta_m = (135 - 60)/\ln(135/60) = 92.5$ deg K

If a is the area per unit length of tube multiplied by the number of tubes, then:

$A = 1.0 \times a$ m^2 and in equation 9.1:

$$G_o C_p(420 - 370) = Ua\ 92.5 \text{ or } (G_o C_p/Ua) = 1.85$$

For the *new cooler*, for the oil: $Q = G_o C_{po}(420 - 350)$

and for the water, $Q = G_w C_{pw}(T - 285)$

where T is the water outlet temperature.

Thus: $\qquad (T - 285) = (G_o C_p / G_w C_p)(420 - 350) = (0.5 \times 70)$

and: $\qquad\qquad\qquad\qquad\qquad T = 320$ K

$\therefore \quad \theta_1 = (420 - 285) = 135 \deg$ K, $\theta_2 = (350 - 320) = 30 \deg$ K, again for co-current flow,

and from equation 9.9: $\theta_m = (135 - 30)/\ln(135/30) = 69.8 \deg$ K

In equation 9.1: $G_o s_o (420 - 350) = Ual69.8$

$\therefore \qquad l = (G_o C_p / Ua) \times 1.003 = (1.85 \times 1.003) = \underline{\underline{1.86 \text{ m}}}$

PROBLEM 9.37

In a countercurrent-flow heat exchanger, 1.25 kg/s of benzene (specific heat 1.9 kJ/kg K and density 880 kg/m^3) is to be cooled from 350 K to 300 K with water which is available at 290 K. In the heat exchanger, tubes of 25 mm external and 22 mm internal diameter are employed and the water passes through the tubes. If the film coefficients for the water and benzene are 0.85 and 1.70 kW/m^2 K respectively and the scale resistance can be neglected, what total length of tube will be required if the minimum quantity of water is to be used and its temperature is not to be allowed to rise above 320 K?

Solution

Heat load:

For the benzene: $Q = 1.25 \times 1.9(350 - 300) = 118.75$ kW.

In order to use the minimum amount, water must leave the unit at the maximum temperature, 320 K. Thus for G kg/s water:

$$118.75 = G \times 4.18(320 - 290) \text{ or } G = 0.947 \text{ kg/s}$$

Temperature driving force

$$\theta_1 = (350 - 320) = 30 \deg \text{ K}, \theta_2 = (300 - 290) = 10 \deg \text{ K}$$

and in equation 9.9: $\theta_m = (30 - 10)/\ln(30/10) = 18.2 \deg$ K. In the absence of further data, it will be assumed that the correction factor is unity.

Overall coefficient

Inside: $h_i = 0.85$ kW/m^2 K or based on the tube o.d., $h_{io} = (0.85 \times 22/25)$ = 0.748 kW/m^2 K.

Outside: $h_o = 1.70$ kW/m^2 K.

Wall: Taking $k_{steel} = 45$ W/m K, $x/k = (0.003/45) = 0.00007$ m^2 K/W or
0.07 m^{2}K/kW.

Thus neglecting any scale resistance: $1/U = (1/0.748) - (1/1.70) + 0.07$
$= 1.995$ m^{2}K/kW

and: $U = 0.501$ kW/m^2 K

Area

In equation 9.1: $A = Q/U\theta_m = 118.75/(0.0501 \times 18.2) = 13.02$ m^2.

Surface area of a 0.025 m o.d. tube $= (\pi \times 0.025 \times 1.0) = 0.0785$ m^2/m and hence total length of tubing required $= (1302/0.0785) = \underline{165.8 \text{ m}}$

PROBLEM 9.38

Calculate the rate of loss of heat from a 6 m long horizontal steam pipe of 50 mm internal diameter and 60 mm external diameter when carrying steam at 800 kN/m^2. The temperature of the surroundings is 290 K.

What would be the cost of steam saved by coating the pipe with a 50 mm thickness of 85% magnesia lagging of thermal conductivity 0.07 W/m K, if steam costs £0.5/100 kg? The emissivity of both the surface of the bare pipe and the lagging may be taken as 0.85, and the coefficient h for the heat loss by natural convection is given by:

$$h = 1.65(\Delta T)^{0.25} \text{ W/m}^2 \text{ K}$$

where ΔT is the temperature difference in deg K. The Stefan-Boltzmann constant is 5.67×10^{-8} W/m^2 K^4.

Solution

For the bare pipe

Steam is saturated at 800 kN/m^2 and 443 K.

Neglecting the inside resistance and that of the wall, it may be assumed that the surface temperature of the pipe is 443 K.

For *radiation* from the pipe, the surface area $= (\pi \times 0.060 \times 6.0) = 1.131$ m^2 and in equation 9.119:

$$q_r = (5.67 \times 10^{-8} \times 0.85 \times 1.131)(443^4 - 290^4) = 1714 \text{ W}.$$

For *convection* from the pipe, the heat loss:

$$q_c = h_c A(T_s - T)$$
$$= 1.65(443 - 290)^{0.25} \times 1.131(443 - 290) = 1.866(443 - 290)^{1.25} = 1004 \text{ W}$$

and the total loss $= 2718$ W or $\underline{\underline{2.71 \text{ kW}}}$

For the insulated pipe

The heat conducted through the lagging q_l must equal the heat lost from the surface $(q_r + q_c)$.

Mean diameter of the lagging $= [(0.060 + 2 \times 0.050) + 0.060]/2 = 0.110$ m

at which the area $= (\pi \times 0.110 \times 6.0) = 2.07 \text{ m}^2$

and in equation 9.12: $q_l = 0.07 \times 2.07(443 - T_s)/0.050 = (1280 - 2.90T_s)$ W

where T_s is the surface temperature.

The outside area $= \pi(0.060 + 2 \times 0.050) \times 6.0 = 3.016 \text{ m}^2$

and from equation 9.119 : $q_r = (5.67 \times 10^{-8} \times 0.85 \times 3.016(T_s^4 - 290^4)$

$$= 1.456 \times 10^{-7} T_s^4 = 1030 \text{ W}$$

and:
$$q_c = 1.65(T_s - 290)^{0.25} \times 3.016(T_s - 290)$$
$$= 4.976(T_s - 290)^{1.25} \text{ W}$$

Making a heat balance:

$$(1280 - 2.90T_s) = (1.456 \times 10^{-7} T_s^4) - 1030 + 4.976(T_s - 290)^{1.25}$$

or:
$$4.976(T_s - 290)^{1.25} + (1.456 \times 10^{-7} T_s^4) + 2.90T_s = 2310$$

Solving by trial and error: $T_s = 305$ K

and hence the heat lost $= (1280 - 2.90 \times 305) = 396$ W.

The heat saved by lagging the pipe $= (2712 - 396) = 2317$ W or 2.317 kW.

At 800 kN/m^2, the latent heat of steam is 2050 kJ/kg

and the reduction in the amount of steam condensed $= (2.317/2050) = 0.00113$ kg/s

or:
$$(0.00113 \times 3600 \times 24 \times 365) = 35{,}643 \text{ kg/year}$$
∴
$$\text{annual saving} = (35{,}643 \times 0.5)/100 = \underline{\underline{£178/\text{year}}}$$

It may be noted that arithmetic mean radius should only be used with thin walled tubes, which is not the case here. If a logarithmic mean radius is used in applying equation 9.8, $T_s = 305.7$ K and the difference is, in this case, negligible.

PROBLEM 9.39

A stirred reactor contains a batch of 700 kg reactants of specific heat 3.8 kJ/kg K initially at 290 K, which is heated by dry saturated steam at 170 kN/m² fed to a helical coil. During the heating period the steam supply rate is constant at 0.1 kg/s and condensate leaves at the temperature of the steam. If heat losses are neglected, calculate the true temperature of the reactants when a thermometer immersed in the material reads 360 K. The bulb of the thermometer is approximately cylindrical and is 100 mm long by 10 mm in diameter with a water equivalent of 15 g, and the overall heat transfer coefficient to the thermometer is 300 W/m² K. What temperature would a thermometer with a similar bulb of half the length and half the heat capacity indicate under these conditions?

Solution

The latent heat of dry saturated steam at 170 kN/m² and 388 K = 2216 kJ/kg.

Therefore heat added to the reactor = $(2216 \times 0.1) = 221.6$ kJ/s = 221.6 kW

which is equal to the increase in enthalpy, dH/dt.

The enthalpy of the contents, neglecting the heat capacity of the reactor and losses = $mC_p dT/dt = (700 \times 3.8) dT/dt$ or $2660 \, dT/dt$ kW

$\therefore$ $$2660 \, dT/dt = 221.6$$

and the rate of temperature rise, $dT/dt = 0.083$ deg K/s.

At time t, the temperature of the reactants is:

$$T = (290 + 0.083t) \text{ K} \tag{i}$$

The increase in enthalpy of the thermometer is equal to the rate of heat transfer from the fluid, or:

$$(mC_p)_t dT_t/dt = UA_t(T - T_t) \tag{ii}$$

where the subscript t refers to the thermometer.

$\therefore$ $\quad (15/1000) \times 4.18(dT_t/dt) = 0.300(\pi \times 0.010 \times 0.100)(T - T_t)$

and: $\quad (dT_t/dt) = 0.0150(T - T_t)$ deg K/s

At time t s, the temperature of the thermometer is therefore:

$$T_t = 290 + [0.0150(T - T_t)]t \text{ K} \tag{iii}$$

When $T_t = 360$ K, then substituting from equation (i):

$$360 = 290 + \{0.0150[290 + 0.083t) - 360]\}t$$

or: $0.00125t^2 - 1.05t - 70 = 0$ and $t = 902$ s

Therefore in (i): $T = (290 + (0.083 \times 902)) = \underline{\underline{364.9 \text{ K}}}$

With half the length, that is 0.050 m, and half the heat capacity, that is 7.5 g water, then in equation (ii):

$$(7.5/1000) \times 4.18(dT_t/dt) = 0.300(\pi \times 0.010 \times 0.050)(T - T_t)$$

or: $$(dT_t/dt) = 0.0150(T - T_t)$$

The same result as before and hence the new thermometer would also read $\underline{\underline{360 \text{ K}}}$.

PROBLEM 9.40

How long will it take to heat 0.18 m^3 of liquid of density 900 kg/m^3 and specific heat 2.1 kJ/kg K from 293 to 377 K in a tank fitted with a coil of area 1 m^2? The coil is fed with steam at 383 K and the overall heat transfer coefficient can be taken as constant at 0.5 kW/m^2 K. The vessel has an external surface of 2.5 m^2, and the coefficient for heat transfer to the surroundings at 293 K is 5 W/m^2 K.

The batch system of heating is to be replaced by a continuous countercurrent heat exchanger in which the heating medium is a liquid entering at 388 K and leaving at 333 K. If the heat transfer coefficient is 250 W/m^2 K, what heat exchange area is required? Heat losses may be neglected.

Solution

Mass of liquid in the tank $= (0.18 \times 900) = 162$ kg

$\therefore$ $mC_p = (162 \times 2100) = 340,200$ J/deg K

Using the argument given in Problem 9.77:

$$340,200 \, dT/dt = (500 \times 1)(383 - T) - (5 \times 2.5)(T - 293)$$

or: $$= 191,500 - 500T - 12.5T + 3663 = 195,163 - 512.5T$$

or: $664 \, dT/dt = 380.8 - T$

The time taken to heat the liquid from 293 to 377 K is:

$$t = 664 \int_{293}^{377} dT/(380.8 - T)$$

$$= 664 \ln[(380.8 - 293)/(380.8 - 377)] = \underline{\underline{2085 \text{ s}}} \ (0.58 \text{ h})$$

For the heat exchanger:

$$\Delta T_1 = (388 - 377) = 11 \deg K, \ \Delta T_2 = (333 - 293) = 40 \deg K$$

and from equation 9.9:

$$\Delta T_m = (40 - 11)/\ln(40/11) = 22.5 \deg K.$$

Mass flow $= (162/2085) = 0.0777$ kg/s

Heat load: $Q = 0.0777 \times 2.1(377 - 293) = 13.71$ kW

In equation 9.1: $U = (250/1000) = 0.250$ kW/m^2 K,

The area required: $A = 13.71/(0.250 \times 22.5) = \underline{\underline{2.44 \ m^2}}$.

PROBLEM 9.41

The radiation received by the earth's surface on a clear day with the sun overhead is 1 kW/m^2 and an additional 0.3 kW/m^2 is absorbed by the earth's atmosphere. Calculate approximately the temperature of the sun, assuming its radius to be 700,000 km and the distance between the sun and the earth to be 150,000,000 km. The sun may be assumed to behave as a black body.

Solution

The total radiation received $= 1.3$ kW/m^2 of the earth's surface. The equivalent surface area of the sun is obtained by comparing the area of a sphere at the radius of the sun, 7×10^5 km and the area of a sphere of radius (radius of sun $+$ distance between sun and earth) or:

$$A_1/A_2 = 4\pi(7 \times 10^5)^2/4\pi(150 \times 10^6 + 7 \times 10^5)^2 = 2.16 \times 10^{-5}.$$

Therefore radiation at the sun's surface $= (1.3 \times 10^3/2.16 \times 10^{-5}) = 6.03 \times 10^7$ W/m^2. For a black body, the intensity of radiation is given by equation 9.112:

$$6.03 \times 10^7 = 5.67 \times 10^{-8}T^4 \quad \text{and} \quad \underline{\underline{T = 5710 \ K}}$$

PROBLEM 9.42

A thermometer is immersed in a liquid which is heated at the rate of 0.05 K/s. If the thermometer and the liquid are both initially at 290 K, what rate of passage of liquid over the bulb of the thermometer is required if the error in the thermometer reading after 600 s is to be no more than 1 deg K? Take the water equivalent of the thermometer as 30 g, the heat transfer coefficient to the bulb to be given by $U = 735 \ u^{0.8}$ W/m^2 K. The area of the bulb is 0.01 m^2 where u is the velocity in m/s.

Solution

If T and T' are the liquid and thermometer temperatures respectively after time t s, then:

$$dT/dt = 0.05 \text{ K/s} \quad \text{and hence} \quad T = 290 + 0.05t$$

When $t = 600$ s, $(T - T') = 1$.

$$\therefore \qquad T = 290 + (600 \times 0.05) = 320 \text{ K} \quad \text{and} \quad T' = 319 \text{ K}$$

Balancing: $G_t C \rho_t \, dT'/dt = UA(T - T')$

$$\therefore \qquad (30/1000)4.18 \, dT'/dt = UA(290 + 0.05t + T')$$

$$\therefore \qquad dT'/dt + 7.98UAT' = 2312UA(1 + 0.000173t)$$

$$\therefore \qquad e^{7.98UAt}T' = 2312UA\left\{(1 + 0.000173t)\frac{e^{7.98UAt}}{7.98UA} - \int 0.000173\frac{e^{7.98UAt}}{7.98UA} \, dt\right\}$$

$$= 290(1 + 0.000173t)e^{7.98UAt} - 0.050\frac{e^{7.98UAt}}{7.98UA} + k$$

When $t = 0$, $T' = 290$ K and $k = 0.00627/UA$.

$$\therefore \qquad T' = 290(1 + 0.000173t) - (0.00627/UA)(1 - e^{-7.98UAt})$$

When $t = 600$ s, $T' = 319$ K.

$$\therefore \qquad 319 = 320 + (0.00627/UA)(1 - e^{-4789UA})$$

$$\therefore \qquad -4789UA = \ln(1 - 159.5UA)$$

and: $\qquad UA = -0.000209\ln(1 - 159.5UA)$

Solving by trial and error: $UA = 0.00627$ kW/K.

$$A = 0.01 \text{ m}^2 \text{ and hence: } U = 0.627 \text{ kW/m}^2 \text{ K or } 627 \text{ W/m}^2\text{K}$$

$$\therefore \qquad 627 = 735u^{0.8} \quad \text{and} \quad \underline{\underline{u = 0.82 \text{ m/s}}}$$

PROBLEM 9.43

In a shell-and-tube type of heat exchanger with horizontal tubes 25 mm external diameter and 22 mm internal diameter, benzene is condensed on the outside of the tubes by means of water flowing through the tubes at the rate of 0.03 m^3/s. If the water enters at 290 K and leaves at 300 K and the heat transfer coefficient on the water side is 850 W/m^2 K, what total length of tubing will be required?

Solution

Mass flow of water $= (0.03 \times 1000) = 30$ kg/s

and hence the heat load $= 30 \times 4.18(300 - 290) = 1254$ kW

At atmospheric pressure, benzene condenses at 353 K and hence:

$$\theta_1 = (353 - 290) = 63 \deg K, \quad \theta_2 = (353 - 300) = 53 \deg K$$

and from equation 9.9:

$$\theta_m = (63 - 53)/\ln(63/53) = 57.9 \deg K$$

No correction factor is required, because of isothermal conductions on the shell side.

For condensing benzene, h_o will be taken as 1750 W/m² K. From Table 9.18: $h_i = 850$ W/m² K or, based on the outside diameter, $h_{io} = (850 \times 22/25) = 748$ W/m² K. Neglecting scale and wall resistances:

$$1/U = (1/1750) + (1/748) = 0.00191 \text{ m}^2 \text{ K/W}$$

and: $\qquad\qquad U = 524$ W/m² K or 0.524 kW/m² K

Therefore, from equation 9.1: $A = 1254/(0.524 \times 57.9) = 41.3$ m².

Outside area of 0.025 m tubing $= (\pi \times 0.025 \times 1.0) = 0.0785$ m²/m

and total length of tubing required $= (41.3/0.0785) = \underline{\underline{526 \text{ m}}}$.

PROBLEM 9.44

In a contact sulphuric acid plant, the gases leaving the first convertor are to be cooled from 845 to 675 K by means of the air required for the combustion of the sulphur. The air enters the heat exchanger at 495 K. If the flow of each of the streams is 2 m³/s at NTP, suggest a suitable design for a shell-and-tube type of heat exchanger employing tubes of 25 mm internal diameter.

(a) Assume parallel co-current flow of the gas streams.
(b) Assume parallel countercurrent flow.
(c) Assume that the heat exchanger is fitted with baffles giving cross-flow outside the tubes.

Solution

Heat load

At a mean temperature of 288 K, the density of air $= (29/22.4)(273/288) = 1.227$ kg/m³, where 29 kg/kmol is taken as the mean molecular mass of air.

$\therefore \qquad\qquad$ mass flow of air $= (2.0 \times 1.227) = 2.455$ kg/s.

If, as a first approximation, the thermal capacities of the two streams can be assumed equal for equal flowrates, then the outlet air temperature $= 495 + (845 - 675) = 665$ K and for a mean specific heat of 1.0 kJ/kg K, the heat load is $Q = 2.455 \times 1.0(665 - 495) = 417.4$ kW

For gas to gas heat transfer, an overall coefficient of $1/(1/60 + 1/60) = 30$ W/m^2 K will be taken using the data in Table 9.17.

(a) Co-current flow

$$\theta_1 = (845 - 495) = 350 \deg K, \quad \theta_2 = (675 - 665) = 10 \deg K$$

and in equation 9.9: $\theta_m = (350 - 10)/\ln(350/10) = 95.6 \deg K$.

Therefore in equation 9.1: $A = (417.4 \times 10^3)/(30 \times 95.6) = 145.5$ m^2.

For 25 mm i.d. tubes an o.d. of 32 mm will be assumed for which the outside area $= (\pi \times 0.032 \times 1.0) = 0.1005$ m^2/m

and total length of tubing $= (145.5/0.1005) = 1447$ m.

At a mean air temperature of 580 K: $\rho = (29/22.4)(273/580) = 0.609$ kg/m^3.

$\therefore$ volume flow of air $= (2.445/0.609) = 4.03$ m^3/s.

For a reasonable gas velocity of say 15 m/s: area for flow $= (4.03/15) = 0.268$ m^2. Cross-sectional area of one tube $= (\pi/4)0.025^2 = 0.00050$ m^2.

$\therefore$ number of tubes/pass $= (0.268/0.00050) = 545$, each of length $= (1447/545) = 2.65$ m.

In practice, the standard length of 2.44 m would be adopted with $(1447/2.44) = 594$ tubes in a single pass.

(b) Countercurrent flow

In this case, $\theta_1 = (845 - 665) = 180 \deg K$, $\theta_2 = (675 - 495) = 180 \deg K$, and $\theta_m = 180 \deg K$

In equation 9.1: $A = (417.4 \times 10^3)/(30 \times 180) = 77.3$ m^2

and total length of tubing $= (77.3/0.1005) = 769$ m.

With a velocity of 15 m/s, each tube would be $(769/545) = 1.41$ m long.

A better arrangement would be the use of $(769/2.44) = 315$ tubes, 2.44 m long, though this would give a higher velocity and hence an increased air side pressure drop. With such an arrangement, 315×32 mm o.d. tubes could be accommodated in a 838 mm i.d. shell on 40 mm triangular pitch.

(c) Cross flow

As in (b), $\theta_m = 180 \deg K$.

From equation 9.213: $X = (t_2 - t_1)/(T_1 - t_1) = (665 - 495)/(845 - 495) = 0.486$

and: $Y = (T_1 - T_2)/(t_2 - t_1) = (845 - 675)/(665 - 495) = 1.0$

Thus, assuming one shell pass, two tube-side passes, from Figure 9.71: $F = 0.82$ and $\theta_m F = (0.82 \times 180) = 147.6$ K

Thus, in equation 9.212: $A = (417.4 \times 10^3)/(30 \times 147.6) = 94.3$ m^2 and:

$$\text{total length of tubing} = (94.3/0.1005) = 938 \text{ m}.$$

Using standard tubes 2.44 m long, number of tubes $= (938/2.44) = 384$ or $(384/2) = 192$ tubes/pass.

The cross-sectional area for flow would then be $(192 \times 0.00050) = 9.61 \times 10^{-2}$ m^2 and the air velocity $= (4.03/9.61 \times 10^{-2}) = 41.9$ m/s.

This is not excessive providing the minimum acceptable pressure drop is not exceeded. The nearest standard size is $\underline{390 \times 32 \text{ mm o.d. tubes, 2.44 m in a 940 mm i.d. shell}}$ arranged on 40 mm triangular pitch in two passes.

PROBLEM 9.45

A large block of material of thermal diffusivity $D_H = 0.0042$ cm^2/s is initially at a uniform temperature of 290 K and one face is raised suddenly to 875 K and maintained at that temperature. Calculate the time taken for the material at a depth of 0.45 m to reach a temperature of 475 K on the assumption of unidirectional heat transfer and that the material can be considered to be infinite in extent in the direction of transfer.

Solution

This problem is identical to Problem 9.2 except for slight variations in temperature, and reference may be made to that solution.

PROBLEM 9.46

A 50% glycerol–water mixture is flowing at a Reynolds number of 1500 through a 25 mm diameter pipe. Plot the mean value of the heat transfer coefficient as a function of pipe length, assuming that: $Nu = 1.62(Re\,Pr\,d/l)^{0.33}$.

Indicate the conditions under which this is consistent with the predicted value $Nu = 4.1$ for fully developed flow.

Solution

For 50% glycerol–water at, say, 290 K: $\mu = 0.007$ N s/m^2, $k = 0.415$ W/m K and $C_p = 3135$ J/kg K.

$\therefore \qquad (h \times 0.025)/0.415 = 1.62[(1500 \times 3135 \times 0.007/0.415)(0.025/l)]^{0.33}$

$\therefore \qquad\qquad h = 26.89(1983/l)^{0.33} = \underline{\underline{330/l^{0.33}}} \text{ W/m}^2 \text{ K}$

h is plotted as a function of l over the range $l = 0$–10 m in Fig. 9d.

When $Nu = 4.1$: $h = 4.1k/d = (4.1 \times 0.415)/0.025 = 68.1$ W/m^2 K.

Taking this as a point value, $l = (330/68.1)^3 = \underline{113.8 \text{ m}}$

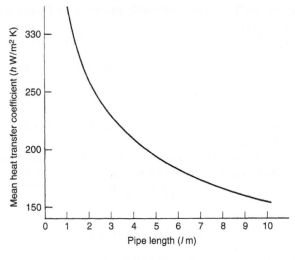

Figure 9d.

which would imply that the flow is fully developed at this point. For further discussion on this point reference should be made to the turbulent flow of gases in Section 9.4.3.

PROBLEM 9.47

A liquid is boiled at a temperature of 360 K using steam fed at 380 K to a coil heater. Initially the heat transfer surfaces are clean and an evaporation rate of 0.08 kg/s is obtained from each square metre of heating surface. After a period, a layer of scale of resistance 0.0003 m^2 K/W, is deposited by the boiling liquid on the heat transfer surface. On the assumption that the coefficient on the steam side remains unaltered and that the coefficient for the boiling liquid is proportional to its temperature difference raised to the power of 2.5, calculate the new rate of boiling.

Solution

When the surface is clean, taking the wall and the inside resistances as negligible, the surface temperature will be 380 K.

Thus:
$$Q = h_0 A (T_s - T)$$

where $Q = GL$, G kg/s is the rate of evaporation of fluid of latent heat L J/kg, $A = 1$ m^2, and T_s and T are the surface and fluid temperature respectively.

∴ $\qquad\qquad 0.08L = h_o \times 1.0(380 - 360)$ or $h_o = 0.004L$

$$h_o \propto (T_s - T)^{2.5}$$

or: $\qquad\qquad h_o = k'(380 - 360)^{2.5} = 1.79 \times 10^3 k'$

∴ $\qquad\qquad k' = 0.004L/(1.79 \times 10^3) = 2.236 \times 10^{-6}L$

When the scale has formed, the total resistance is:

$$0.0003 + 1/[2.236 \times 10^{-6}L(T_s - 360)^{2.5}] = 0.0003 + 4.472 \times 10^5 L^{-1}(T_s - 360)^{-2.5}$$

For conduction through the scale:

$$GL = (380 - T_s)/0.0003 = 3.33 \times 10^3 (380 - T_s) \qquad \text{(i)}$$

For transfer through the outside film:

$$GL = (t - 360)/[4.472 \times 10^5 L^{-1}(T_s - 360)^{-2.5}] - 2.236 \times 10^{-6}L(T_s - 360)^{3.5} \qquad \text{(ii)}$$

and for overall transfer:

$$GL = (380 - 360)/[0.0003 + 4.472 \times 10^5 L^{-1}(T_s - 360)^{-2.5}] \qquad \text{(iii)}$$

Inspection of these equations shows that the rate of evaporation G is a function not only of the surface temperature T_s but also of the latent heat of the fluid L. Using equations (i) and (ii) and selecting values of T in the range 360 to 380 K, the following results are obtained:

Surface temperature T_s (K)	Mass rate of evaporation G (kg/s)	Latent heat of vaporisation L (kJ/kg)
362	0.000025	2,400,000
364	0.00029	186,000
366	0.0012	39,600
368	0.0033	12,200
370	0.0071	4710
372	0.013	1990
374	0.023	869
376	0.036	364
378	0.055	121
380	0.080	0

At a boiling point of 360 K it is likely that the liquid is organic with a latent heat of, say, 900 kJ/kg. This would indicate a surface temperature of 374 K and an evaporation rate of 0.023 kg/s. A precise result requires more specific data on the latent heat.

PROBLEM 9.48

A batch of reactants of specific heat 3.8 kJ/kg K and of mass 1000 kg is heated by means of a submerged steam coil of area 1 m^2 fed with steam at 390 K. If the overall heat transfer coefficient is 600 W/m^2 K, calculate the time taken to heat the material from 290 to 360 K if heat losses to the surroundings are neglected.

If the external area of the vessel is 10 m^2 and the heat transfer coefficient to the surroundings at 290 K is 8.5 W/m^2 K, what will be the time taken to heat the reactants over the same temperature range and what is the maximum temperature to which the reactants can be raised?

What methods would you suggest for improving the rate of heat transfer?

Solution

Use is made of equation 9.209:

$$\ln[(T_s - T_1)]/(T_s - T_2) = UAt/GC_p$$

$\therefore$ $\ln[(390 - 290)/(390 - 360)] = 600 \times 1.0t/(1000 \times 3.8 \times 10^3)$

or: $\ln 3.33 = 0.000158t$ and $\underline{t = 7620 \text{ s}}$ (2.12 h)

The heat lost from the vessel: $Q_L = hA_v(T - T_a)$, where T_a is the ambient temperature.

$\therefore$ $Q_L = 8.5 \times 10.0(T - 290) = (85.0T - 24650) \text{ W}$

Heat from the steam = heat to the reactants + heat losses

$\therefore$ $UA(T_s - T) = GC_p \, dT/dt + 85.0T - 24650$

$$600 \times 1.0(390 - T) = (1000 \times 3.8 \times 10^3) \, dT/dt + 85.0T - 24650$$

$$\int_0^t dt = 5548 \int_{T_1}^{T_2} dT/(3777.6 - T)$$

$\therefore$ $t = 5548 \ln[(377.6 - T_1)/(337.6T_2)]$

$$= 5548 \ln[(377.6 - 290)/(377.6 - 360)] = \underline{8904 \text{ s}} \text{ (2.47 h)}$$

The maximum temperature of the reactants is attained when the heat transferred from the steam is equal to the heat losses, or:

$$UA(T_s - T) = hA_v(T - T_a)$$

Thus: $600 + 1.0(390 - T) = 8.5 \times 10.0(T - 290)$ and $\underline{T = 378 \text{ K}}$

The heating-up time could be reduced by improving the rate of heat transfer to the fluid, by agitation of the fluid for example, and by reducing heat losses from the vessel by insulation. In the case of a large vessel there is a limit to the degree of agitation and circulation of the fluid through an external heat exchanger is an attractive alternative.

PROBLEM 9.49

What do you understand by the terms "black body" and "grey body" when applied to radiant heat transfer?

Two large parallel plates with grey surfaces are situated 75 mm apart; one has an emissivity of 0.8 and is at a temperature of 350 K and the other has an emissivity of 0.4 and is at a temperature of 300 K. Calculate the net rate of heat exchange by radiation per square metre taking the Stefan–Boltzmann constant as 5.67×10^{-8} W/m^2 K^4. Any formula (other than Stefan's law) which you use must be proved.

Solution

The terms "black body" and "grey body" are discussed in Sections 9.5.2 and 9.5.3.

For two large parallel plates with grey surfaces, the heat transfer by radiation between them is given by putting $A_1 = A_2$ in equation 150 to give:

$$q = [e_1 e_2 \sigma / (e_1 + e_2 - e_1 e_2)](T_1^4 - T_2^4) \text{ W/m}^2$$

In this case: $q = [(0.8 \times 0.4 \times 5.67 \times 10^{-8})/(0.8 + 0.4 - 0.8 \times 0.4)](350^4 - 300^4)$

$$= (0.367 \times 5.67 \times 10^{-8} \times 6.906 \times 10^9) = \underline{\underline{143.7 \text{ W/m}^2}}$$

PROBLEM 9.50

A longitudinal fin on the outside of a circular pipe is 75 mm deep and 3 mm thick. If the pipe surface is at 400 K, calculate the heat dissipated per metre length from the fin to the atmosphere at 290 K if the coefficient of heat transfer from its surface may be assumed constant at 5 W/m^2 K. The thermal conductivity of the material of the fin is 50 W/m K and the heat loss from the extreme edge of the fin may be neglected. It should be assumed that the temperature is uniformly 400 K at the base of the fin.

Solution

The heat lost from the fin is given by equation 9.254:

$$Q_f = \sqrt{(hbkA)}\theta_1 \tan hmL$$

where h is the coefficient of heat transfer to the surroundings = 5 W/m^2 K, b is the fin perimeter = $(2 \times 0.075 + 0.003) = 0.153$ m, k is the thermal conductivity of the fin = 50 W/mK, A is the cross-sectional area of the fin = $(0.003 \times 1.0) = 0.003$ m^2, θ_1 is the temperature difference at the root = $(T_1 - T_G) = (400 - 290) = 100$ deg K, $m = \sqrt{(hb/kA)} = \sqrt{((5 \times 0.153)/(50 \times 0.003))} = 2.258$ and L is the length of the fin = 0.075 m.

$\therefore$ $$Q_f = \sqrt{(5 \times 0.153 \times 50 \times 0.003)}[110 \tanh(2.258 - 0.075)]$$

$$= (0.339 \times 100 \tanh 0.169) = \underline{\underline{6.23 \text{ W/m}}}$$

PROBLEM 9.51

Liquid oxygen is distributed by road in large spherical insulated vessels, 2 m internal diameter, well lagged on the outside. What thickness of magnesia lagging, of thermal conductivity 0.07 W/m K, must be used so that not more than 1% of the liquid oxygen evaporates during a journey of 10 ks (2.78 h) if the vessel is initially 80% full? Latent heat of vaporisation of oxygen = 215 kJ/kg. Boiling point of oxygen = 90 K. Density of liquid oxygen = 1140 kg/m^3. Atmospheric temperature = 288 K. Heat transfer coefficient from the outside surface of the lagging surface to atmosphere = 4.5 W/m^2 K.

Solution

For conduction through the lagging:

$$Q = 4\pi k (T_1 - T_2)/(1/r_1 - 1/r_2) \qquad \text{(equation 9.25)}$$

where T_1 will be taken as the temperature of boiling oxygen $= 90$ K and the tank radius, $r_1 = 1.0$ m.

In this way, the resistance to heat transfer in the inside film and the walls is neglected. r_2 is the outer radius of the lagging.

$$\therefore \qquad Q = 4\pi \times 0.07(90 - T_2)/(1/1.0 - 1/r_2) \text{ W} \qquad \text{(i)}$$

For heat transfer from the outside of the lagging to the surroundings, $Q = hA(T_2 - T_a)$ where $h = 4.5$ W/m² K, $A = 4\pi r_2^2$ and T_a, ambient temperature $= 288$ K.

$$\therefore \qquad Q = 4.5 \times 4\pi r_2^2 (T_2 - 288) = 18\pi r_2^2 (T_2 - 288) \text{ W} \qquad \text{(ii)}$$

The volume of the tank $= 4\pi r_1^3/3 = (4\pi \times 1.0^3/3) = 4.189$ m³.

$$\therefore \qquad \text{volume of oxygen} = (4.189 \times 80/100) = 3.351 \text{ m}^3$$

$$\text{and mass of oxygen} = (3.351 \times 1140) = 3820 \text{ kg}$$

$$\therefore \qquad \text{mass of oxygen which evaporates} = (3820 \times 1/100) = 38.2 \text{ kg}$$

or: $\qquad 38.2/(10 \times 10^3) = 0.00382$ kg/s

$\therefore$ heat flow into vessel: $\qquad Q = (215 \times 10^3 \times 0.00382) = 821$ W

$\therefore$ In (ii) $\qquad 821 = 18\pi r_2^2 (T_2 - 288)$ and $T_2 = 288 - (14.52/r_2^2)$

Substituting in (i):

$$821 = 4\pi \times 0.07[90 - 288 + (14.52/r_2^2)]/(1 - 1/r_2)$$

or: $\qquad r_2^2 - 1.27r_2 + 0.0198 = 0$ and $r_2 = 1.25$ m

Thus the thickness of lagging $= (r_2 - r_1) = \underline{\underline{0.25 \text{ m}}}$.

PROBLEM 9.52

Benzene is to be condensed at the rate of 1.25 kg/s in a vertical shell and tube type of heat exchanger fitted with tubes of 25 mm outside diameter and 2.5 m long. The vapour condenses on the outside of the tubes and the cooling water enters at 295 K and passes through the tubes at 1.05 m/s. Calculate the number of tubes required if the heat exchanger is arranged for a single pass of the cooling water. The tube wall thickness is 1.6 mm.

Solution

Preliminary calculation

At 101.3 kN/m^2, benzene condenses at 353 K at which the latent heat $= 394$ kJ/kg.

$\therefore$ heat load: $Q = (1.25 \times 394) = 492$ kW

The maximum water outlet temperature to minimise scaling is 320 K and a value of 300 K will be selected. Thus the water flow is given by:

$$492 = G \times 4.18(300 - 295)$$

or: $\qquad G = 23.5$ kg/s [or $(23.5/1000) = 0.0235$ m^3/s]

$\therefore$ area required for a velocity of 1.05 m/s $= (0.0235/1.05) = 0.0224$ m^2

The cross-sectional area of a tube of $(25 - 2 \times 1.6) = 21.8$ mm i.d. is:

$$(\pi/4) \times 0.0218^2 = 0.000373 \text{ m}^2$$

and hence number of tubes required $= (0.0224/0.000373) = 60$ tubes.

The outside area $= (\pi \times 0.025 \times 2.5 \times 60) = 11.78$ m^2

$$\theta_1 = (353 - 295) = 58 \deg K, \quad \theta_2 = (353 - 300) = 53 \deg K$$

and in equation 9.9: $\quad \theta_m = (58 - 53)/\ln(58/53) = 55.5 \deg K$

$\therefore$ $\qquad U = 492/(55.5 \times 11.78) = 0.753$ kW/m^2 K

This is quite reasonable as it falls in the middle of the range for condensing organics as shown in Table 9.17. It remains to check whether the required overall coefficient will be attained with this geometry.

Overall coefficient

Inside:

The simplified equation for water in tubes may be used:

$$h_i = 4280(0.00488T - 1)u^{0.8}/d_i^{0.2} \text{ W/m}^2 \text{ K.} \qquad \text{(equation 9.221)}$$

where $\quad T = 0.5(300 + 295) = 297.5$ K

$\quad u = 105$ m/s and $d_i = 0.0218$ m

$\therefore$ $\quad h_i = 4280(0.00488 \times 297.5 - 1)1.05^{0.8}/0.218^{0.2} = 4322$ W/m^2 K or 4.32 kW/m^2 K

Based on the outside diameter:

$$h_{io} = (4.32 \times 0.218/0.025) = 3.77 \text{ kW/m}^2 \text{ K}$$

Wall:

For steel, $k = 45$ W/m K, $x = 0.0016$ m and hence:

$$x/k = (0.0016/45) = 0.000036 \text{ m}^2 \text{ K/W} \quad \text{or} \quad 0.036 \text{ m}^2 \text{ K/kW}$$

Outside:
For condensation on vertical tubes:

$$h_o(\mu^2/k^3\rho^2 g)^{0.33} = 1.47(4M/\mu)^{-0.33} \qquad \text{(equation 9.174)}$$

The wall temperature is approximately $0.5(353 + 297.5) = 325$ K, and the benzene film temperature will be taken as $0.5(353 + 325) = 339$ K.

At 339 K: $k = 0.15$ W/m K, $\rho = 880$ kg/m^3, and $\mu = 0.35 \times 10^{-3}$ Ns/m^2.

With 60 tubes, the mass flow of benzene per tube, $G' = (1.25/60) = 0.0208$ kg/s.

For vertical tubes, $M = G'/\pi d_o = 0.0208/(\pi \times 0.025) = 0.265$ kg/ms

$$\therefore \quad h_o[(0.35 \times 10^{-3})^2/0.15^2 \times 880^2 \times 9.8]^{0.33} = 1.47[4 \times 0.0208/(0.35 \times 10^{-3})]^{-0.33}$$

$$\therefore \qquad\qquad 1.699 \times 10^{-4} h_o = (1.47 \times 1.62 \times 10^{-1})$$

and: $\qquad\qquad\qquad\qquad h_o = 1399$ W/m^2 K or 1.40 kW/m^2 K

Overall:
Neglecting scale resistances:

$$1/U = 1/h_{io} + x/k + 1/h_o = 0.265 + 0.036 + 0.714 = 1.015 \text{ m}^2 \text{ K/kW}$$

and: $\qquad U = 0.985$ kW/m^2 K

This is in excess of the value required and would allow for a reasonable scale resistance. If this were negligible, the water throughput could be reduced.

On the basis of these calculations, 60 tubes are required.

PROBLEM 9.53

One end of a metal bar 25 mm in diameter and 0.3 m long is maintained at 375 K and heat is dissipated from the whole length of the bar to surroundings at 295 K. If the coefficient of heat transfer from the surface is 10 W/m^2 K, what is the rate of loss of heat? Take the thermal conductivity of the metal as 85 W/m K.

Solution

Use is made of equation 9.254:

$$Q_f = \sqrt{(hbkA)}\theta_1 \tanh mL$$

where the coefficient of heat transfer from the surface, $h = 10$ W/m^2 K; the perimeter, $b = (\pi \times 0.025) + 0.0785$ m; the cross-sectional area, $A = (\pi/4) \times 0.025^2 = 0.000491$ m^2; the thermal conductivity of the metal, $k = 85$ W/m K; the temperature difference at the root, $\theta_1 = (375 - 295) = 80$ deg K; the value of $m = \sqrt{(hb/kA)} = \sqrt{[(10 \times 0.0785)/(85 \times 0.000491)]} = 4.337$, and the length of the rod, $L = 0.3$ m.

$$\therefore \qquad Q_f = \sqrt{(10 \times 0.0785 \times 85 \times 0.000491)}[80 \tanh(4.337 \times 0.3)]$$

$$= \sqrt{(0.0328)}(80 \tanh 1.3011)$$

$$= 14.49(e^{1.301} - e^{-1.301})/(e^{1.301} + e^{-1.301})$$

$$= 14.49(3.673 - 0.272)/(3.673 + 0.272) = \underline{\underline{12.5 \text{ W}}}$$

PROBLEM 9.54

A shell-and-tube heat exchanger consists of 120 tubes of internal diameter 22 mm and length 2.5 m. It is operated as a single-pass condenser with benzene condensing at a temperature of 350 K on the outside of the tubes and water of inlet temperature 290 K passing through the tubes. Initially there is no scale on the walls, and a rate of condensation of 4 kg/s is obtained with a water velocity of 0.7 m/s through the tubes. After prolonged operation, a scale of resistance 0.0002 m^2 K/W is formed on the inner surface of the tubes. To what value must the water velocity be increased in order to maintain the same rate of condensation on the assumption that the transfer coefficient on the water side is proportional to the velocity raised to the 0.8 power, and that the coefficient for the condensing vapour is 2.25 kW/m^2 K, based on the inside area? The latent heat of vaporisation of benzene is 400 kJ/kg.

Solution

Area for heat transfer, based on the tube i.d. $= (\pi \times 0.022 \times 1.0) = 0.0691$ m^2/m or: $(120 \times 2.5 \times 0.0691) = 20.74$ m^2.

With no scale

Heat load: $Q = (4 \times 400) = 1600$ W.
Cross-sectional area of one tube $= (\pi/4)0.022^2 = 0.00038$ m^2
and hence area for flow per pass $= (120 \times 0.00038) = 0.0456$ m^2.

$\therefore$ volume of flow of water $= (0.0456 \times 0.7) = 0.0319$ m^3/s

and: mass flow of water $= (0.0319 \times 1000) = 31.93$ kg/s

The water outlet temperature is given by, $1600 = 31.93 \times 4.18(T - 290)$ or $T = 302$ K

$$\theta_1 = (350 - 290) = 60 \deg \text{K}, \theta_2 = (350 - 302) = 48 \deg \text{K}$$

and in equation 9.9, $\theta_m = (60 - 48)/\ln(60/48) = 53.8 \deg \text{K}$.

In equation 9.1, $U = Q/A\theta_m = 1600/(20.74 \times 53.8) = 1.435$ kW/m^2K
Neglecting the wall resistance, $1/U = 1/h_i + 1/h_{0i}$

$$(1/1.435) = 1/h_i + (1/2.25) \text{ and } h_i = 3.958 \text{ kW/m}^2 \text{ K}$$

$$h_i \text{ is proportional to } u^{0.8} \text{ or } 3.958 = k(0.7)^{0.8} \text{ and } k = 5.265$$

With scale

$$h_i = 5.265u^{0.8} \text{ kW/m}^2 \text{ K, scale resistance} = 0.20 \text{ m}^2 \text{ K/kW}$$

and: $1/U = 1/(5.265u^{0.8}) + 0.20 + (1/2.25)$

∴ $U = u^{0.8}/(0.190 + 0.644u^{0.8}) \text{ kW/m}^2 \text{ K}$

that is: $Q = 1600 \text{ kW as before.}$

The mass flow of water is: $(u \times 0.0456 \times 1000) = 45.6u$ kg/s and the outlet water temperature is given by:

$$1600 = 45.6u \times 4.18(T - 290)$$

or: $T = (290 + 8.391/u) \text{ K}$

$\theta_1 = (350 - 290) = 60 \deg \text{K}, \theta_2 = (350 - 290 - 8.391)/u = (60 - 8.391)/u$

and: $\theta_m = (60 - (60 - 8.391)/u)/\ln[60/(60 - 8.391)/u]$

$= 8.391/\{u\ln[60u/(60u - 8.391)]\}$

In equation 9.1:

$$1600 = [u^{0.8}/(0.190 + 0.644u^{0.8})] \times 20.74 \times 8.391/\{u\ln[60u/(60u - 8.391)]\}$$

or: $1/\{u^{0.2}(0.190 + 0.644u^{0.8})\ln[60u/(60u - 8.391)]\} = 9.194$

The left-hand side of this equation is plotted against u in Fig. 9e and the function equals 9.194 when $u = 2.06$ m/s.

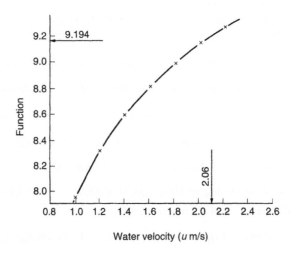

Figure 9e.

PROBLEM 9.55

Derive an expression for the radiant heat transfer rate per unit area between two large parallel planes of emissivities e_1 and e_2 and at absolute temperatures T_1 and T_2 respectively.

Two such planes are situated 2.5 mm apart in air. One has an emissivity of 0.1 and is at a temperature of 350 K, and the other has an emissivity of 0.05 and is at a temperature of 300 K. Calculate the percentage change in the total heat transfer rate by coating the first surface so as to reduce its emissivity to 0.025. Stefan–Boltzmann constant = 5.67×10^{-8} W/m^2 K^4. Thermal conductivity of air = 0.026 W/m K.

Solution

The theoretical derivation is laid out in Section 9.5.5 and the heat transfer by radiation is given by putting $A_1 = A_2$ in equation 9.150 to give:

$$q_r = [(e_1 e_2 \sigma)/(e_1 + e_2 - e_1 e_2)](T_1^4 - T_2^4)$$

For *conduction* between the two planes:

$$q_c = kA(T_1 - T_2)/x \qquad \text{(equation 9.12)}$$
$$= 0.026 \times 1.0(350 - 300)/0.0025 = 520 \text{ W/m}^2$$

For radiation between the two planes:

$$q_r = [(e_1 e_2 \sigma)/(e_1 + e_2 - e_1 e_2)](T_1^4 - T_2^4)$$
$$= [(0.1 \times 0.05 \times 5.67 \times 10^{-8})/(0.1 + 0.05 - 0.1 \times 0.05)](350^4 - 300^4)$$
$$= 13.5 \text{ W/m}^2$$

Thus neglecting any convection in the very narrow space, the total heat transferred is 533.5 W/m^2. When $e_1 = 0.025$, the heat transfer by radiation is:

$$q_r = [(0.025 \times 0.05 \times 5.67 \times 10^{-8})/(0.025 + 0.05 - 0.025 \times 0.05)]$$
$$\times (350^4 - 300^4) = 6.64 \text{ W/m}^2$$

and: $(q_r + q_c) = 526.64$ W/m^2

Thus, although the heat transferred by radiation is reduced to $(100 \times 6.64)/13.5 = 49.2\%$ of its initial value, the total heat transferred is reduced to $(100 \times 526.64)/533.5 = \underline{98.7\%}$ of the initial value.

PROBLEM 9.56

Water flows at 2 m/s through a 2.5 m length of a 25 mm diameter tube. If the tube is at 320 K and the water enters and leaves at 293 and 295 K respectively, what is the value of the heat transfer coefficient? How would the outlet temperature change if the velocity was increased by 50%?

Solution

The cross-sectional area of 0.025 m tubing $= (\pi/4)0.025^2 = 0.000491 \text{ m}^2$.

Volume flow of water $= (2 \times 0.000491) = 0.000982 \text{ m}^3/\text{s}$

Mass flow of water $= (1000 \times 0.000982) = 0.982 \text{ kg/s}$

$\therefore$ Heat load, $Q = 0.982 \times 4.18(295 - 293) = 8.21 \text{ kW}$

Surface area of 0.025 m tubing $= (\pi \times 0.025 \times 1.0) = 0.0785 \text{ m}^2/\text{m}$

and: $A = (0.0785 \times 2.5) = 0.196 \text{ m}^2$

         $\theta_1 = (320 - 293) = 27 \text{ deg K}, \theta_2 = (320 - 295) = 25 \text{ deg K}$

and: $\theta_m = (27 - 25)/\ln(27/25) = 25.98 \text{ say } 26 \text{ deg K}$

In equation 9.1: $U = 8.21/(0.196 \times 26) = \underline{\underline{1.612 \text{ kW/m}^2 \text{ K}}}$

An estimate may be made of the inside film coefficient from equation 9.221, where T, the mean water temperature, is 294 K.

Thus: $h_i = 4280(0.00488 \times 294 - 1)2.0^{0.8}/0.025^{0.2}$

         $= (4280 \times 0.435 \times 1.741/0.478) = 6777 \text{ W/m}^2 \text{ K or } 6.78 \text{ kW/m}^2 \text{ K}$

The scale resistance is therefore given by:

$$(1/1.612) = (1/6.78) + R \text{ or: } R = 0.473 \text{ m}^2 \text{ K/kW}$$

With a water velocity of $(2.0 \times 150/100) = 3.0 \text{ m/s}$, assuming a mean water temperature of 300 K, then:

         $h_i = 4280(0.00488 \times 300 - 1)3.0^{0.8}/0.025^{0.2}$

         $= (4280 \times 0.464 \times 2.408/0.478) = 10004 \text{ or } 10.0 \text{ kW/m}^2 \text{ K}$

$\therefore$ $1/U = (0.473 + 1/10.0)$ and $U = 1.75 \text{ kW/m}^2 \text{ K}$

For an outlet water temperature of T K: $\theta_1 = 27 \text{ deg K}$, $\theta_2 = (320 - T) \text{ deg K}$

and, taking an arithmetic mean: $\theta_m = 0.5(27 + 320 - T) = (173.5 - 0.5T) \text{ deg K}$.

The mass flow of water $= (0.982 \times 150)/100 = 1.473 \text{ kg/s}$,

and the heat load, $Q = 1.473 \times 4.18(T - 293) = (6.157T - 1804) \text{ kW}$

$\therefore$ $(6.157T - 1804) = [1.75 \times 0.196(173.5 - 0.5T)]$

from which: $\underline{\underline{T = 294.5 \text{ K}}}$

The use of 300 K as a mean water temperature has a minimal effect on the result and recalculation is not necessary.

PROBLEM 9.57

A liquid hydrocarbon is fed at 295 K to a heat exchanger consisting of a 25 mm diameter tube heated on the outside by condensing steam at atmospheric pressure. The flowrate of hydrocarbon is measured by means of a 19 mm orifice fitted to the 25 mm feed pipe. The reading on a differential manometer containing hydrocarbon-over-water is 450 mm and the coefficient of discharge of the meter is 0.6.

Calculate the initial rate of rise of temperature (deg K/s) of the hydrocarbon as it enters the heat exchanger.

The outside film coefficient = 6.0 W/m^2 K.

The inside film coefficient h is given by:

$$hd/k = 0.023(ud\rho/\mu)^{0.8}(C_p\mu/k)^{0.4}$$

where: u = linear velocity of hydrocarbon (m/s). d = tube diameter (m), ρ = liquid density (800 kg/m^3), μ = liquid viscosity (9×10^{-4} N s/m^2), C_p = specific heat of liquid (1.7×10^3 J/kgK), and k = thermal conductivity of liquid (0.17 W/mK).

Solution

The effective manometer fluid density, is 200 kg/m^3.

The pressure difference across the orifice = 450 mm water

or: $(450 \times 800/200) = 1800$ mm hydrocarbon

that is: $H = 1.80$ m

The area of the orifice = $(\pi/4)0.019^2 = 2.835 \times 10^{-4}$ m^2

In equation 6.21: $G = (0.6 \times 2.835 \times 10^{-4} \times 800)\sqrt{(2 \times 9.81 \times 1.80)}$

$$= 1.36\sqrt{(35.3)} = 0.808 \text{ kg/s}$$

The volume flow = $(0.808/800) = 0.00101$ m^3/s.

Cross-sectional area of a 0.025 m diameter pipe = $(\pi/4)0.025^2 = 0.000491$ m^2 and hence the velocity, $u = (0.00101/0.000491) = 2.06$ m/s.

The inside film coefficient is given by:

$$(h_i \times 0.025/0.17) = 0.023((2.06 \times 0.025 \times 800)/9 \times 10^{-4})^{0.8}$$

$$\times ((1.7 \times 10^3 \times 9 \times 10^{-4})/0.17)^{0.4}$$

or: $h_i = 0.1564(4.58 \times 10^4)^{0.8}(9.0)^{0.4} = 2016$ W/m^2 K or 2.02 kW/m^2 K

Neglecting scale and wall resistances:

$$1/U = (1/6.0) + (1/2.02) \text{ and } U = 1.511 \text{ kW/m}^2 \text{ K}$$

For steam at atmospheric pressure, the saturation temperature = 373 K and at the inlet the temperature driving force = $(373 - 295) = 78$ deg K.

The heat flux is: $(1.511 \times 78) = 117.9$ kW/m^2.

For a small length of tube, say 0.001 m, the area for heat transfer $= (\pi \times 0.025 \times 0.001) = 7.854 \times 10^{-5}$ m^2

and the heat transfer rate $= (117.9 \times 7.854 \times 10^{-5} \times 1000) = 9.27$ W.

In the small length (0.001 m) of tube, mass of material $= (0.000491 \times 0.001 \times 800) = 3.93 \times 10^{-4}$ kg and hence temperature rise $= [9.27/(3.93 \times 10^{-4} \times 1.7 \times 10^3)] = 13.9$ deg K/s

PROBLEM 9.58

Water passes at a velocity of 1.2 m/s through a series of 25 mm diameter tubes 5 m long maintained at 320 K. If the inlet temperature is 290 K, at what temperature would it leave?

Solution

Assuming an outlet water temperature of T K, the mean water temperature is therefore:

$$= 0.5(T + 290) = (0.5T + 145) \text{ K.}$$

The coefficient may be calculated from:

$$h = 4280(0.00488T - 1)u^{0.8}/d^{0.2} \qquad \text{(equation 9.221)}$$

$$= 4280[0.00488(0.5T + 145) - 1]1.2^{0.8}/0.025^{0.2}$$

$$= (25.28T - 3028.1) \text{ W/m}^2 \text{ K}$$

Area for heat transfer $= (\pi \times 0.025 \times 5.0) = 0.393$ m^2

and the heat load, $Q = [1.2(\pi/4)0.025^2 \times 1000 \times 4.18 \times 10^3(T - 290)]$

$$= (2462T - 714,045) \text{ W}$$

Therefore neglecting any scale resistance:

$$(2462T - 714,045) = (25.28T - 3028.1)0.393[320 - (0.5T + 145)]$$

from which: $\qquad\qquad T^2 + 25.98T - 101,851 = 0$

and: $\qquad\qquad\qquad T = \underline{306.4 \text{ K}}$

[*An alternative approach is as follows:*
The heat transferred per unit time in length dL of pipe,

$$= h \times \pi \times 0.025\text{d}L(320 - T_k) \text{ W}$$

where T_k is the water temperature at L m from the inlet.

The rate of increase in the heat content of the water is:

$$(\pi/4) \times 0.025^2 \times 1.2 \times 1000 \times 4.18 \times 10^3 \text{ d}T = 2462 \text{ d}T$$

The outlet temperature T' is then given by:

$$\int_{290}^{T'} \frac{dT}{(320-T)} = 0.0000319h \int_0^5 dL$$

or: $\ln(320 - T') = \ln 30 - 0.0001595h = 3.401 - 0.0001595h$

At a mean temperature of say 300 K, in equation 9.221:

$$h = 4280(0.00488 \times 300 - 1)1.2^{0.8}/0.025^{0.2} = 4805 \text{ W/m}^2 \text{ K}$$

Thus: $\ln(320 - T') = 3.401 - (0.0001595 \times 4805)$

and: $T' = \underline{306.06 \text{ K}}$

PROBLEM 9.59

Heat is transferred from one fluid stream to a second fluid across a heat transfer surface. If the film coefficients for the two fluids are, respectively, 1.0 and 1.5 kW/m^2 K, the metal is 6 mm thick (thermal conductivity 20 W/m K) and the scale coefficient is equivalent to 850 W/m^2 K, what is the overall heat transfer coefficient?

Solution

From equation 9.201:

$$1/U = 1/h_o + x_w/k_w + R + 1/h_i$$
$$= (1/1000) + (10.006/20) + (1/850) + (1/1500)$$
$$= (0.001 + 0.00030 + 0.00118 + 0.00067) = 0.00315 \text{ m}^2 \text{ K/W}$$

$\therefore$ $U = 317.5 \text{ W/m}^2 \text{ K or } \underline{0.318 \text{ kW/m}^2 \text{ K}}$

PROBLEM 9.60

A pipe of outer diameter 50 mm carries hot fluid at 1100 K. It is covered with a 50 mm layer of insulation of thermal conductivity 0.17 W/m K. Would it be feasible to use magnesia insulation, which will not stand temperatures above 615 K and has a thermal conductivity of 0.09 W/m K for an additional layer thick enough to reduce the outer surface temperature to 370 K in surroundings at 280 K? Take the surface coefficient of transfer by radiation and convection as 10 W/m^2 K.

Solution

The solution is presented as Problem 9.8.

PROBLEM 9.61

A jacketed reaction vessel containing 0.25 m^3 of liquid of density 900 kg/m^3 and specific heat 3.3 kJ/kg K is heated by means of steam fed to a jacket on the walls. The contents of the tank are agitated by a stirrer rotating at 3 Hz. The heat transfer area is 2.5 m^2 and the steam temperature is 380 K. The outside film heat transfer coefficient is 1.7 kW/m^2 K and the 10 mm thick wall of the tank has a thermal conductivity of 6.0 W/m K. The inside film coefficient was 1.1 kW/m^2 K for a stirrer speed of 1.5 Hz and proportional to the two-thirds power of the speed of rotation. Neglecting heat losses and the heat capacity of the tank, how long will it take to raise the temperature of the liquid from 295 to 375 K?

Solution

For a stirrer speed of 1.5 Hz, $h_i = 1.1$ kW/m^2 K.

$$\therefore \qquad\qquad 1.1 = k'1.5^{0.67} \text{ and } k' = 0.838$$

Thus at a stirrer speed of 3 Hz, $h_i = (0.838 \times 3.0^{0.67}) = 1.75$ kW/m^2 K.
The overall coefficient is given by:

$$1/U = (1/1750) + (0.010/6.0) + (1/1700) + 0.00283 \text{ (equation 9.201)}$$

and: $\qquad\qquad U = 353.8$ W/m^2 K neglecting scale resistances.

The time for heating the liquid is given by:

$$\ln[(T_s - T_1)/(T_s - T_2)] = UAt/mC_p \qquad\qquad \text{(equation 9.209)}$$

In this case: $m = (0.25 \times 900) = 225$ kg and $C_p = 3300$ J/kg K.

$$\therefore \qquad\qquad \ln[(380 - 295)/(380 - 375)] = 353.8 \times 2.5t/(225 \times 3300)$$

$$2.833 = 0.00119t \text{ and } t = \underline{\underline{2381 \text{ s}}} \text{ (40 min)}$$

PROBLEM 9.62

By dimensional analysis, derive a relationship for the heat transfer coefficient h for natural convection between a surface and a fluid on the assumption that the coefficient is a function of the following variables:

k = thermal conductivity of the fluid, C_p = specific heat of the fluid, ρ = density of the fluid, μ = viscosity of the fluid, βg = the product of the coefficient of cubical expansion of the fluid and the acceleration due to gravity, l = a characteristic dimension of the surface, and ΔT = the temperature difference between the fluid and the surface.

Indicate why each of these quantities would be expected to influence the heat transfer coefficient and explain how the orientation of the surface affects the process.

Under what conditions is heat transfer by natural convection important in Chemical Engineering?

Solution

If the heat transfer coefficient h can be expressed as a product of powers of the variables, then:

$$h = k'(k^a C_p^b \rho^c \mu^d (\beta g)^e l^f \Delta T^g) \text{ where } k' \text{ is a constant.}$$

The dimensions of each variable in terms of **M, L, T, Q,** and θ are:

$$\begin{aligned}
\text{heat transfer coefficient,} \quad & h = \mathbf{Q/L^2 T\theta} \\
\text{thermal conductivity,} \quad & k = \mathbf{Q/LT\theta} \\
\text{specific heat,} \quad & C_p = \mathbf{Q/M\theta} \\
\text{viscosity,} \quad & \mu = \mathbf{M/LT} \\
\text{density,} \quad & \rho = \mathbf{M/L^3} \\
\text{the product,} \quad & \beta g = \mathbf{L/T^2\theta^{-1}} \\
\text{length,} \quad & l = \mathbf{L} \\
\text{temperature difference,} \quad & \Delta T = \theta
\end{aligned}$$

Equating indices:

$$\begin{aligned}
\mathbf{M:} \quad & 0 = -b + c + d \\
\mathbf{L:} \quad & -2 = -a - 3c - d + e + f \\
\mathbf{T:} \quad & -1 = -a - d - 2e \\
\mathbf{Q:} \quad & 1 = a + b \\
\theta: \quad & -1 = -a - b - e + g
\end{aligned}$$

Solving in terms of b and c:

$$a = (1 - b), d = (b - c), e = (c/2), f = (3c/2 - 1), g = (c/2)$$

and hence:

$$h = k' \left(\frac{k}{k^b} C_p^b \rho^c \frac{\mu^b}{\mu^c} (\beta g)^{c/2} \frac{l^{3c/2}}{l} \Delta T^{c/2} \right) = k' \left(\frac{k}{l} \right) \left(\frac{C_p \mu}{k} \right)^b \left(\frac{l^{3/2} \rho (\beta g)^{1/2} T^{1/2}}{\mu} \right)^c$$

or:

$$\frac{hl}{k} = k' \left(\frac{C_p \mu}{k} \right)^b \left(\frac{l^3 \rho^2 \beta g T}{\mu^2} \right)^{c/2}$$

where $(C_p \mu / k)$ is the Prandtl number and $(l^3 \rho^2 \beta g \Delta T / \mu^2)$ the Grashof number. A full discussion of the significance of this result and the importance of free of natural convection is presented in Section 9.4.7.

PROBLEM 9.63

A shell-and-tube heat exchanger is used for preheating the feed to an evaporator. The liquid of specific heat 4.0 kJ/kg K and density 1100 kg/m^3 passes through the inside of tubes and is heated by steam condensing at 395 K on the outside. The exchanger heats liquid at 295 K to an outlet temperature of 375 K when the flowrate is 1.75×10^{-4} m^3/s and to 370 K when the flowrate is 3.25×10^{-4} m^3/s. What is the heat transfer area and the value of the overall heat transfer coefficient when the flow rate is 1.75×10^{-4} m^3/s?

Assume that the film heat transfer coefficient for the liquid in the tubes is proportional to the 0.8 power of the velocity, that the transfer coefficient for the condensing steam remains constant at 3.4 kW/m^2 K and that the resistance of the tube wall and scale can be neglected.

Solution

i) *For a flow of* 1.75×10^{-4} m^3/s:

Density of the liquid $= 1100$ kg/m^3

Mass flow $\quad\quad\quad = (1.75 \times 10^{-4} \times 1100) = 0.1925$ kg/s.

Heat load $\quad\quad\quad = 0.1925 \times 4.0(373 - 295) = 61.6$ kW

$\theta_1 = (395 - 295) = 100$ deg K, $\quad \theta_2 = (395 - 375) = 20$ deg K

and in equation 9.9:

$$\theta_m = (100 - 20)/\ln(100/20) = 49.7 \text{ deg K}$$

Thus, in equation 9.1:

$$U_1 A = (61.6/49.7) = 1.239 \text{ kW/K}$$

ii) *For a flow of* $3.25 \times 10^{-4} m^3/s$:

Mass flow $= (3.25 \times 10^{-4} \times 1100) = 0.3575$ kg/s

Heat load $= 0.3575 \times 4.0(370 - 295) = 107.3$ kW

$\theta_1 = (395 - 295) = 100$ deg K, $\quad \theta_2 = (395 - 370) = 25$ deg K

and in equation 9.9:

$$\theta_m = (100 - 25)/\ln(100/25) = 54.1 \text{ deg K}$$

Thus in equation 9.1:

$$U_2 A = (107.3/54.1) = 1.983 \text{ kW/K}$$

$$\therefore \quad\quad U_2/U_1 = (1.983/1.239) = 1.60$$

The velocity in the tubes is proportional to the volumetric flowrate, v cm^3/s and hence

$$h_i \propto v^{0.8} \text{ or } h_i = k'v^{0.8}, \text{ where } k' \text{ is a constant.}$$

Neglecting scale and wall resistances:

$1/U = 1/h_0 + 1/h_i$

$\quad = (1/3.4) + (1/k'v^{0.8})$

and: $\quad U = 3.4k'v^{0.8}/(3.4 + k'v^{0.8})$

$$\therefore \quad U_1 = (3.4k' \times 175^{0.8})/(3.4 + k' \times 175^{0.8}) = 211.8k'/(3.4 + 62.3k')$$

$$\text{and: } U_2 = (3.4k' \times 325^{0.8})/(3.4 + k' \times 325^{0.8}) = 347.5k'/(3.4 + 102.2k')$$

$$\therefore \qquad [347.5k'/(3.4 + 102.2k')]/[211.18k'/(3.4 + 62.3k')] = 1.60$$

$$\therefore \quad k' = 0.00228$$

$$\therefore \quad U_1 = (3.4 \times 0.00228 \times 175^{0.8})/(3.4 + 0.00228 \times 175^{0.8}) = \underline{\underline{0.136 \text{ kW/m}^2 \text{ K}}}$$

and the heat transfer area, $A = (1.239/0.136) = \underline{\underline{9.09 \text{ m}^2}}$.

PROBLEM 9.64

0.1 m^3 of liquid of specific heat capacity 3 kJ/kg K and density 950 kg/m^3 is heated in an agitated tank fitted with a coil, of heat transfer area 1 m^2, supplied with steam at 383 K. How long will it take to heat the liquid from 293 to 368 K, if the tank, of external area 20 m^2 is losing heat to the surroundings at 293 K? To what temperature will the system fall in 1800 s if the steam is turned off? Overall heat transfer coefficient in coil = 2000 W/m^2 K. Heat transfer coefficient to surroundings = 10 W/m^2 K.

Solution

If T K is the temperature of the liquid at time t s, then:

heat input from the steam $= UA(T_s - T) = (2000 \times 1)(383 - T)$ or $2000(383 - T)$ W

Similarly, heat losses to the surroundings $= (10 \times 20)(T - 293) = 200(T - 293)$ W

and, net heat input to the liquid $= 2000(383 - T) - 200(T - 293) = (824,600 - 2200T)$ W

This is equal to: $\qquad\qquad Q = mC_p(\mathrm{d}T/\mathrm{d}t)$

where $m = (0.1 \times 950) = 95$ kg and $C_p = 3000$ J/kg K.

$$\therefore \qquad\qquad (95 \times 3000)\mathrm{d}T/\mathrm{d}t = (824,600 - 2200T)$$

or: $\qquad\qquad 129.6\,\mathrm{d}T/\mathrm{d}t = (374.8 - T)$

Thus the time taken to heat from 293 to 368 K is:

$$t = 129.6 \int_{293}^{368} \mathrm{d}T/(374.8 - T)$$

$$= 129.6 \ln((374.8 - 293)/(374 - 368)) = \underline{\underline{1559 \text{ s } (0.43 \text{ h})}}$$

The steam is turned off for 1800s and, during this time, a heat balance gives:

$$(95 \times 3000)\mathrm{d}T/\mathrm{d}t = -(10 \times 20)(T - 293)$$

$$\therefore \qquad\qquad 285,000\,\mathrm{d}T/\mathrm{d}t = (58600 - 200T) \text{ or } 1425\,\mathrm{d}T/\mathrm{d}t = (293 - T)$$

The change in temperature is then given by:

$$\int_{368}^{T} dT/(293 - T) = (1/1425) \int_{0}^{1800} dt$$

$$\ln((293 - 368)/(293 - T)) = (1800/1425) = 1.263 \text{ and } \underline{\underline{T = 311.8 \text{ K}}}.$$

PROBLEM 9.65

The contents of a reaction vessel are heated by means of steam at 393 K supplied to a heating coil which is totally immersed in the liquid. When the vessel has a layer of lagging 50 mm thick on its outer surfaces, it takes one hour to heat the liquid from 293 to 373 K. How long will it take if the thickness of lagging is doubled? Outside temperature = 293 K. Thermal conductivity of lagging = 0.05 W/mK. Coefficient for heat loss by radiation and convection from outside surface of vessel = 10 W/m^2 K. Outside area of vessel = 8 m^2. Coil area = 0.2 m^2. Overall heat transfer coefficient for steam coil = 300 W/m^2 K.

Solution

If T K is the temperature of the liquid at time t s and $T_1 K$ the temperature at the outside surface of the vessel, then heat flowing through the insulation is equal to the heat lost by convection and radiation to the surroundings or:

$$(kA/x)(T - T_1) = h_c A(T_1 - T_0)$$

where h_c is the coefficient for heat loss, A the outside surface area of the vessel and T_0 the ambient temperature. Thus:

$$(0.05 \times 8/0.050)(T - T_1) = (10 \times 8)(T_1 - 293)$$

and: $T_1 = (0.0909T + 266.4)$ K

$\therefore$ Heat loss to the surroundings $= (10 \times 8)(0.0909T + 266.4 - 293)$

$$= (7.272T - 2128) \text{ W}$$

Heat input from the coil $= (300 \times 0.2)(393 - T) = (23580 - 60T)$ W

and net heat input $= (23580 - 60T) - (7.272T - 2128) = 25708 - 67.3T$ W

which is equal to: $Q = mC_p \, dT/dt$

or: $mC_p \, dT/dt = 25708 - 67.3T$

and: $0.0149mC_p \, dT/dt = 382 - T$

It takes $t = 3600$ s to heat the contents from 293 to 373 K, or:

$$3600 = 0.0149mC_p \int_{293}^{373} dT/(382 - T)$$

$\therefore$ $241610 = mC_p \ln((382 - 293)/(382 - 373)) = 2.291mC_p$

and: $mC_p = 105442$ J/K

If the thickness of the lagging is doubled to 0.100 m, then:

$$(0.05 \times 8/0.100)(T - T_1) = (10 \times 8)(T_1 - 293)$$

and: $\qquad\qquad\qquad\qquad\qquad T_1 = (0.0476T + 279.1) \text{ K}$

∴ Heat loss to the surroundings $\qquad = (10 \times 8)(0.0476T + 279.1 - 293)$

$$= (3.808T - 1112) \text{ W}$$

Heat input from the coil $= (300 \times 0.2)(393 - T) = (23580 - 60T) \text{ W}$

and net heat input $= (23580 - 60T) - (3.808T - 1112) = 24692 - 63.808T$.

∴ $\qquad\qquad\qquad\qquad mC_p \, dT/dt = (24{,}692 - 63.808T)$

$$105442 \, dT/dt = (24{,}692 - 63.808T)$$

or: $\qquad\qquad\qquad\qquad 1652.5 \, dT/dt = 387 - T$

Thus, the time taken to heat the contents from 293 to 373 K is:

$$t = 1652.5 \int_{293}^{373} dT/(387 - T)$$

$$= 1625.5 \ln[(387 - 293)/(387 - 373)] = (1652.5 \times 1.904)$$

$$= \underline{\underline{3147 \text{ s } (0.87 \text{ h})}}$$

PROBLEM 9.66

A smooth tube in a condenser which is 25 mm internal diameter and 10 m long is carrying cooling water and the pressure drop over the length of the tube is 2×10^4 N/m². If vapour at a temperature of 353 K is condensing on the outside of the tube and the temperature of the cooling water rises from 293 K at inlet to 333 K at outlet, what is the value of the overall heat transfer coefficient based on the inside area of the tube? If the coefficient for the condensing vapour is 15,000 W/m² K, what is the film coefficient for the water? If the latent heat of vaporisation is 800 kJ/kg, what is the rate of condensation of vapour?

Solution

From equation 3.23: $(R/\rho u^2) Re^2 = -\Delta P_f d^3 \rho/(4l/\mu^2)$

Taking the viscosity of water as 1 mN s/m² $\equiv 0.001$ Ns/m², then:

$$-\Delta P_f d^3 \rho/(4l/\mu^2) = 20{,}000(0.025)^3 1000/(4 \times 10(0.001)^2) = 7{,}812{,}500$$

From Fig. 3.8, for a smooth pipe, $Re = 57{,}000$

∴ $\qquad\qquad\qquad dup/\mu = (0.025u1000)/0.001 = 57{,}000$

and: $\qquad\qquad\qquad u = 2.28 \text{ m/s}$

∴ Volume flow of water $= \pi(0.025^2/4)2.28 = 0.00112 \text{ m}^3/\text{s}$

Mass flow of water $= (1000 \times 0.00112) = 1.12$ kg/s

Heat removed by water $= 1.12 \times 4.187(333 - 293) = 187.6$ kW

Surface area of tube, based on inside diameter $= (\pi \times 0.025 \times 10) = 0.785$ m^2

Vapour temperature $= 353$ K $\quad \therefore \Delta T_1 = (353 - 293) = 60$ deg K

$$\Delta T_2 = (353 - 333) = 20 \text{ deg K}$$

and from equation 9.9, $\Delta T_m = (60 - 20)/\ln(60/20) = 36.4$ deg K

From equation 9.1, $187.6 = (U \times 0.785 \times 36.4)$

and the overall coefficient based on the inside diameter is: $\underline{U = 6.57 \text{ kW/m}^2 \text{ K}}$.

In equation 9.201, neglecting the wall and scale resistances:

$$1/U = 1/h_o + 1/h_i$$

$$(1/6.57) = (1/15.0) + 1/h_i \text{ and } \underline{h_i = 11.68 \text{ kW/m}^2 \text{ K}}$$

If the latent heat of condensation is 800 kJ/kg, then assuming the vapour enters and the condensate leaves at the boiling point:

$$\text{rate of condensation} = (187.6/800) = \underline{0.235 \text{ kg/s}}$$

PROBLEM 9.67

A chemical reactor, 1 m in diameter and 5 m long, operates at a temperature of 1073 K. It is covered with a 500 mm thickness of lagging of thermal conductivity 0.1 W/m K. The heat loss from the cylindrical surface to the surroundings is 3.5 kW. What is the heat transfer coefficient from the surface of the lagging to the surroundings at a temperature of 293 K? How would the heat loss be altered if the coefficient were halved?

Solution

From equation 9.20, the heat flow at any radius r is given by:

$$Q = -k(2\pi + 1)\mathrm{d}T/\mathrm{d}r \text{ W}$$

$$\therefore \qquad \mathrm{d}r/r = (-2\pi kl/Q)\mathrm{d}T$$

Integrating between the limits r_1 and r_2 at which the temperatures are T_1 and T_2 respectively:

$$\int_{r_1}^{r_2} (\mathrm{d}r/r) = (-2\pi kl/Q) \int_{T_1}^{T_2} \mathrm{d}T$$

$$\therefore \qquad \ln(r_2/r_1) = (-2\pi kl/Q)(T_1 - T_2)$$

In this case: $r_1 = (1/2) = 0.50$ m, $r_2 = (0.50 + 500/1000) = 1.0$ m, $l = 5$ m, $k = 0.1$ W/mK, $Q = 3500$ W and $T_1 = 1073$ K.

$$\therefore \qquad \ln(1.0/0.50) = (2\pi \times 0.1 \times 5/3500)(1073 - T_2)$$

and: $\qquad 0.693 = 0.00090(1073 - T_2)$ and $\underline{T_2 = 301 \text{ K}}$

The heat flow to the surroundings is: $Q = h_o A_o (T_2 - T_o)$

$\therefore$ $\qquad$ $3500 = h_o (\pi \times 2.0 \times 5.0)(301 - 293)$ and $\underline{h_o = 14.05 \text{ W/m}^2 \text{ K}}$

If this value is halved, that is $h_{o2} = 7.02 \text{ W/m}^2 \text{ K}$, then:

$$Q_2 = 7.02(2\pi \times 2.0 \times 5.0)(T_2 - 293) = 220.5(T_2 - 293)$$

and: $\qquad$ $T_2 = (Q_2/220.5) + 293 \ K.$

But: $\qquad$ $(Q_2/Q_1) = (T_1 - T_2)_2/(T_1 - T_2)_1$

$\therefore$ $\qquad$ $(Q_2/3500) = [1073 - (Q_2/220.5) - 293]/(1073 - 301)$

$\qquad$ $0.000286Q_2 = 0.00130(780 - 0.00454Q_2)$

and: $\qquad$ $\underline{\underline{Q_2 = 3473 \text{ W}}}$ — a very slight reduction in the heat loss.

In this case, $T_2 = (3473/220.5) + 293 = \underline{\underline{308.7 \text{ K}}}.$

PROBLEM 9.68

An open cylindrical tank 500 mm diameter and 1 m deep is three-quarters filled with a liquid of density 980 kg/m^3 and of specific heat capacity 3 kJ/kg K. If the heat transfer coefficient from the cylindrical walls and the base of the tank is 10 W/m^2 K and from the surface is 20 W/m^2 K, what area of heating coil, fed with steam at 383 K, is required to heat the contents from 288 K to 368 K in a half hour? The overall heat transfer coefficient for the coil may be taken as 100 W/m^2 K. The surroundings are at 288 K. The heat capacity of the tank itself may be neglected.

Solution

The rate of heat transfer from the steam to the liquid is:

$$U_c A_c (383 - T) = 100 A_c (383 - T) \ W$$

where A_c is the surface area of the coil.

The rate of heat transfer from the tank to the surroundings $= U_T A_T (T - 288)$
where U_T is the effective overall coefficient and A_T the surface area of the tank and liquid surface. In this case:

$$U_T A_T = 10((\pi \times 0.5 \times 1) + (\pi/4)0.5^2) + 20(\pi/4)0.5^2 = 21.6 \text{ W/K}.$$

$\therefore$ rate of heat loss $= 21.6(T - 288) \text{ W}$

$\therefore$ net rate of heat input to the tank $= 100 A_c (383 - T) - 21.6(T - 288) \text{ W}.$

This is equal to $mC_p \, dT/dt$, where the mean specific heat, $C_p = 3000$ J/kgK.

Volume of liquid $= (75/100)(\pi/4)0.5^2 \times 1 = 0.147 \text{ m}^3$

Mass of liquid: $m = (0.147 \times 980) = 144.3$ kg

and: $\qquad\qquad mC_p = (144.3 \times 3000) = 432{,}957$ J/K.

$\therefore \qquad 432{,}957\, dT/dt = 100A_c(383 - T) - 21.6(T - 288)$

$$= (38{,}300A_c + 6221) - (100A_c + 21.6)T$$

$$\therefore \int_{288}^{368} dT/((38{,}300A_c + 6221)/(100A_c + 21.6) - T)$$

$$= ((100A_c + 21.6)/432{,}957) \int_0^{1800} dt$$

$\therefore \qquad \ln\{[(38{,}300A_c + 6221)/(100A_c + 21.6)) - 288]/$

$$[(38{,}300A_c + 6221)/(100A_c + 21.6) - 368)]\}$$

$$= 0.00416(100A_c + 21.6)$$

This equation is solved by trial and error to give: $\underline{\underline{A_c = 5.0 \text{ m}^2}}$.

PROBLEM 9.69

Liquid oxygen is distributed by road in large spherical vessels, 1.82 m in internal diameter. If the vessels were unlagged and the coefficient for heat transfer from the outside of the vessel to the atmosphere were 5 W/m^2 K, what proportion of the contents would evaporate during a journey lasting an hour? Initially the vessels are 80% full.

What thickness of lagging would be required to reduce the losses to one tenth? Atmospheric temperature = 288 K. Boiling point of oxygen = 90 K. Density of oxygen = 1140 kg/m^3. Latent heat of vaporisation of oxygen = 214 kJ/kg. Thermal conductivity of lagging = 0.07 W/m K.

Solution

Volume of the vessel $\qquad = \pi d^3/6 = (\pi \times 1.82^3/6) = 3.16$ m^3
Volume of liquid oxygen $\qquad = (80/100)3.16 = 2.53$ m^3
Mass of liquid oxygen $\qquad = (2.53 \times 1140) = 2879$ kg
Surface area of unlagged vessel $= (\pi \times 1.82^2) = 10.41$ m^2
Heat leakage into the vessel $\qquad = h_cA(T_1 - T_2) = 5.0 \times 10.41(288 - 90)$
$\qquad\qquad\qquad\qquad\qquad\qquad = 10{,}302$ W or 10.3 kW

$\therefore \qquad\qquad$ Evaporation rate of oxygen = $(10.3/214) = 0.048$ kg/s

$\therefore \qquad$ Evaporation taking place during 1 h = $(0.048 \times 3600) = 173.3$ kg

which is $(100 \times 173.3/2879) = \underline{6.02\% \text{ of the contents}}$

In order to reduce the losses to one tenth, the heat flow into the vessel must be 1.03 kW and this will be achieved by reducing the temperature driving force to:

$$(288 - 90)/10 = 19.8 \deg \text{K}.$$

In this case the outside temperature of the lagging will be $(288 - 19.8) = 268.2$ K and the temperature drop through the lagging will be $(268.2 - 90) = 178.2 \deg$ K. Thus, the heat flow through the lagging is:

$$1030 = (kA/x)\Delta T_{\text{lagging}} = (0.07 \times 10.41/x)178.2$$

from which the thickness of the lagging, $x = 0.126$ m or 126 mm

This calculation does not take into account the increase in the surface area at the lagging surface since it was assumed to be that of the tank, 10.41 m². In practice, it will be larger than this and, if this is taken into account, the reasoning is as follows:

Radius of the tank $= (1.82/2) = 0.91$ m

∴ For a lagging thickness of x m, the new radius is $(0.91 + x)$m and the surface area is:

$$4\pi(0.91 + x)^2 \text{ m}^2$$

∴ convective heat gain $= 5.0 \times 4\pi(0.91 + x)^2(288 - T) = 1030$ W
and the outside temperature of the lagging:

$$T = 288 - 16.39/(0.91 + x)^2 \text{ K} \tag{i}$$

The heat flow through the lagging (taking an arithmetic mean area and neglecting the curvature),

$$1030 = (0.07/x)4\pi(0.91 + x/2)^2(T - 90)$$

Substituting for T from (i) into (ii):

$$1030 = (0.07/x)4\pi(0.91 + x/2)^2(198 - 16.39)/(0.91 + x)^2)$$

Solving by trial and error: $x = 0.151$ m or 151 mm

PROBLEM 9.70

Water at 293 K is heated by passing it through a 6.1 m coil of 25 mm internal diameter pipe. The thermal conductivity of the pipe wall is 20 W/m K and the wall thickness is 3.2 mm. The coil is heated by condensing steam at 373 K for which the film coefficient is 8 kW/m² K. When the water velocity in the pipe is 1 m/s, its outlet temperature is 309 K. What will the outlet temperature be if the velocity is increased to 1.3 m/s, if the coefficient of heat transfer to the water in the tube is proportional to the velocity raised to the 0.8 power?

Solution

The surface area of the coil $= \pi d_o l = \pi((25 + 2 \times 3.2)/1000)6.1 = 0.602$ m²

i) *When the water velocity is 1 m/s:*

$$\text{Area for flow} = \pi d_i^2/4 = \pi(25/1000)^2/4 = 0.00049 \text{ m}^2$$

$$\text{Volume flow} = (0.00049 \times 1.0) = 0.00049 \text{ m}^3/\text{s}$$

Taking the density as 1000 kg/m^3,

$$\text{Mass flow of water} = (1000 \times 0.00049) = 0.491 \text{ kg/s}$$

and taking the mean specific heat as 4.18 kJ/kg K,

$$\text{Heat load} = (0.491 \times 4.18)(309 - 293) = 32.83 \text{ kW}$$

With steam at 373 K, $\Delta T_1 = (373 - 293) = 80 \deg \text{K}$,

$$\Delta T_2 = (373 - 309) = 64 \deg \text{K}$$

and from equation 9.9, $\Delta T_m = (80 - 64)/\ln(80/64) = 71.7 \deg \text{K}$

Therefore from equation 9.1, the overall coefficient, $U = 32.83/(0.602 \times 71.7) = 0.761$ kW/m^2 K or 761 W/m^2 K.

From equation 9.201, and neglecting any scale resistance:

$$1/U = 1/h_i + 1/h_0 + x/k$$

In this case, $h_o = 8$ kW/m^2 K $= 8000$ W/m^2 K, $k = 20$ W/mK, $x = 3.2$ mm or 0.0032 m and $h_i = Ku^{0.8}$ where $u = 1$ m/s and K is a constant.

$\therefore$ $\qquad\qquad 1/761 = 1/K1^{0.8} + (1/8000) + (0.0032/20)$

$$0.00131 = (1/K + (0.000125 + 0.00016)) \text{ and } \underline{K = 976}$$

ii) *When the velocity is 1.3 m/s*

Volume flow of water $= (0.00049 \times 1.3) = 0.000637$ m^3/s

Mass flow of water $\quad= (1000 \times 0.000637) = 0.637$ kg/s

$\therefore$ Heat load $\qquad\quad= (0.637 \times 4.18)(T - 293)$

$\qquad\qquad\qquad\quad= 2.663(T - 293)$ kW or $2663(T - 293)$ W

The inside coefficient, $h_i = 976 \times 1.3^{0.8} = 1204$ W/m^2 K
and the overall coefficient, U is given by:

$$1/U = (1/1204 + 1/8000 + 0.0032/20)$$

$\therefore$ $\qquad\qquad U = 896.4$ W/m^2 K

$$\Delta T_1 = (373 - 293) = 80 \deg \text{K and } \Delta T_2 = (373 - T) \deg \text{K}.$$

Thus, from equation 9.9:

$$\Delta T_m = (80 - 373 + T)/\ln[80/(373 - T)] = (T - 293)/\ln[80/(373 - T)] \deg \text{K}$$

and in equation 9.1:

$$2663(T - 293) = (896.4 \times 0.602)(T - 293)/\ln[80/(373 - T)]$$

$$\ln 80/(373 - T) = 0.2026 \text{ and: } \underline{T = 307.7 \text{ K}}.$$

PROBLEM 9.71

Liquid is heated in a vessel by means of steam which is supplied to an internal coil in the vessel. When the vessel contains 1000 kg of liquid it takes half an hour to heat the contents from 293 to 368 K if the coil is supplied with steam at 373 K. The process is modified so that liquid at 293 K is continuously fed to the vessel at the rate of 0.28 kg/s. The total contents of the vessel are always being maintained at 1000 kg. What is the equilibrium temperature which the contents of the vessel will reach, if heat losses to the surroundings are neglected and the overall heat transfer coefficient remains constant?

Solution

Use is made of equation 9.209:

$$\ln((T_s - T_1)/(T_s - T_2)) = (UA/mC_p)t$$

In this case: $T_s = 373$ K, $T_1 = 293$ K, $T_2 = 368$ K, $m = 1000$ kg and $t = 0.5$ h or 1800 s

$$\therefore \quad \ln((373 - 293)/(373 - 368)) = (UA/C_p)(1800/1000) = 2.773$$

and: $$UA/C_p = 1.54 \text{ kg/s}$$

For continuous heating, assuming (UA/C_p) is constant and losses are negligible then:

$$Q = UA(T_s - T) = mC_p(T - T_1)$$

where T is the temperature of the contents.

$$\therefore \quad UA(373 - T) = 0.28C_p(T - 293)$$

$$(UA/C_p)(373 - T) = 0.28T - 82.04$$

Substituting for (UA/C_p):

$$(1.54 \times 373) - 1.54T = 0.28T - 82.04$$

and: $$\underline{\underline{T = 360.7 \text{ K}}}$$

PROBLEM 9.72

The heat loss through a firebrick furnace wall 0.2 m thick is to be reduced by addition of a layer of insulating brick to the outside. What is the thickness of insulating brick necessary to reduce the heat loss to 400 W/m^2? The inside furnace wall temperature is 1573 K, the ambient air adjacent to the furnace exterior is at 293 K and the natural convection heat transfer coefficient at the exterior surface is given by $h_o = 3.0\Delta T^{0.25}$ W/m^2 K, where ΔT is the temperature difference between the surface and the ambient air. Thermal conductivity of firebrick = 1.5 W/m K. Thermal conductivity of insulating brick = 0.4 W/m K.

Solution

The conduction through the firebrick is given by:

$$Q = 400 = (1.5 \times 1.0/0.2)(1573 - T_2) \qquad \text{(equation 9.12)}$$

and T_2, the temperature at the firebrick/insulating brick interface, is:

$$T_2 = 1579.7 \text{ K}.$$

For the natural convection to the surroundings:

$$Q = h_o A(T_3 - T_a)$$

or: $$400 = 3.0 \Delta T^{0.25} \times 1.0(T_3 - 293)$$

but $(T_3 - 293) = \Delta T$ and:

$$400 = 3.0 \Delta T^{1.25}$$

$$\therefore \qquad \Delta T = 133.3^{0.8} = 50.1 \deg \text{K}$$

and the temperature at the outer surface of the insulating brick, $T_3 = (293 + 50.1) = 343.1$ K. Thus, applying equation 9.12 to the insulating brick:

$$400 = (0.4 \times 1.0/x)(1519.7 - 343.1)$$

and the thickness of the brick, $\underline{x = 1.18 \text{ m}}$

PROBLEM 9.73

2.8 kg/s of organic liquid of specific heat capacity 2.5 kJ/kg K is cooled in a heat exchanger from 363 to 313 K using water whose temperature rises from 293 to 318 K flowing countercurrently. After maintenance, the pipework is wrongly connected so that the two streams, flowing at the same rates as previously, are now in co-current flow. On the assumption that overall heat transfer coefficient is unaffected, show that the new outlet temperatures of the organic liquid and the water will be 320.6 K and 314.5 K, respectively.

Solution

i) *Countercurrent flow*

Heat load, $Q = (2.8 \times 2.5)(363 - 313) = 350$ kW

$\therefore$ water flow $= 350/4.18(318 - 293) = 3.35$ kg/s

$\Delta T_1 = (363 - 318) = 35 \deg \text{K}, \ \Delta T_2 = (313 - 293) = 20 \deg \text{K}$
and, from equation 9.9, $\Delta T_m = (45 - 20)/\ln(45/20) = 30.83 \deg \text{K}.$

From equation 9.1: $350 = UA \times 30.83$

and: $UA = 11.35$ kW/K.

ii) *Co-current flow*

Heat load, $Q = (2.8 \times 2.5)(363 - T) = 7.0(363 - T)$ kW for the organic
and for the water, $Q = (3.35 \times 4.18)(T' - 293) = 14.0(T')$ kW
where T and T' are the outlet temperatures of the organic and water respectively.

$\therefore$ $(363 - T) = (14.0/7.0)(T' - 293)$

and: $T' = (474.5 - 0.5T)$ K.

$$\Delta T_1 = (363 - 293) = 70 \deg \text{K}, \Delta T_2 = (T - T')$$

and from equation 9.9, $\Delta T_m = 70 - (T - T')/\ln[70/(T - T')] \deg \text{K}$.
In equation 9.1:

$$70(360 - T) = 11.35(70 - T + T')/\ln[70/(T - T')]$$

$\therefore$ $7.0(360 - T) = 11.35(70 - T + 474.5 - 0.5T)/\ln[70/(T - 474.5 + 0.5T)]$

or: $0.617(360 - T) = (544.5 - 1.5T)/\ln[70/(1.5T - 474.5)]$

Solving by trial and error, $T = \underline{319.8 \text{ K}}$ which is very close to the value suggested,
320.6 K.
$\therefore$ The outlet temperature of the water is:

$$T' = 474.5 - (0.5 \times 319.8) = \underline{314.6 \text{ K}}$$

which agrees almost exactly with the given value.
Thus for co-current flow: $Q = 7.0(363 - 319.8) = 302.4$ kW

$$\Delta T_1 = 70 \deg \text{K (as before)}, \Delta T_2 = (319.8 - 314.6) = 5.2 \deg \text{K}$$

and from equation 9.9, $\Delta T_m = (70 - 5.2)/\ln(70/5.2) = 24.92 \deg \text{K}$
$\therefore$ in equation 9.1: $302.4 = UA \times 24.92$

and: $UA = 12.10$ kW/K

which is in relatively close agreement with the counter-current value.

PROBLEM 9.74

An organic liquid is cooled from 353 to 328 K in a single-pass heat exchanger. When the
cooling water of initial temperature 288 K flows countercurrently its outlet temperature
is 333 K. With the water flowing co-currently, its feed rate has to be increased in order
to give the same outlet temperature for the organic liquid, the new outlet temperature of
the water is 313 K. When the cooling water is flowing countercurrently, the film heat
transfer coefficient for the water is 600 W/m^2 K.

What is the coefficient for the water when the exchanger is operating with cocurrent flow if its value is proportional to the 0.8 power of the water velocity?

Calculate the film coefficient from the organic liquid, on the assumptions that it remains unchanged, and that heat transfer resistances other than those attributable to the two liquids may be neglected.

Solution

i) *For countercurrent flow:*

$$\Delta T_1 = (353 - 333) = 20 \deg \text{ K and } \Delta T_2 = (328 - 288) = 40 \deg \text{K}.$$

∴ From equation 9.9: $\Delta T_m = (40 - 20)/\ln(40/20) = 28.85 \deg \text{K}$.

ii) *For co-current flow:*

$$\Delta T_1 = (353 - 288) = 65 \deg \text{K and } \Delta T_2 = (328 - 313) = 15 \deg \text{K}.$$

∴ $\Delta T_m = (65 - 15)/\ln(65/15) = 34.1 \deg \text{K}.$

Taking countercurrent flow as state 1 and co-current flow as state 2, then, in equation 9.1:

$$Q = U_1 A \Delta T_{m_1} = U_2 A \Delta T_{m_2}$$

or: $U_1/U_2 = (28.85/34.1) = 0.846$ (i)

The water velocity, $u \propto 1/\Delta T$, where ΔT is the rise in temperature of the water,

or: $u = K/\Delta T$ where K is a constant.

∴ $u_1/u_2 = \Delta T_2/\Delta T_1 = (313 - 288)/(333 - 288) = 0.556.$

But: $h_i \propto u^{0.8}$ or: $h_i = k'u^{0.8}$

Thus: $h_{i_1}/h_{i_2} = (u_1/u_2)^{0.8} = (0.556)^{0.8} = 0.625$

∴ $600/h_{i_2} = 0.625$ and $h_{i_2} = \underline{\underline{960 \text{ W/m}^2 \text{ K}}}.$

From equation 9.201, ignoring scale and wall resistances:

$$1/U = 1/h_o + 1/h_i$$

∴ $1/U_1 = 1/h_o + 1/600$ and $U_1 = 600h_o/(600 + h_o)$

and $1/U_2 = 1/h_o + 1/960$ and $U_2 = 960h_o/(960 + h_o)$

∴ $U_1/U_2 = (600/960)(960 + h_o)/(600 + h_o) = 0.846$ (from (i))

∴ $0.625(960 + h_o) = 0.846(600 + h_o)$ and $h_o = \underline{\underline{417 \text{ W/m}^2 \text{ K}}}.$

PROBLEM 9.75

A reaction vessel is heated by steam at 393 K supplied to a coil immersed in the liquid in the tank. It takes 1800 s to heat the contents from 293 K to 373 K when the outside temperature is 293 K. When the outside and initial temperatures are only 278 K, it takes 2700 s to heat the contents to 373 K. The area of the steam coil is 2.5 m^2 and of the external surface is 40 m^2. If the overall heat transfer coefficient from the coil to the liquid in the vessel is 400 W/m^2 K, show that the overall coefficient for transfer from the vessel to the surroundings is about 5 W/m^2 K.

Solution

Using the argument in Section 9.8.3, the net rate of heating is given by:

$$mC_p \, dT/dt = U_c A_c(T_s - T) - U_o A_o(T - T_o)$$

where U_c and U_o are the overall coefficients from the coil and the outside of the vessel respectively, A_c and A_o are the areas of the coil and the outside of the vessel and T_s, T and T_o are the temperature of the steam, the contents and the surroundings respectively. Writing $U_c A_c = a$ and $U_o A_o = b$:

$$mC_p \, dT/dt = a(393 - T) - b(T - T_a) = 393a + bT_a - (a + b)T$$

Integrating: $t = (mC_p/(a + b)) \left[\ln(1/[393a + bT_a - (a + b)T])\right]_{T_a}^{373}$ s

$$= (mC_p/(a + b)) \ln[(a(393 - T_a)/(20a - 373b + bT_a)] \text{ s}$$

When $T_a = 293$ K: $1800 = (mC_p/(a + b)) \ln[(100a)/(20a - 80b)]$ (i)

When $T_a = 278$ K: $2700 = (mC_p/(a + b)) \ln[(115a)/(20a - 95b)]$ (ii)

Dividing (i) by (ii): $0.667 = \ln[5a/(a - 4b)]/\ln[23a/(4a - 19b)]$ (iii)

But $U_c = 400$ W/m^2 K, $A_c = 2.5$ m^2 and hence:

$$a = U_c A_c = 1000 \text{ W/K or } 1 \text{ kW/K.}$$

Substituting in (iii): $1.5 = \ln[23/(4 - 19b)]/\ln[5/(1 - 4b)]$
Solving by trial and error: $b = 0.2$ kW/K or 200 W/K.

∴ $(U_o \times 40) = 200$ and $U_o = (200/40) = \underline{\underline{5 \text{ W/m}^2 \text{ K}}}$.

PROBLEM 9.76

Steam at 403 K is supplied through a pipe of 25 mm outside diameter. Calculate the heat loss per metre to surroundings at 293 K, on the assumption that there is a negligible drop in temperature through the wall of the pipe. The heat transfer coefficient h from the outside of the pipe of the surroundings is given by:

$$h = 1.22(\Delta T/d)^{0.25} \text{W/m}^2 \text{ K}$$

where d is the outside diameter of the pipe (m) and ΔT is the temperature difference (deg K) between the surface and surroundings.

The pipe is then lagged with a 50 mm thickness of lagging of thermal conductivity 0.1 W/m K. If the outside heat transfer coefficient is given by the same equation as for the bare pipe, by what factor is the heat loss reduced?

Solution

For 1 m length of pipe: surface area $= \pi dl = \pi(25/1000) \times 1.0 = 0.0785$ m^2

With a negligible temperature drop through the wall, the wall is at the steam temperature, 403 K, and $\Delta T = (403 - 293) = 110 \deg$ K.

Thus, the coefficient of heat transfer from the pipe to the surroundings is:

$$h = 1.22[110/(25/1000)]^{0.25} = 9.94 \text{ W/m}^2 \text{ K.}$$

and the heat loss: $Q = hA(T_w - T_s)$

$$= (9.94 \times 0.0785)(403 - 293) = \underline{\underline{85.8 \text{ W/m}}}$$

With the lagging:

$$Q = k(2\pi r_m l)(T_1 - T_2)/(r_2 - r_1) \qquad \text{(equation 9.22)}$$

In this case: $k = 0.1$ W/mK, $T_1 = 403$ K and T_2 is the temperature at the surface of the lagging.

$$r_1 = (25/1000)/2 = 0.0125 \text{ m}$$

$$r_2 = 0.0125 + (50/1000) = 0.0625 \text{ m}$$

and: $r_m = (0.0625 - 0.0125)/\ln(0.0625/0.0125) = 0.0311$ m

Thus: $Q = (0.1 \times 2\pi \times 0.0311 \times 1)(403 - T_2)/(0.0625 - 0.0125) = 0.391(403 - T_2)$(i)

But: $Q = 1.22[(T_2 - 293)/(2 \times 0.0625)]^{0.25}(\pi \times 2 \times 0.0625 \times 1)(T_2 - 293)$

$$= 0.806(T_2 - 293)^{1.25} \qquad \text{(ii)}$$

From (i) and (ii):

$$0.391(403 - T_2) = 0.806(T_2 - 293)^{1.25}$$

Solving by trial and error: $T_2 = 313.5$ K

and hence:

$$Q = 0.39/(403 - 313.5) = 35.0 \text{ W/m,}$$

a reduction of: $\qquad\qquad\qquad (85.8 - 35.0) = 50.8$ W/m

or: $\qquad\qquad\qquad (50.8 \times 100/85.8) = \underline{\underline{59.2\%}}$

PROBLEM 9.77

A vessel contains 1 tonne of liquid of specific heat capacity 4.0 kJ/kg K. It is heated by steam at 393 K which is fed to a coil immersed in the liquid and heat is lost to the surroundings at 293 K from the outside of the vessel. How long does it take to heat the liquid from 293 to 353 K and what is the maximum temperature to which the liquid can be heated? When the liquid temperature has reached 353 K, the steam supply is turned off for two hours and the vessel cools. How long will it take to reheat the material to 353 K? Coil: Area 0.5 m². Overall heat transfer coefficient to liquid, 600 W/m² K. Outside of vessel: Area 6 m². Heat transfer coefficient to surroundings, 10 W/m² K.

Solution

If T K is the temperature of the liquid at time t s, then the net rate of heat input to the vessel, $U_c A_c(T_s - T) - U_s A_s(T - T_a) = m C_p \, dT/dt$ W

where the coefficient at the coil, $U_c = 600$ W/m² K, the coefficient at the outside of the vessel, $U_s = 10$ W/m² K, the areas are: coil, $A_c = 0.5$ m², vessel, $A_s = 6.0$ m², the temperatures are: steam, $T_s = 393$ K, ambient, $T_a = 293$ K, the mass of liquid, $m = 1000$ kg and the specific heat capacity, $C_p = 4.0$ kJ/kg K or 4000 J/kg K.

Thus: $(1000 \times 4000)dT/dt = (600 \times 0.5)(393 - T) - (10 \times 6)(T - 293)$

and: $\quad 11{,}111 \, dT/dt = 376.3 - T$ \hfill (i)

$$\therefore \quad t = 11{,}111 \int_{T_1}^{T_2} dT/(376.3 - T) = 11{,}111 \ln[(376.3 - T_1)/(376.3 - T_2)] \quad \text{(ii)}$$

When $T_1 = 293$ K and $T_2 = 353$ K then: $t = 11{,}111 \ln(83.3/23.3) = \underline{\underline{14{,}155 \text{ s}}}$ (3.93 h)

The maximum temperature to which the liquid can be heated is obtained by putting $dT/dt = 0$ in (i) to give: $\underline{\underline{T = 376.3 \text{ K}}}$.

During the time the steam is turned off (for a period of 7200 s) a heat balance gives:

$$m C_p \, dT/dt = -U_s A_s(T - T_a)$$

or: $\quad (1000 \times 4000)dT/dt = -(10 \times 6)(T - 293)$

$\therefore \quad\quad\quad 66{,}700 \, dT/dt = (293 - T)$

Integrating: $\displaystyle\int_{353}^{T} dT/(293 - T) = 0.000015 \int_{0}^{7200} dt$

$\therefore \quad \ln((293 - 353)/(293 - T)) = (0.000015 \times 7200) = 0.108$ and $T = 346.9$ K.

The time taken to reheat the liquid to 353 K is then given by (ii):

$$t = 11{,}111 \int_{346.9}^{353} dT/(376.3 - T)$$

$$= 11{,}111 \ln[(376.3 - 346.9)/(376.3 - 353)] = \underline{\underline{2584 \text{ s}}} \ (0.72 \text{ h})$$

PROBLEM 9.78

A bare thermocouple is used to measure the temperature of a gas flowing through a hot pipe. The heat transfer coefficient between the gas and the thermocouple is proportional to the 0.8 power of the gas velocity and the heat transfer by radiation from the walls to the thermocouple is proportional to the temperature difference.

When the gas is flowing at 5 m/s the thermocouple reads 323 K. When it is flowing at 10 m/s it reads 313 K, and when it is flowing at 15.0 m/s it reads 309 K. Show that the gas temperature is about 298 K and calculate the approximate wall temperature. What temperature will the thermocouple indicate when the gas velocity is 20 m/s?

Solution

If the gas and thermocouple temperatures are T_g and T_k respectively, then the rate of heat transfer from the thermocouple to the gas:

$$Q_1 = Ku^{0.8}(T_g - T) \tag{i}$$

where K is a constant and u the gas velocity.

Similarly, the rate of heat transfer from the walls to the thermocouple is:

$$Q_2 = k'(T_w - T) \text{ W} \tag{ii}$$

where k' is a constant and T_w is the wall temperature.

At equilibrium: $\quad Q_1 = Q_2$ and $u^{0.8} = (k'/k)(T_w - T)/(T - T_g) \tag{iii}$

When $u = 5$ m/s, $\quad T = 323$ K and in (iii):

$$5^{0.8} = (k'/k)(T_w - 323)/(323 - T_g) = 3.624$$

$\therefore \qquad (k'/k) = 3.624(323 - T_g)/(T_w - 323) \tag{iv}$

When $u = 10$ m/s, $T = 313$ K and in (iii):

$$10^{0.8} = (k'/k)(T_w - 313)/(313 - T_g) = 6.31$$

Substituting for (k'/k) from (iv):

$$6.31 = 3.624(323 - T_g)(T_w - 313)/[(T_w - 323)(313 - T_g)] \tag{v}$$

When $u = 15$ m/s, $T = 309$ K and in (iii):

$$15^{0.8} = (k'/k)(T_w - 309)/(309 - T_g) = 8.73$$

Substituting for (k'/k) from (iv):

$$8.73 = 3.624(323 - T_g)(T_w - 309)/[(T_w - 323)(309 - T_g)] \tag{vi}$$

If $T_g = 298$ K, then in (v):

$$1.741 = [(323 - 298)(T_w - 313)]/[(T_w - 323)(313 - 298)]$$

or: $\qquad (T_w - 313)/(T_w - 323) = 1.045$ and $T_w = 533$ K.

If $T_g = 298$ K, then in (vi): $2.409 = (323 - 298)(T_w - 309)/[(T_w - 323)(309 - 298)]$

or: $\qquad (T_w - 309)/(T_w - 323) = 1.060$ and $T_w = 556$ K

This result agrees fairly well and a mean value of $T_w = \underline{\underline{545 \text{ K}}}$ is indicated.

In equation (iv): $(k'/k) = 3.624(323 - 298)/(545 - 323) = 0.408$

$\therefore$ in equation (iii) $u^{0.8} = 0.408(545 - T)/(T - 298)$

When $u = 20$ m/s: $10.99 = 0.408(545 - T)/(T - 298)$ and $\underline{\underline{T = 306.8 \text{ K}}}$.

PROBLEM 9.79

A hydrocarbon oil of density 950 kg/m^3 and specific heat capacity 2.5 kJ/kg K is cooled in a heat exchanger from 363 to 313 K by water flowing countercurrently. The temperature of the water rises from 293 to 323 K. If the flowrate of the hydrocarbon is 0.56 kg/s, what is the required flowrate of water?

After plant modifications, the heat exchanger is incorrectly connected so that the two streams are in co-current flow. What are the new outlet temperatures of hydrocarbon and water, if the overall heat transfer coefficient is unchanged?

Solution

Heat lost by the oil $= 0.56 \times 2.5(363 - 313) = 70.0$ kW

For a flow of water of G kg/s, heat gained by the water is:

$$70.0 = (G \times 4.18)(323 - 293) \text{ and } \underline{\underline{G = 0.56 \text{ kg/s}}}$$

i) *For countercurrent flow:*

$$\Delta T_1 = (363 - 323) = 40 \deg K, \ \Delta T_2 = (313 - 293) = 20 \deg K$$

and from equation 9.9:

$$\Delta T_m = (40 - 20)/\ln(40/20) = 28.85 \deg K$$

In equation 9.1:

$$70.0 = UA \times 28.85 \text{ and } UA = 2.43 \text{ kW/K}.$$

ii) *For co-current flow:*
If T_o and T_w are the outlet temperature of the oil and water respectively, the heat load is:

$$Q = (0.56 \times 2.5)(363 - T_o) = (0.56 \times 4.18)(T_w - 293) \qquad \text{(i)}$$

$$508.2 - 1.4T_o = 2.34T_w - 685.9$$

and: $$T_w = 510.3 - 0.60T_o \qquad \qquad \text{(ii)}$$

$$\Delta T_1 = (363 - 293) = 70 \text{ K}$$

$$\Delta T_1 = (T_o - T_w) = (T_o - 510.3 + 0.60T_o) = (1.60T_o - 510.3) \text{ K}$$

and in equation 9.9: $\Delta T_m = (70 - 1.60T_o + 510.3)/\ln[70/(1.60T_o - 510.3)]$

$$= (580.3 - 1.60T_o)/\ln[70/1.60T_o - 510.3)]\deg K$$

and: $(508.2 - 1.4T_o) = 2.43(580.3 - 1.60T_o)/\ln[70/(1.60T_o - 510.3)]$

$$(508.2 - 1.4T_o) = 2.126(508.2 - 1.40T_o)/\ln[70/(1.60T_o - 510.3)]$$

∴ $70/(1.60T_o - 510.3) = e^{2.126} = 8.38$ and $T_o = \underline{\underline{324.2 \text{ K}}}$.

and in equation $T(ii) = 510.3 - (0.60 \times 324.2) = \underline{\underline{315.8 \text{ K}}}$.

PROBLEM 9.80

A reaction mixture is heated in a vessel fitted with an agitator and a steam coil of area 10 m^2 fed with steam at 393 K. The heat capacity of the system is equal to that of 500 kg of water. The overall coefficient of heat transfer from the vessel of area 5 m^2 is 10 W/m^2 K. It takes 1800 s to heat the contents from ambient temperature of 293 to 333 K. How long will it take to heat the system to 363 K and what is the maximum temperature which can be reached? Specific heat capacity of water $= 4200 \text{ J/kgK}$.

Solution

Following the argument of Problem 9.77 and taking ambient temperature as the initial temperature of the mixture, 293 K, then:

net rate of heating $= (500 \times 4200)dT/dt = U_c \times 10(393 - T) - (10 \times 5)(T - 293)$ W

∴ $2,100,000\, dT/dt = 3930U_c - 10U_cT - 50T + 14,650$

$$= (3930U_c + 14,650) - (10U_c + 50)T.$$

∴ $(2,100,000/(10U_c + 50))dT/dt = ((3930U_c + 14,650)/(10U_c + 50)) - T.$

∴ $t = [2,100,000/(10U_c + 50)] \int_{293}^{333} dT/[(3930U_c + 14,650)/(10U_c + 50) - T]$

In heating from 293 to 333 K, the time taken is 1800 s and:

$$1800 = [2,100,000/(10U_c + 50)] \ln\{[(3930U_c + 14,650)/(10U_c + 50) - 293]/$$
$$[(3930U_c + 14,650)/(10U_c + 50) - 333]\}$$

Solving by trial and error: $U_c = 61.0 \text{ W/m}^2$ K.
Thus, net rate of heating is:

$2,100,000\, dT/dt = (61.0 \times 10)(393 - T) - (10 \times 5)(T - 293) = 254,380 - 660\, T$ W

or: $3182\, dT/dt = 385.4 - T$ W   (i)

∴ time for heating, $t = 3182 \int_{293}^{363} dT/(385.4 - T)$

$$= 3182 \ln[(385.4 - 293)/(385.4 - 363)] = 3182 \ln(92.4/22.4) = \underline{\underline{4509 \text{ s}}} \ (1.25 \text{ h})$$

The maximum temperature which can be attained is obtained by putting $dT/dt = 0$ in (i) which gives:

$$T_{max} = \underline{\underline{385.5 \text{ K}}}.$$

PROBLEM 9.81

A pipe, 50 mm outside diameter, is carrying steam at 413 K and the coefficient of heat transfer from its outer surface to the surroundings at 288 K is 10 W/m^2 K. What is the heat loss per unit length?

It is desired to add lagging of thermal conductivity 0.03 W/m K as a thick layer to the outside of the pipe in order to cut heat losses by 90%. If the heat transfer from the outside surface of the lagging is 5 W/m^2 K, what thickness of lagging is required?

Solution

Outside area of pipe: $\pi dl = \pi(50/1000) \times 1.0 = 0.157$ m^2/m. Assuming the pipe wall is at the temperature of the steam, that is the resistance of the wall is negligible, then the heat loss is:

$$Q = hA(T_w - T_a) = (10 \times 0.157)(413 - 288) = \underline{\underline{196 \text{ W/m}}}$$

With the addition of lagging of thickness, x mm, the required heat flow is 19.6 W/m The diameter of the lagging $= (50 + 2x)/1000$ m

and the surface area of the lagging $= [\pi(50 + 2x)/1000] \times 1.0$
$$= (0.157 + 0.00628x) \text{ m}^2\text{/m}.$$

The heat transferred to the surroundings,

$$19.6 = 5(0.157 + 0.00628x)(T_2 - 288) \text{ W}$$

and: $\qquad T_2 = (49.14 + 1.809x)/(0.157 + 0.00628x) \text{ K}$ \hfill (i)

For conduction through the lagging:

$$Q = k(2\pi r_m l)(T_1 - T_2)/(r_2 - r_1) \text{ W} \qquad \text{(equation 9.22)}$$

where $Q = 19.6$ W/m, $k = 0.03$ W/mK, $l = 1.0$ m and the steam temperature, $T_1 = 413$ K.

$$r_1 = 25 \text{ mm or } 0.025 \text{ m},$$

$$r_2 = (25 + x) \text{ mm or } (25 + x)/1000 = (0.025 + 0.001x) \text{ m}$$

∴ $\qquad r_m = (0.025 + 0.001x - 0.025)/\ln[(0.025 + 0.001x)/0.025]$

$$= 0.001x/\ln(1 + 0.04x) \text{ m}$$

Thus, in equation 9.22:

$$19.6 = 0.03(2\pi(0.001x/\ln(1 + 0.04x)) \times 1.0(413 - T_2)/(0.025 + 0.001x - 0.025)$$

∴ $\qquad 103,981 = [x/\ln(1 + 0.04x)](413 - T_2)/0.001x$

Substituting for T_2 from (i):

$$104 = [413 - (49.14 + 1.809x)/(0.157 + 0.00628x)]/\ln(1 + 0.04x)$$

Solving by trial and error, $\underline{\underline{x = 52.5 \text{ mm}}}$.

PROBLEM 9.82

It takes 1800 s (0.5 h) to heat a tank of liquid from 293 to 323 K using steam supplied to an immersed coil when the steam temperature is 383 K. How long will it take when the steam temperature is raised to 393 K? The overall heat transfer coefficient from the steam coil to the tank is 10 times the coefficient from the tank to surroundings at a temperature of 293 K, and the area of the steam coil is equal to the outside area of the tank.

Solution

Using the argument in Problem 9.77,

$$mC_p \, dT/dt = U_c A_c(T_s - T) - U_s A_s(T - T_a)$$

In this case, $T_s = 383$ K, $T_a = 293$ K, $U_s = U_c/10$ and $A_c = A_s = A$ (say).

$\therefore$ $\qquad\qquad mC_p \, dT/dt = U_c A_c(383 - T) - (U_c A_c/10)(T - 293)$

$\therefore$ $\qquad\qquad (U_c A_c/mC_p)dt = dT/(412.3 - 1.1T)$

On integration:

$$(U_c A_c/mC_p)t = \int_{293}^{323} dT/(412.3 - 1.1T) = (1/1.1)\left[\ln(1/(412.3 - 1.1T)\right]_{293}^{323}$$

Since it takes 1800 s to heat the liquid from 293 to 333 K, then:

$$1800(U_c A_c/mC_p) = 0.909 \ln[(412.3 - (1.1 \times 293)]/[412.3 - (1.1 \times 323)]$$

$$= 0.909 \ln(90/57)$$

and: $\qquad (U_c A_c/mC_p) = 0.00023071 \text{ s}^{-1}$

On increasing the steam temperature to 393 K,

Heat transferred from the steam $= U_c A_c(393 - T)$ W

Heat lost to the surroundings $= (U_c A_c/10)(T - 293)$ W

and: $\qquad\qquad mC_p \, dT/dt = U_c A_c(393 - T) - (0.1 U_c A_c)(T - 293)$ W

$\therefore$ $\qquad\qquad dT/(422.3 - 1.1T) = (U_c A_c/mC_p)dt = 0.0002307 \, dt$

On integration:

$$0.0002307t = \int_{293}^{323} dT/(422.3 - 1.1T) = (1/1.1)[\ln(1/(422.3 - 1.1T)]_{293}^{333}$$

Thus, on heating from 293 to 323 K:

$$0.0002307t = 0.909 \ln\{[422.3 - (1.1 \times 293)]/[422.3 - (1.1 \times 323)]\} = 0.909 \ln(100/67)$$

and: $t = \underline{\underline{1578 \text{ s}}}$ (0.44h)

PROBLEM 9.83

A thermometer is situated in a duct in an air stream which is at a constant temperature. The reading varies with the gas flowrate as follows:

air velocity (m/s)	thermometer reading (K)
6.1	553
7.6	543
12.2	533

The wall of the duct and the gas stream are at somewhat different temperatures. If the heat transfer coefficient for radiant heat transfer from the wall to the thermometer remains constant, and the heat transfer coefficient between the gas stream and thermometer is proportional to the 0.8 power of the velocity, what is the true temperature of the air stream? Neglect any other forms of heat transfer.

Solution

As with Problem 9.78, a heat balance on the thermometer gives: $h_w(T_w - T)$ $= h_g(T - T_g)$ where h_w and h_g are the coefficients for radiant heat transfer from the wall and for convection to the gas respectively and T_w, T and T_g are the temperatures of the wall, thermometer and gas, respectively, *above a datum of 533 K*.

When $u = 12.2$ m/s, $\quad h_w(T_w - 0) = h_g(0 + T_g)$ (i)

When $u = 7.6$ m/s, since $h_g \propto u^{0.8}$, $\quad h_w(T_w - 10) = h_g(7.6/12.2)^{0.8}(-10 + T_g)$ (ii)

When $u = 6.1$ m/s, $\quad h_w(T_w - 20) = h_g(6.1/12.2)^{0.8}(-20 + T_g)$ (iii)

Dividing equation (i) by equation (ii): $T_w/(T_w - 10) = (12.2/7.6)^{0.8}T_g/(T_g - 10)$

$$= 1.46T_g/(T_g - 10) \quad \text{(iv)}$$

and dividing equation (i) by equation (iii): $T_w/(T_w - 20) = (12.2/6.1)^{0.8}T_g/(T_g - 20)$

$$= 1.741T_g/(T_g - 20) \quad \text{(v)}$$

From (v): $(T_w T_g - 20 T_w) = 1.741 T_w T_g - 34.82 T_g$

$\therefore$ $T_w = 34.82 T_g / (20 + 0.741 T_g)$ K   (vi)

Substituting for T_w from (vi) into (iv):

$$34.82 T_g / (27.4 T_g - 200) = 1.46 T_g / (T_g - 10) \text{ and } T_g = -11.20 \text{ K}$$

and hence, the temperature of the gas is $(533 - 11.2) = \underline{\underline{521.8 \text{ K}}}$.

SECTION 10

Mass Transfer

PROBLEM 10.1

Ammonia gas is diffusing at a constant rate through a layer of stagnant air 1 mm thick. Conditions are fixed so that the gas contains 50% by volume of ammonia at one boundary of the stagnant layer. The ammonia diffusing to the other boundary is quickly absorbed and the concentration is negligible at that plane. The temperature is 295 K and the pressure atmospheric, and under these conditions the diffusivity of ammonia in air is 0.18 cm^2/s. Calculate the rate of diffusion of ammonia through the layer.

Solution

See Volume 1, Example 10.1.

PROBLEM 10.2

A simple rectifying column consists of a tube arranged vertically and supplied at the bottom with a mixture of benzene and toluene as vapour. At the top, a condenser returns some of the product as a reflux which flows in a thin film down the inner wall of the tube. The tube is insulated and heat losses can be neglected. At one point in the column, the vapour contains 70 mol% benzene and the adjacent liquid reflux contains 59 mol% benzene. The temperature at this point is 365 K. Assuming the diffusional resistance to vapour transfer to be equivalent to the diffusional resistance of a stagnant vapour layer 0.2 mm thick, calculate the rate of interchange of benzene and toluene between vapour and liquid. The molar latent heats of the two materials can be taken as equal. The vapour pressure of toluene at 365 K is 54.0 kN/m^2 and the diffusivity of the vapours is 0.051 cm^2/s.

Solution

In this solution, subscripts 1 and 2 refer to the liquid surface and vapour side of the stagnant layer respectively and subscripts B and T refer to benzene and toluene.

If the latent heats are equal and there are no heat losses, there is no net change of phase across the stagnant layer.

This is an example of equimolecular counter diffusion and:

$$N_A = -D(P_{A2} - P_{A1})/\mathbf{R}TL \qquad \text{(equation 10.23)}$$

where L = thickness of the stagnant layer = 0.2 mm = 0.0002 m.

As the vapour pressure of toluene $= 54$ kN/m^2, the partial pressure of toluene from Raoult's law $= (1 - 0.59) \times 54 = 22.14$ kN/m$^2 = P_{T1}$ and:

$$P_{T2} = (1 - 0.70) \times 101.3 = 30.39 \text{ kN/m}^2$$

For toluene: $N_T = -(0.051 \times 10^{-4})(30.39 - 22.14)/(8.314 \times 365 \times 0.0002)$

$$= -6.93 \times 10^{-5} \text{ kmol/m}^2\text{s}$$

For benzene: $P_{B1} = 101.3 - 22.14 = 79.16$ kN/m^2

$$P_{B2} = 101.3 - 30.39 = 70.91 \text{ kN/m}^2$$

Hence, for benzene: $N_B = -(0.051 \times 10^{-4})(70.91 - 79.16)/(8.314 \times 365 \times 0.0002)$

$$= 6.93 \times 10^{-5} \text{ kmol/m}^2\text{s}$$

Thus the rate of interchange of benzene and toluene is equal but opposite in direction.

PROBLEM 10.3

By what percentage would the rate of absorption be increased or decreased by increasing the total pressure from 100 to 200 kN/m^2 in the following cases?

  (a) The absorption of ammonia from a mixture of ammonia and air containing 10% of ammonia by volume, using pure water as solvent. Assume that all the resistance to mass transfer lies within the gas phase.
  (b) The same conditions as (a) but the absorbing solution exerts a partial vapour pressure of ammonia of 5 kN/m^2.

The diffusivity can be assumed to be inversely proportional to the absolute pressure.

Solution

(a) The rates of diffusion for the two pressures are given by:

$$N_A = -(D/\mathbf{R}TL)(P/P_{BM})(P_{A2} - P_{A1}) \qquad \text{(equation 10.34)}$$

where subscripts 1 and 2 refer to water and air side of the layer respectively and subscripts A and B refer to ammonia and air.

Thus: $P_{A2} = (0.10 \times 100) = 10$ kN/m^2 and $P_{A1} = 0$ kN/m^2

$$P_{B2} = (100 - 10) = 90 \text{ kN/m}^2 \text{ and } P_{B1} = 100 \text{ kN/m}^2$$

$$P_{BM} = (100 - 90)/\ln(100/90) = 94.91 \text{ kN/m}^2$$

∴ $P/P_{BM} = (100/94.91) = 1.054$

Hence: $N_A = -(D/\mathbf{R}TL)1.054(10 - 0) = -10.54D/\mathbf{R}TL$

If the pressure is doubled to 200 kN/m^2, the diffusivity is halved to 0.5D (from equation 10.18) and:

$$P_{A2} = (0.1 \times 200) = 20 \text{ kN/m}^2 \text{ and } P_{A1} = 0 \text{ kN/m}^2$$

$$P_{B2} = (200 - 20) = 180 \text{ kN/m}^2 \text{ and } P_{B1} = 200 \text{ kN/m}^2$$

∴ $$P_{BM} = (200 - 180)/\ln(200/180) = 189.82 \text{ kN/m}^2$$

$$P/P_{BM} = (200/189.82) = 1.054 \text{ i.e. unchanged}$$

Hence: $N_A = -(0.5D/\mathbf{R}TL)1.054(20 - 0) = -10.54D/\mathbf{R}TL$, that is the rate is unchanged

(b) If the absorbing solution now exerts a partial vapour pressure of ammonia of 5 kN/m^2, then at a total pressure of 100 kN/m^2:

$$P_{A2} = 10 \text{ kN/m}^2 \text{ and } P_{A1} = 5 \text{ kN/m}^2$$

$$P_{B2} = 90 \text{ kN/m}^2 \text{ and } P_{B1} = 95 \text{ kN/m}^2$$

$$P_{BM} = (95 - 90)/\ln(95/90) = 92.48 \text{ kN/m}^2$$

∴ $$P/P_{BM} = (100/92.48) = 1.081$$

$$N_A = -(D/\mathbf{R}TL) \times 1.081(10 - 5) = -5.406D/\mathbf{R}TL$$

At 200 kN/m^2, the diffusivity = 0.5D and:

$$P_{A2} = 20 \text{ kN/m}^2 \text{ and } P_{A1} = 5 \text{ kN/m}^2$$

$$P_{B2} = 180 \text{ kN/m}^2 \text{ and } P_{B1} = 195 \text{ kN/m}^2$$

∴ $$P_{BM} = (195 - 180)/\ln(195/180) = 187.4 \text{ kN/m}^2$$

$$P/P_{BM} = 1.067$$

$$N_A = -(0.5D/\mathbf{R}TL)1.067(20 - 5) = -8.0D/\mathbf{R}TL$$

Thus the rate of diffusion has been increased by $100(8 - 5.406)/5.406 = \underline{\underline{48\%}}$.

PROBLEM 10.4

In the Danckwerts' model of mass transfer it is assumed that the fractional rate of surface renewal s is constant and independent of surface age. Under such conditions the expression for the surface age distribution function is se^{-st}.

If the fractional rate of surface renewal were proportional to surface age (say $s = bt$, where b is a constant), show that the surface age distribution function would then assume the form:

$$(2b/\pi)^{1/2}e^{-bt^2/2}$$

Solution

From equation 10.117: $f'(t) = sf(t) = 0$

In this problem: $s = bt$ and $e^{bt^2/2} f(t) = \text{constant} = k$

$$\therefore \qquad\qquad f(t) = ke^{-bt^2/2}$$

The total area of surface considered is unity and:

$$\therefore \qquad\qquad \int_0^\infty f(t)\,dt = 1$$

$$\therefore \qquad\qquad \int_0^\infty ke^{-bt^2/2}\,dt = 1$$

and by substitution as in equation 10.120:

$$k(\pi/2b)^{0.5} = 1$$

$$k = (2b/\pi)^{0.5} \text{ and } f(t) = \underline{\underline{(2b/\pi)^{1/2} e^{-bt^2/2}}}$$

PROBLEM 10.5

By consideration of the appropriate element of a sphere show that the general equation for molecular diffusion in a stationary medium and in the absence of a chemical reaction is:

$$\frac{\partial C_A}{\partial t} = D\left(\frac{\partial^2 C_A}{\partial r^2} + \frac{1}{r^2}\frac{\partial^2 C_A}{\partial \beta^2} + \frac{1}{r^2 \sin^2 \beta}\frac{\partial^2 C_A}{\partial \phi^2} + \frac{2}{r}\frac{\partial C_A}{\partial r} + \frac{\cot\beta}{r^2}\frac{\partial C_A}{\partial \beta} \right)$$

where C_A is the concentration of the diffusing substance, D the molecular diffusivity, t the time, and r, β, ϕ are spherical polar coordinates, β being the latitude angle.

Solution

The basic equation for unsteady state mass transfer is:

$$\frac{\partial C_A}{\partial t} = D\left[\left(\frac{\partial^2 C_A}{\partial x^2}\right)_{yz} + \left(\frac{\partial^2 C_A}{\partial y^2}\right)_{zx} + \left(\frac{\partial^2 C_A}{\partial z^2}\right)_{xy} \right] \qquad \text{(equation 10.67) (i)}$$

This equation may be transformed into other systems of orthogonal coordinates, the most useful being the spherical polar system. (Carslaw and Jaeger, *Conduction of Heat in Solids*, gives details of the transformation.) When the operation is performed:

$$x = r \sin\beta \cos\phi$$

$$y = r \sin\beta \sin\phi$$

$$z = r \cos\beta$$

and the equation for C_A becomes:

$$\frac{\partial C_A}{\partial t} = \frac{D}{r^2}\left[\frac{\partial}{\partial r}\left(r^2 \frac{\partial C_A}{\partial r}\right) + \frac{1}{\sin\beta}\frac{\partial}{\partial \beta}\left(\sin\beta\frac{\partial C_A}{\partial \beta}\right) + \frac{1}{\sin^2\beta}\frac{\partial^2 C_A}{\partial \phi^2} \right] \qquad \text{(ii)}$$

which may be written as:

$$\frac{\partial C_A}{\partial t} = D\left[\frac{\partial^2 C_A}{\partial r^2} + \frac{2}{r}\frac{\partial C_A}{\partial r} + \frac{1}{r^2}\frac{\delta}{\partial \mu}\left((1-\mu^2)\frac{\partial C_A}{\partial \mu}\right) + \frac{1}{r^2(1-\mu^2)}\frac{\partial^2 C_A}{\partial \phi^2}\right] \quad \text{(iii)}$$

where: $\mu = \cos\beta$. (iv)

In this problem $\partial C_A/\partial t$ is given by:

$$\frac{\partial C_A}{\partial t} = D\left(\frac{\partial^2 C_A}{\partial r^2} + \frac{1}{r^2}\frac{\partial^2 C_A}{\partial \beta^2} + \frac{1}{r^2 \sin^2 \beta}\frac{\partial^2 C_A}{\partial \phi^2} + \frac{2}{r}\frac{\partial C_A}{\partial r} + \frac{\cot\beta}{r^2}\frac{\partial C_A}{\partial \beta}\right) \quad \text{(v)}$$

Comparing equations (iii) and (v) is necessary to prove that:

$$\frac{1}{r^2}\frac{\partial}{\partial \mu}\left((1-\mu^2)\frac{\partial C_A}{\partial \mu}\right) + \frac{1}{r^2(1-\mu^2)}\frac{\partial^2 C_A}{\partial \phi^2} = \frac{1}{r^2}\frac{\partial^2 C_A}{\partial \beta^2} + \frac{1}{r^2 \sin^2 \beta}\frac{\partial^2 C_A}{\partial \phi^2} + \frac{\cot\beta}{r^2}\frac{\partial C_A}{\partial \beta}$$

$$\mu = \cos\beta, \, 1 - \mu^2 = 1 - \cos^2\beta = \sin^2\beta$$

$$\therefore \qquad\qquad \frac{1}{r^2(1-\mu^2)}\frac{\partial^2 C_A}{\partial \phi^2} = \frac{1}{r^2 \sin^2 \beta}\frac{\partial^2 C_A}{\partial \phi^2}$$

It now becomes necessary to prove that:

$$\frac{1}{r^2}\frac{\partial^2 C_A}{\partial \beta^2} + \frac{\cot\beta}{r^2}\frac{\partial C_A}{\partial \beta} = \frac{1}{r^2}\frac{\partial}{\partial \mu}\left((1-\mu^2)\frac{\partial C_A}{\partial \mu}\right) \quad \text{(vi)}$$

From equation (iv): $\mu = \cos\beta$

$$\therefore \qquad\qquad \partial\mu/\partial\beta = -\sin\beta \qquad\qquad \text{(vii)}$$

and: $\partial^2\mu/\partial\beta^2 = -\cos\beta$ (viii)

$$\frac{1}{r^2}\frac{\partial}{\partial \mu}\left((1-\mu^2)\frac{\partial C_A}{\partial \mu}\right) = \frac{1}{r^2}\frac{\partial}{\partial \beta}\left((1-\mu^2)\frac{\partial C_A}{\partial \beta}\frac{\partial \beta}{\partial \mu}\right)\frac{\partial \beta}{\partial \mu}$$

Substituting from equation (iv) for μ from equation (vii) for $\partial\beta/\partial\mu$ gives:

$$= \frac{1}{r^2}\frac{\partial}{\partial \beta}\left((1-\cos^2\beta)\frac{\partial C_A}{\partial \beta}\frac{1}{-\sin\beta}\right)\frac{1}{-\sin\beta} = \frac{1}{r^2}\frac{\partial}{\partial \beta}\left(-\sin\beta\frac{\partial C_A}{\partial \beta}\right)\frac{1}{-\sin\beta}$$

$$= \frac{1}{r^2}\left[-\sin\beta\frac{\partial^2 C_A}{\partial \beta^2} + \frac{\partial C_A}{\partial \beta}(-\cos\beta)\right]\frac{1}{-\sin\beta} = \frac{1}{r^2}\frac{\partial^2 C_A}{\partial \beta^2} + \frac{\cot\beta}{r^2}\frac{\partial C_A}{\partial \beta}$$

PROBLEM 10.6

Prove that for equimolecular counter diffusion from a sphere to a surrounding stationary, infinite medium, the Sherwood number based on the diameter of the sphere is equal to 2.

Solution

If the particle has a radius r, and is surrounded by a spherical shell of radius s then, Moles per unit time diffusing through the shell, M is given by:

$$M = 4\pi s^2 \left(-D\frac{dC_A}{ds} \right)$$

At steady state, M is constant and:

$$M \int_{s_1}^{s_2} \frac{ds}{s^2} = 4\pi(-D) \int_{C_{A_1}}^{C_{A_2}} dC_A$$

$$M \left(\frac{1}{s_1} - \frac{1}{s_2} \right) = 4\pi D(C_{A_1} - C_{A_2})$$

If C_{A_1} is the concentration at $s_1 = r$ and C_{A_2} is the concentration at $s_2 = \infty$, then:

$$M/r = 4\pi D(-\Delta C_A)$$

The mass transfer coefficient: $h_d = \dfrac{M}{A(-\Delta C_A)} = \dfrac{M}{4\pi r^2(-\Delta C_A)}$

$$h_D 4\pi r^2(-\Delta C_A)/r = 4\pi D(-\Delta C_A)$$

$$h_D = D/r = 2D/d$$

$$\underline{\underline{h_D d/D = Sh = 2}}$$

PROBLEM 10.7

Show that the concentration profile for unsteady-state diffusion into a bounded medium of thickness L, when the concentration at the interface is suddenly raised to a constant value C_{Ai} and kept constant at the initial value of C_{Ao} at the other boundary is:

$$\frac{C_A - C_{Ao}}{C_{Ai} - C_{Ao}} = 1 - \frac{z}{L} - \frac{2}{\pi} \left[\sum_{n=1}^{n=\infty} \frac{1}{n} \exp(-n^2\pi^2 Dt/L^2) \sin(nz\pi/L) \right].$$

Assume the solution to be the sum of the solution for infinite time (steady-state part) and the solution of a second unsteady-state part, which simplifies the boundary conditions for the second part.

Solution

The system is shown in Fig. 10a.
    The boundary conditions are:

At time, $t = 0$ $C_A = C_{Ao}$ $0 < y < L$
          $t > 0$ $C_A = C_{Ai}$ $y = 0$
          $t > 0$ $C_A = C_{Ao}$ $y = L$

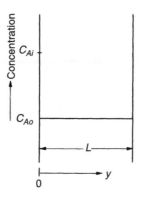

Figure 10a.

Replacing C_{Ai} by C_i' and C_A by C' where: $C_A = C' + C_{Ao}$ and $C_{Ai} = C_i' + C_{Ao}$, then using these new variables:

At: $\quad t = 0 \quad C' = 0 \quad 0 < y < L$

$\qquad t > 0 \quad C' = C_i' \quad y = 0$

$\qquad t > 0 \quad C' = 0 \quad\quad y = L$

The problem states that the solution of the one dimensional diffusion equation is:

$$C' = (\text{steady state solution}) + \sum_{0}^{\infty} \exp(-n^2\pi^2 Dt/L^2) A_n \sin(n\pi y/L)$$

where the steady state solution $= C_i' - C_i' y/L$.

(A derivation of the analogous equation for heat transfer may be found in *Conduction of Heat in Solids* by H. S. Carslaw and J. C. Jaeger, Oxford, 1960.)

$$A_n = \frac{2}{L} \int_0^L (\text{initial concentration profile} - \text{steady state}) \sin(n\pi y/L) \, dy$$

$$= \frac{2}{L} \int_0^L [0 + (C_i' y/L) - C_i'] \sin(n\pi y/L) \, dy$$

$$= -2C_i'/n\pi \text{ (this proof is given at the end of this problem)}.$$

Hence: $\qquad C' = C_i' - C_i' y/L - \frac{2C_i'}{\pi} \sum_{n=0}^{\infty} \frac{1}{n} \exp(-n^2\pi^2 Dt/L^2) \sin(n\pi y/L)$

$\therefore \qquad C = C_o + (C_i - C_o)\left[1 - \frac{y}{L} - \frac{2}{\pi} \sum_{n=0}^{\infty} \frac{1}{n} \exp(-n^2\pi^2 Dt/L^2) \sin(n\pi y/L) \right]$

$$A_n = \frac{2}{L} \int_0^L [(C_i' y/L) - C_i'] \sin(n\pi y/L) \, dy$$

$$= \frac{2C_i'}{L^2} \int_0^L y \sin(n\pi y/L)\, dy - \frac{2C_i'}{L} \int_0^L \sin(n\pi y/L)\, dy$$

$$= \frac{2C_i'}{L^2} \int_0^L ① - \frac{2C_i'}{L} \int_0^L ②$$

$$\int_0^L ① = \left[-\frac{Ly}{n\pi} \cos \frac{n\pi y}{L} \right]_0^L + \int_0^L \frac{L}{n\pi} \cos \frac{n\pi y}{L}\, dy$$

Putting $u = y$, $du = dy$

and: $\qquad dv = \sin(n\pi y/L)\, dy, \qquad v = -\frac{L}{n\pi} \cos \frac{n\pi y}{L}$

$$\therefore \qquad \int_0^L ① = \left(-\frac{Ly}{n\pi} \cos \frac{n\pi y}{L} \right)_0^L + \left(\frac{L^2}{n^2\pi^2} \sin \frac{n\pi y}{L} \right)_0^L$$

$$= -\frac{L^2}{n\pi} \cos n\pi + \frac{L^2}{n^2\pi^2} \sin n\pi = -\frac{L^2}{n\pi}(-1)^n$$

$$\int_0^L ② = \left(-\frac{L}{n\pi} \cos \frac{n\pi y}{L} \right)_0^L = -\frac{L}{n\pi} \cos n\pi + \frac{L}{n\pi}$$

$$= -\frac{L}{n\pi} \cos n\pi + \frac{L}{n\pi}$$

$$= -\frac{L}{n\pi}(-1)^n + \frac{L}{n\pi}$$

$$A_n = \frac{2C_i'}{L^2} ① - \frac{2C_i'}{L} ② = \frac{2C_i'}{L^2} \left(-\frac{L^2}{n\pi}(-1)^n \right) - \frac{2C_i'}{L} \left(-\frac{L}{n\pi}(-1)^n + \frac{L}{n\pi} \right)$$

$$= -2C_i'/n\pi$$

PROBLEM 10.8

Show that under the conditions specified in Problem 10.7 and assuming the Higbie model of surface renewal, the average mass flux at the interface is given by:

$$(N_A)_t = (C_{Ai} - C_{Ao})D/L \left\{ 1 + (2L^2/\pi^2 Dt) \sum_{n=1}^{n=\infty} \left[\frac{\pi^2}{6} - \frac{1}{n^2} \exp(-n^2\pi^2 Dt/L^2) \right] \right\}$$

Use the relation $\sum_{n=1}^{\infty} \frac{1}{n^2} = \pi^2/6$.

Solution

The rate of transference across the phase boundary is given by:

$$N_A = -D(\partial C_A/\partial y)_{y=0}$$

According to the Higbie model, if the element is exposed for a time t_e, the average rate of transfer is given by:

$$N_A = \frac{1}{t_e} \int_0^{t_e} -D(\partial C/\partial z)_{z=0} \, dt$$

From Problem 10.7, the concentration C is:

$$C = C_{Ao} + (C_{Ai} - C_{Ao})\left[1 - \frac{y}{L} - \frac{2}{\pi}\sum_{n=0}^{\infty}\frac{1}{n}\exp(-n^2\pi^2 Dt_e/L^2)\sin n\pi y/L\right]$$

$$\frac{\partial C}{\partial y} = (C_{Ai} - C_{Ao})\left[-\frac{1}{L} - \frac{2}{\pi}\sum_{n=0}^{\infty}\frac{\pi}{L}\exp(-n^2\pi^2 Dt_e/L^2)\cos n\pi y/L\right]$$

$$\left(\frac{\partial C}{\partial y}\right)_{y=0} = (C_{Ai} - C_{Ao})\left[-\frac{1}{L} - \frac{2}{\pi}\sum_{0}^{\infty}\frac{\pi}{L}\exp(-n^2\pi^2 Dt_e/L^2)\right]$$

$$N_A = -\frac{D(C_{Ai} - C_{Ao})}{t_e}\int_0^{t_e}\left[-\frac{1}{L} - \frac{2}{\pi}\sum_{0}^{\infty}\frac{\pi}{L}\exp(-n^2\pi^2 Dt_e/L^2)\right]dt$$

$$= -\frac{D(C_{Ai} - C_{Ao})}{t_e}\left[-\frac{t_e}{L} - \frac{2}{\pi}\sum_{0}^{\infty}\frac{\pi}{L}\left(-\frac{L^2}{n^2\pi^2 D}\right)\exp(-n^2\pi^2 Dt_e/L^2)\right]_0^{t_e}$$

$$= -\frac{D(C_{Ai} - C_{Ao})}{t_e}$$

$$\left[-\frac{t_e}{L} - \frac{2}{\pi}\sum_{0}^{\infty}\left(-\frac{L}{n^2\pi D}\right)\exp(-n^2\pi^2 Dt_e/L^2) + \frac{2}{\pi}\sum_{0}^{\infty}\frac{\pi}{L}\left(-\frac{L^2}{n^2\pi^2 D}\right)\right]$$

$$N_A = \frac{D}{L}(C_{Ai} - C_{Ao})\left\{1 + \frac{2L^2}{\pi^2 Dt_e}\left[\sum_{0}^{\infty}\frac{-1}{n^2}\exp(-n^2\pi^2 Dt_e/L^2) + \sum_{0}^{\infty}\frac{1}{n^2}\right]\right\}$$

$$\sum_{0}^{\infty}\frac{-1}{n^2}\exp(-n^2\pi^2 Dt_e/L^2) + \sum_{0}^{\infty}\frac{1}{n^2}$$

$$= \sum_{0}^{1}\frac{-1}{n^2}\exp(-n^2\pi^2 Dt_e/L^2) + \sum_{1}^{\infty}\frac{-1}{n^2}\exp(-n^2\pi^2 Dt_e/L^2) + \sum_{0}^{\infty}\frac{1}{n^2} + \sum_{1}^{\infty}\frac{1}{n^2}$$

$$= -\exp(-\pi^2 Dt_e/L^2) + \sum_{1}^{\infty}\frac{-1}{n^2}\exp(-n^2\pi^2 Dt_e/L^2) + 1 + \pi^2/6$$

$$= \sum_{1}^{\infty}\left[\frac{\pi^2}{6} - \frac{1}{n^2}\exp(-n^2\pi^2 Dt_e/L^2) + 1 - \exp(-\pi^2 Dt_e/L^2)\right]$$

Considering the terms $1 - \exp(-\pi^2 Dt_e/L^2)$ and Dt_e/L^2 to be very small so that $(-\pi^2 Dt_e/L^2)$ is small and $\exp(-\pi^2 Dt_e/L^2) \to 1$. Therefore, $1 - \exp(-\pi^2 Dt_e/L^2)$ is approximately zero and:

$$N_A = \frac{D}{L}(C_{Ai} - C_{Ao})\left\{1 + \frac{2L^2}{\pi^2 Dt_e}\sum_{n=1}^{\infty}\left[\frac{\pi^2}{6} - \frac{1}{n^2}\exp(-n^2\pi^2 Dt_e/L^2)\right]\right\}$$

PROBLEM 10.9

According to the simple penetration theory the instantaneous mass flux:

$$(N_A)_t = (C_{Ai} - C_{Ao})\left(\frac{D}{\pi t}\right)^{0.5}$$

What is the equivalent expression for the instantaneous heat flux under analogous conditions?

Pure sulphur dioxide is absorbed at 295 K and atmospheric pressure into a laminar water jet. The solubility of SO_2, assumed constant over a small temperature range, is 1.54 kmol/m^3 under these conditions and the heat of solution is 28 kJ/kmol.

Calculate the resulting jet surface temperature if the Lewis number is 90. Neglect heat transfer between the water and the gas.

Solution

The heat flux at any time, $f = -k(\partial\theta/\partial x)$ where k is the thermal conductivity, θ the temperature, and y the distance in the direction of transfer.

The flux satisfies the same differential equation as θ, that is:

$$D_H(\partial^2 f/\partial y^2) = (\partial f/\partial t) \quad y > 0, t > 0$$

where D_H = thermal diffusivity = $k/\rho C_p$.

This last equation is analogous to the mass transfer equation 10.66:

$$(\partial C/\partial t) = D(\partial^2 C/\partial y^2)$$

The solution of the heat transfer equation with $f = F_o$ (constant) at $y = 0$ when $t > 0$ is:

$$f = F_o \text{ erfc }\frac{y}{2\sqrt{D_H t}}$$

The temperature rise is due to the heat of solution H_S. Heat is liberated at the jet surface at a rate $\mathcal{H}(t) = N_A^o H_S$,

or:
$$\mathcal{H}(t) = (C_{Ai} - C_{Ao})H_S(D/\pi t)^{0.5}$$

The temperature rise, T, due to the heat flux $\mathcal{H}(t)$ into the surface is:

$$T = \frac{1}{\rho C_p\sqrt{\pi D_H}}\int_0^L \frac{\mathcal{H}(t - \theta)\,d\theta}{\sqrt{\theta}}$$

and:
$$T = \frac{(C_{Ai} - C_{Ao})H_S\sqrt{D/D_H}}{\rho C_p}$$

The Lewis number $= h/C_p \rho h_D = Pr/Sc = DC_p \rho/k$.

$$D/D_H = DC_p \rho/k = 90$$

$$(C_{Ai} - C_{Ao}) = 1.54 \text{ kmol/m}^3$$

$$H_S = 28 \text{ kJ/kmol}$$

$\therefore \qquad T = (1.54 \times 28\sqrt{90})/(1000 \times 4.186) = \underline{\underline{0.1 \text{ deg K}}}$

PROBLEM 10.10

In a packed column, operating at approximately atmospheric pressure and 295 K, a 10% ammonia-air mixture is scrubbed with water and the concentration is reduced to 0.1%. If the whole of the resistance to mass transfer may be regarded as lying within a thin laminar film on the gas side of the gas-liquid interface, derive from first principles an expression for the rate of absorption at any position in the column. At some intermediate point where the ammonia concentration in the gas phase has been reduced to 5%, the partial pressure of ammonia in equilibrium with the aqueous solution is 660 N/m² and the transfer rate is 10^{-3} kmol/m²s. What is the thickness of the hypothetical gas film if the diffusivity of ammonia in air is 0.24 cm²/s?

Solution

The equation for the rate of absorption is derived in Section 10.2.2 as:

$$N_A = -(D/\mathbf{R}TL)(P_{A2} - P_{A1}) \qquad \text{(equation 10.23)}$$

If subscripts 1 and 2 refer to the water and air side of the stagnant film and subscripts A and B refer to ammonia and air, then:

$$P_{A1} = 66.0 \text{ kN/m}^2 \text{ and } P_{A2} = (0.05 \times 101.3) = 5.065 \text{ kN/m}^2$$

$$D = 0.24 \times 10^{-4} \text{ m}^2/\text{s}, \quad \mathbf{R} = 8.314 \text{ kJ/kmol K},$$

$$T = 295 \text{ K and } N_A = 1 \times 10^{-3} \text{ kmol/m}^2\text{s}$$

$\therefore \qquad L = -(D/N_A \mathbf{R}T)(P_{A2} - P_{A1})$

$\qquad = -(0.24 \times 10^{-4}/10^{-3} \times 8.314 \times 295)(66.0 - 5.065) = -0.000043 \text{ m}$

The negative sign indicates that the diffusion is taking place in the opposite direction and the thickness of the gas film is $\underline{\underline{0.043 \text{ mm.}}}$

PROBLEM 10.11

An open bowl, 0.3 m in diameter, contains water at 350 K evaporating into the atmosphere. If the air currents are sufficiently strong to remove the water vapour as it is formed and if the resistance to its mass transfer in air is equivalent to that of a 1 mm layer for conditions of molecular diffusion, what will be the rate of cooling due to evaporation? The water can be considered as well mixed and the water equivalent of the system is equal to 10 kg. The diffusivity of water vapour in air may be taken as 0.20 cm^2/s and the kilogram molecular volume at NTP as 22.4 m^3.

Solution

If subscripts 1 and 2 refer to the water and air side of the stagnant layer and subscripts A and B refer to water vapour and air, then the rate of diffusion through a stagnant layer is:

$$N_A = -(D/\mathbf{R}TL)(P/P_{BM})(P_{A2} - P_{A1}) \qquad \text{(equation 10.34)}$$

where, P_{A1} is the vapour pressure of water at 350 K $= 41.8$ kN/m^2.

$P_{A2} = 0$ (since the air currents remove the vapour as it is formed.)

$P_{B1} = (101.3 - 41.8) = 59.5$ kN/m^2 and $P_{B2} = 101.3$ kN/m^2.

$\therefore \quad P_{BM} = (101.3 - 59.5)/\ln(101.3/59.5) = 78.17$ kN/m^2.

and: $P/P_{BM} = (101.3/78.17) = 1.296$.

$\therefore$
$$N_A = -(0.2 \times 10^{-4}/8.314 \times 350 \times 10^{-3})1.296(0 - 41.8)$$

$$= 3.72 \times 10^{-4} \text{ kmol/m}^2\text{s}$$

$$= (3.72 \times 10^{-4} \times 18) = 6.70 \times 10^{-3} \text{ kg water/m}^2\text{s}$$

Area of bowl $= (\pi/4)0.3^2 = 0.0707$ m^2

Therefore the rate of evaporation $= (6.70 \times 10^{-3} \times 0.0707) = 4.74 \times 10^{-4}$ kg/s

Latent heat of vaporisation $= 2318$ kJ/kg

Specific heat capacity of water $= 4.187$ kJ/kg K

Rate of heat removal $= (4.74 \times 10^{-4} \times 2318) = 1.10$ kW

If the rate of cooling $= d\theta/dt$ K/s, then:

$$\text{(water equivalent)} \times \text{(specific heat capacity)} \times (d\theta/dt) = 0.0617$$

or: $\qquad 10 \times 4.187 \times (d\theta/dt) = 1.10$ and $d\theta/dt = \underline{\underline{0.026 \text{ deg K/s}}}$

PROBLEM 10.12

Show by substitution that when a gas of solubility C^+ is absorbed into a stagnant liquid of infinite depth, the concentration at time t and depth y is:

$$C^+ \text{ erfc } \frac{y}{2\sqrt{Dt}}$$

Hence, on the basis of the simple penetration theory, show that the rate of absorption in a packed column will be proportional to the square root of the diffusivity.

Solution

The first part of this question is discussed in Section 10.5.2 and the required equation is presented as equation 10.108.

In Section 10.5.2 the analysis leads to equation 10.113 which expresses the instantaneous rate of mass transfer when the surface element under consideration has an age t, or:

$$(N_A)_t = (C_{Ai} - C_{Ao})\sqrt{(D/\pi t)}$$

The simple penetration theory assumes that each element is exposed for the same time interval t_e before returning to the bulk solution. The average rate of mass transfer is then:

$$N_A = \frac{1}{t_e} \int_0^{t_e} (N_A)_t \, dt = \frac{(C_{Ai} - C_{Ao})}{t_e} \int_0^{t_e} (D/\pi t)^{0.5} \, dt$$

$$= 2(C_{Ai} - C_{Ao})\sqrt{D/\pi t_e}$$

and the rate of absorption is proportional to $\sqrt{D}$.

PROBLEM 10.13

Show that in steady-state diffusion through a film of liquid, accompanied by a first-order irreversible reaction, the concentration of solute in the film at depth y below the interface is:

$$\frac{C_A}{C_{Ai}} = \sinh \frac{\sqrt{\dfrac{k}{D}}(L - y)}{\sinh \sqrt{\dfrac{k}{D}}L} \cdot C_i$$

if $C_A = 0$ at $y = L$ and $C_A = C_{Ai}$ at $y = 0$, corresponding to the interface.

Hence show that according to the "film theory" of gas-absorption, the rate of absorption per unit area of interface, N_A is given by:

$$N_A = k_L C_{Ai} \frac{\beta}{\tanh \beta}$$

where $\beta = \sqrt{Dk/k_L}$, D is the diffusivity of the solute, k the rate constant of the reaction, K_L the liquid film mass transfer coefficient for physical absorption, C_{Ai} the concentration of solute at the interface, y the distance normal to the interface and y_L the liquid film thickness.

Solution

The basic equation for diffusion through a film of liquid accompanied by a first-order irreversible reaction is:

$$D(d^2C_A/dy^2) = k C_A \quad \text{or} \quad (d^2C_A/dy^2) = a^2 C_A \qquad \text{(equation 10.171) (i)}$$

where $a^2 = \sqrt{k/D}$.

The general solution of equation (i) is:

$$C_A = A \cosh ay + B \sinh ay \tag{ii}$$

where A and B are constants.

The boundary conditions are:

$$\text{At } y = L, \ C_A = 0 \tag{iii}$$

$$\text{At } y = 0, \ C_A = C_{Ai} \tag{iv}$$

Substituting equation (iii) in equation (ii):

$$0 = A \cosh aL + B \sinh aL$$

and substituting equation (iv) in equation (ii):

$$C_{Ai} = A + 0 \text{ and } A = C_{Ai} \text{ and}$$

$$B = -C_{Ai} \cosh aL / \sinh aL$$

$$\therefore \quad C_A = C_{Ai} \cosh ay - C_{Ai} \frac{\cosh aL}{\sinh aL} \sinh ay$$

$$= \frac{C_{Ai}}{\sinh aL} (\cosh ay \sinh aL - \cosh aL \sinh ay)$$

$$= \frac{C_{Ai} \sinh a(L - y)}{\sinh aL}$$

$$= C_{Ai} \frac{\sinh a(L - y)}{\sinh aL} = \underline{\underline{C_{Ai} \frac{\sinh \sqrt{k/D}(L - y)}{\sinh \sqrt{k/D}L}}}$$

Rate of absorption:
$$N_A = -D \left(\frac{dC_A}{dy} \right)_{y=0}$$

Assuming C_A to be small so that bulk flow can be neglected, then:

$$N_A = -D \frac{d}{dy} \left(\frac{\sinh a(L - y)}{\sinh aL} \right)$$

$$= \frac{DC_{Ai} a \cosh aL}{\sinh aL}$$

$$= DC_{Ai} a / \tanh aL = DC_{Ai} aL / L \tanh aL$$

$$k_L = D/L$$

$$\beta = \sqrt{Dk/k_L} = \sqrt{k/D} L = aL$$

$$\therefore \quad N_A = \underline{\underline{\frac{k_L C_{Ai} \beta}{\tanh \beta}}}$$

PROBLEM 10.14

The diffusivity of the vapour of a volatile liquid in air can be conveniently determined by Winkelmann's method, in which liquid is contained in a narrow diameter vertical tube maintained at a constant temperature, and an air stream is passed over the top of the tube sufficiently rapidly to ensure the partial pressure of the vapour there remains approximately zero. On the assumption that the vapour is transferred from the surface of the liquid to the air stream by molecular diffusion, calculate the diffusivity of carbon tetrachloride vapour in air at 321 K and atmospheric pressure from the following experimentally obtained data:

Time from commencement of experiment (ks)	Liquid level (cm)
0	0.00
1.6	0.25
11.1	1.29
27.4	2.32
80.2	4.39
117.5	5.47
168.6	6.70
199.7	7.38
289.3	9.03
383.1	10.48

The vapour pressure of carbon tetrachloride at 321 K is 37.6 kN/m^2, and the density of the liquid is 1540 kg/m^3. The kilogram molecular volume is 22.4 m^3.

Solution

Equations 10.37 and 10.38 state that:

$$N_A = -D\frac{(C_{A2} - C_{A1})}{y_2 - y_1}\frac{C_T}{C_{Bm}}$$

In this problem, the distance through which the gas is diffusing will be taken as h and $C_{A2} = 0$.

$$\therefore \qquad N_A = D(C_A/h)(C_T/C_{Bm}) \text{ kmol/m}^2\text{s}$$

where C_A is the concentration at the interface.

If the liquid level falls by a distance dh in time dt, the rate of evaporation is:

$$N_A = (\rho_L/M)dh/dt \text{ kmol/m}^2\text{s}$$

Hence: $\qquad (\rho_L/M)dh/dt = D(C_A/h)(C_T/C_{Bm})$

If this equation is integrated, noting that when $t = 0$, $h = h_0$, then:

$$h^2 - h_0^2 = (2MD/\rho_L)(C_A C_T/C_{Bm})t$$

or:

$$t/(h - h_0) = (\rho_L/2MD)(C_{Bm}/C_A C_T)(h - h_0) + (\rho_L C_{Bm}/MDC_A C_T)h_0$$

Thus a plot of $t/(h - h_0)$ against $(h - h_0)$ will be a straight line of slope s where:

$$s = \rho_L C_{Bm}/2MDC_A C_T \quad \text{or} \quad D = \rho_L C_{Bm}/2MC_A C_T s$$

The following table may be produced:

t (ks)	1.6	11.1	27.4	80.2	117.5	168.6	199.7	289.3	383.1
$(h - h_0)$ (mm)	2.5	12.9	23.2	43.9	54.7	67.0	73.8	90.3	104.8
$t/(h - h_0)$ (s/m $\times 10^{-6}$)	0.64	0.86	1.18	1.83	2.15	3.52	2.71	3.20	3.66

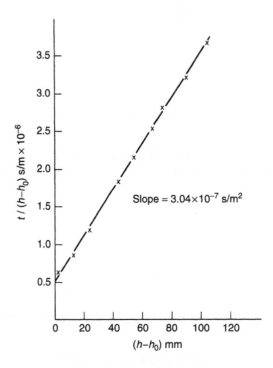

Figure 10b.

These data are plotted as Fig. 10b and the slope is:

$$s = (3.54 - 0.5)10^{-6}/(100 \times 10^{-3}) = 3.04 \times 10^{-7} \text{ s/m}^2$$

$$C_T = (1/22.4)(273/321) = 0.0380 \text{ kmol/m}^3$$

$$M = 154 \text{ kg/kmol}$$

$$C_A = (37.6/101.3)(1/22.4)(273/321) = 0.0141 \text{ kmol/m}^3$$

$$\rho_L = 1540 \text{ kg/m}^3$$

$$C_{B1} = 0.0380 \text{ kmol/m}^3$$

$$C_{B2} = (0.0380 - 0.0141) = 0.0239 \text{ kmol/m}^3$$

$$\therefore \quad C_{Bm} = (0.0380 - 0.0239)/\ln(0.0380/0.0239) = 0.0304 \text{ kmol/m}^3$$

Hence: $\quad D = (1540 \times 0.0304)/(2 \times 154 \times 0.0141 \times 0.0380 \times 3.04 \times 10^{-7})$

$$= \underline{\underline{9.33 \times 10^{-6} \text{ m}^2/\text{s}}}$$

PROBLEM 10.15

Ammonia is absorbed in water from a mixture with air using a column operating at atmospheric pressure and 295 K. The resistance to transfer can be regarded as lying entirely within the gas phase. At a point in the column the partial pressure of the ammonia is 6.6 kN/m^2. The back pressure at the water interface is negligible and the resistance to transfer can be regarded as lying in a stationary gas film 1 mm thick. If the diffusivity of ammonia in air is 0.236 cm^2/s, what is the transfer rate per unit area at that point in the column? If the gas were compressed to 200 kN/m^2 pressure, how would the transfer rate be altered?

Solution

See Volume 1, Example 10.3.

PROBLEM 10.16

What are the general principles underlying the two-film penetration and film-penetration theories for mass transfer across a phase boundary? Give the basic differential equations which have to be solved for these theories with the appropriate boundary conditions.

According to the penetration theory, the instantaneous rate of mass transfer per unit area $(N_A)_t$ at some time t after the commencement of transfer is given by:

$$(N_A)_t = \Delta C_A \sqrt{\frac{D}{\pi t}}$$

where ΔC_A is the concentration driving force and D is the diffusivity.

Obtain expressions for the average rates of transfer on the basis of the Higbie and Danckwerts assumptions.

Solution

The various theories for the mechanism of mass transfer across a phase boundary are discussed in Section 10.5.

The basic equation for unsteady state equimolecular counter-diffusion is:

$$\frac{\partial C_A}{\partial t} = D \left[\left(\frac{\partial^2 C_A}{\partial x^2} \right)_{yz} + \left(\frac{\partial^2 C_A}{\partial y^z} \right)_{xz} + \left(\frac{\partial^2 C_A}{\partial z^2} \right)_{xy} \right] \qquad \text{(equation 10.67)}$$

Considering the diffusion of solute **A** away from the interface in the y-direction this equation becomes:

$$\frac{\partial C_A}{\partial t} = D\frac{\partial^2 C_A}{\partial y^2}$$

The boundary conditions are:

$$
\begin{array}{llll}
t = 0 & 0 < y < \infty & C_A = C_{Ao} \\
t > 0 & y = 0 & C_A = C_{Ai} \\
t > 0 & y = \infty & C_A = C_{Ao}
\end{array}
$$

where C_{Ao} is the concentration in the bulk of the phase and C_{Ai} is the equilibrium concentration at the interface.

The instantaneous rate of mass transfer per unit area N_A at time t is given by:

$$(N_A)_t = \Delta C_A \sqrt{D/\pi t}$$

Higbie assumed that every element of surface is exposed to the gas for the same length of time θ before being replaced by liquid of the bulk composition.

Amount absorbed in time θ:

$$Q = \int_0^\theta (N_A)_t \, d\theta = \int_0^\theta \Delta C_A \sqrt{D/\pi\theta} \, d\theta = 2\Delta C_A \sqrt{D\theta/\pi}$$

The average rate of absorption: $Q/\theta = \left(2\Delta C_A \sqrt{D\theta/\pi}\right)/\theta = \underline{\underline{2\Delta C_A \sqrt{D/\pi\theta}}}$

Danckwerts suggested that each element would not be exposed for the same time but that a random distribution of ages would exist. It is shown in Section 10.5.2 that this age distribution may be expressed $f(t) = se^{-st}$. The average rate of absorption is the value of $(N_A)_t$ averaged over all elements of the surface having ages between 0 and ∞ is then given by:

$$N_A = s\int_0^\infty (N_A)_t e^{-s\theta} \, d\theta = \Delta C_A \sqrt{D/\pi} \int_0^\infty (e^{-s\theta}/\sqrt{\theta}) \, d\theta = \underline{\underline{\Delta C_A \sqrt{Ds}}}$$

PROBLEM 10.17

A solute diffuses from a liquid surface at which its molar concentration is C_{Ai} into a liquid with which it reacts. The mass transfer rate is given by Fick's law and the reaction is first order with respect to the solute. In a steady-state process, the diffusion rate falls at a depth L to one half the value at the interface. Obtain an expression for the concentration C_A of solute at a depth y from the surface in terms of the molecular diffusivity D and the reaction rate constant $\boldsymbol{k}$. What is the molar flux at the surface?

Solution

As in Problem 10.13, the basic equation is: $d^2C_A/dy^2 = a^2C_A$ (i)
where $a = \sqrt{k/D}$

Then: $C_A = A\cosh ay + B\sinh ay$ (ii)

The first boundary condition is at $y = 0$, $C_A = C_{Ai}$, and $C_{Ai} = A$.

Hence:
$$C_A = C_{Ai} \cosh ax + B \sinh ax \qquad \text{(iii)}$$

The second boundary condition is that when $y = L$ and:

$$N_A = -D(dC_A/dy)_{y=0} = -2D(dC_A/dy)_{y=L}$$

Differentiating equation (iii):

$$dC_A/dy = C_{Ai}a \sinh ay + Ba \cosh ay$$

and: $(dC_A/dy)_{y=0} = Ba$

and: $(dC_A/dy)_{y=L} = aB/2 = C_{Ai}a \sinh aL + Ba \cosh aL$

so that:
$$B = \frac{2C_{Ai} \sinh aL}{1 - 2 \cosh aL} \qquad \text{(iv)}$$

Substituting equation (iv) into equation (iii):

$$C_A = C_{Ai} \cosh ay + \frac{2C_{Ai} \sinh aL \sinh ay}{1 - 2 \cosh aL}$$
$$= C_{Ai}[\cosh ay - 2(\cosh ay \cosh aL + \sinh aL \sinh ay)]/(1 - 2 \cosh aL)$$
$$= C_{Ai}[\cosh ay - 2 \cosh a(y + L)]$$

The molar flux at the surface $= N_A = -D(dC_A/dy)_{y=0}$.

$$\frac{dC_A}{dy} = C_{Ai}[a \sinh ay - 2a \sinh a(a + L)]$$

$$(dC_A/dy)_{y=0} = -2C_{Ai}a^2 \sinh aL$$

$$N_A = 2DC_{Ai}a^2 \sinh aL$$

$$a = \sqrt{k/D}$$

$$N_A = 2DC_{Ai}(k/D) \sinh L\sqrt{k/D}$$

$$= \underline{\underline{2C_{Ai}k \sinh(L\sqrt{k/D})}}$$

PROBLEM 10.18

4 cm^3 of mixture formed by adding 2 cm^3 of acetone to 2 cm^3 of dibutyl phthalate is contained in a 6 mm diameter vertical glass tube immersed in a thermostat maintained at 315 K. A stream of air at 315 K and atmospheric pressure is passed over the open top of the tube to maintain a zero partial pressure of acetone vapour at that point. The liquid level is initially 11.5 mm below the top of the tube and the acetone vapour is transferred to the air stream by molecular diffusion alone. The dibutyl phthalate can be regarded as completely non-volatile and the partial pressure of acetone vapour may be calculated from Raoult's law on the assumption that the density of dibutyl phthalate is sufficiently greater than that of acetone for the liquid to be completely mixed.

Calculate the time taken for the liquid level to fall to 5 cm below the top of the tube, neglecting the effects of bulk flow in the vapour. 1 kmol occupies 22.4 m^3. Molecular weights of acetone, dibutyl phthalate $= 58$ and 279 kg/kmol respectively. Liquid densities of acetone, dibutyl phthalate $= 764$ and 1048 kg/m^3 respectively. Vapour pressure of acetone at 315 K $= 60.5$ kN/m^2. Diffusivity of acetone vapour in air at 315 K $= 0.123$ cm^2/s.

Solution

Considering the situation when the liquid has fallen to a depth h cm below the top of the tube, volume of acetone evaporated $= (\pi/4)(0.6)^2(h - 1.15) = 0.283(h - 1.15)$ cm^3.

At this time, the amount of dibutyl phthalate is:

$$(2 \times 1.048/278) = 0.00754 \text{ mol}$$

and the amount of acetone $= [2 - 0.283(h - 1.15)]0.764/58 = (0.0306 - 0.00372h)$

$$\therefore \qquad \text{Mole fraction of acetone} = \frac{0.0306 - 0.00372h}{(0.00754 + 0.0306 - 0.00372h)} = \frac{(8.23 - h)}{10.24 - h}$$

Partial pressure of acetone $= 60.5 \left(\dfrac{8.23 - h}{10.24 - h} \right)$ kN/m^2

Molar concentration of acetone vapour at the liquid surface

$$= \left(\frac{60.5}{101.3} \right) \times \left(\frac{273}{315} \right) \times \left(\frac{1}{22400} \right) \left(\frac{8.23 - h}{10.24 - h} \right)$$

$$= 2.31 \times 10^{-5} \left(\frac{8.23 - h}{10.24 - h} \right) \text{ mol/cm}^3$$

Rate of evaporation of acetone: $N_A = \dfrac{dh}{dt} \times \dfrac{0.764}{58} = 0.0132(dh/dt)$ mol/cm^{2}s

$$= (D/h) \times \text{molar concentration at surface} = (0.123/h)2.31 \times 10^{-5} \left(\frac{8.23 - h}{10.24 - h} \right)$$

$$\therefore \qquad 0.0132\frac{dh}{dt} = \frac{1}{h} \times 2.84 \times 10^{-6} \left(\frac{8.23 - h}{10.24 - h} \right)$$

and: $\qquad \left(\dfrac{10.24 - h}{8.23 - h} \right) h \, dh = \dfrac{dt}{4650}$

The time for the liquid level to fall from 1.15 cm to 5 cm below the top of the tube is obtained by integrating this equation:

$$\int_{1.15}^{5} \left(\frac{10.24 - h}{8.23 - h} \right) h \, dh = \frac{1}{4650} \int_0^t dt = \int_{1.15}^5 \left(h - 2.02 - \frac{16.6}{h - 8.23} \right) dh = \frac{1}{4650} \int_0^t dt$$

and: $\qquad\qquad\qquad t = 79500 \text{ s} \equiv \underline{\underline{79.5 \text{ ks}}} \ (\approx 22 \text{ h})$

PROBLEM 10.19

A crystal is suspended in fresh solvent and 5% of the crystal dissolves in 300 s. How long will it take before 10% of the crystal has dissolved? Assume that the solvent can be regarded as infinite in extent, that the mass transfer in the solvent is governed by Fick's second law of diffusion and may be represented as a unidirectional process, and that changes in the surface area of the crystal may be neglected. Start your derivations using Fick's second law.

Solution

The mass transfer process is governed by Fick's second law:

$$\frac{\partial C_A}{\partial t} = D \frac{\partial^2 C_A}{\partial y^2} \qquad \text{(equation 10.66)}$$

and discussed in Section 10.5.2

The boundary conditions for the crystal dissolving are:

$$
\begin{array}{lll}
\text{When } t = 0 & 0 < y < \infty & C_A = 0 \\
t > 0 & y = \infty & C_A = 0 \\
t > 0 & y = 0 & C_A = C_{As} \text{ (the saturation value)}
\end{array}
$$

These boundary conditions allow the solution of equation 10.66 using Laplace transforms as the most convenient method:

$$\frac{\overline{\partial C_A}}{\mathrm{d}t} = \int_0^\infty \mathrm{e}^{-pt} \frac{\partial C_A}{\partial t}\, \mathrm{d}t \qquad \text{(equation 10.102)}$$

$$= \left[\mathrm{e}^{-pt} C_A\right]_0^\infty + p \int_0^\infty \mathrm{e}^{-pt} C_A\, \mathrm{d}t = 0 + p\bar{C}_A \quad \text{(equation 10.103)}$$

Taking Laplace transforms of both sides of equation 10.66:

$$p\bar{C}_A = D \frac{\partial^2 \bar{C}_A}{\partial y^2}$$

$$\therefore \qquad \frac{\partial^2 \bar{C}_A}{\partial y^2} - \frac{p}{D}\bar{C}_A = 0$$

and: $$\bar{C}_A = A\mathrm{e}^{\sqrt{(p/D)}y} + B\mathrm{e}^{-\sqrt{(p/D)}y} \qquad \text{(equation 10.105)}$$

$$
\begin{array}{llll}
\text{When } y = \infty, & C_A = 0 & \therefore \quad \bar{C}_A = 0 & \text{and } A = 0 \\
\text{When } y = 0, & C_A = C_{As} & \text{and } \bar{C}_A = C_{As}/p_o, & B = C_{As}/p
\end{array}
$$

$$\therefore \qquad \bar{C}_A = \frac{C_{As}}{p}\mathrm{e}^{-\sqrt{(p/D)}y}$$

Inverting: $$C_A = C_{As}\, \mathrm{erfc}(y/2\sqrt{Dt}) \qquad \text{(See Volume 1, Appendix Table 13)}$$

Mass transfer rate at the surface $= -D \left(\dfrac{\partial C_A}{\partial y} \right)_{y=0}$

$$\frac{\partial C_A}{\partial y} = C_{As} \frac{\partial}{\partial y} \left\{ \frac{2}{\sqrt{\pi}} \int_{(y/(2\sqrt{Dt}))}^{\infty} e^{-y^2/4Dt} \, d \left(\frac{y}{2\sqrt{Dt}} \right) \right\}$$

$$= C_{As} \frac{2}{\sqrt{\pi}} \left(-\frac{1}{2\sqrt{Dt}} \right) e^{-y^2/4Dt} \qquad \text{(equation 10.111)}$$

$$\therefore \qquad \left(\frac{\partial C_A}{dy} \right)_{y=0} = -\frac{C_{As}}{\sqrt{\pi Dt}}$$

$$(N_A)_t = -D \left(\frac{\partial C_A}{\partial t} \right)_{y=0} = C_{As} \sqrt{\frac{D}{\pi t}}$$

The mass transfer in time $t = \displaystyle\int_0^t \sqrt{\frac{D}{\pi t}} \, dt = 2\sqrt{\frac{D}{\pi}} \sqrt{t}$

and the mass transfer is proportional to $\sqrt{t}$

Thus: $$\frac{M_1}{M_2} = \sqrt{\frac{t_1}{t_2}}$$

$$M_1 = 5\%, \ M_2 = 10\%, \text{ and } t_1 = 300 \text{ s}$$

and: $$0.5 = \sqrt{300/t_2} \text{ and } t_2 = \underline{\underline{1200 \text{ s}}}$$

PROBLEM 10.20

In a continuous steady state reactor, a slightly soluble gas is absorbed into a liquid in which it dissolves and reacts, the reaction being second-order with respect to the dissolved gas. Calculate the reaction rate constant on the assumption that the liquid is semi-infinite in extent and that mass transfer resistance in the gas phase is negligible. The diffusivity of the gas in the liquid is 10^{-8} m^2/s, the gas concentration in the liquid falls to one half of its value in the liquid over a distance of 1 mm, and the rate of absorption at the interface is 4×10^{-6} kmol/m^{2}s.

Solution

The equation for mass transfer with chemical reaction is:

$$\frac{\partial C_A}{\partial t} = D \frac{\partial^2 C_A}{\partial y^2} - k C_A^n \qquad \text{(equation 10.170)}$$

For steady state second order reaction where $n = 2$:

$$D \frac{d^2 C_A}{dy^2} - k C_A^2 = 0$$

Putting $dC_A/dy = q$:
$$\frac{d^2C_A}{dy^2} = \frac{dq}{dy} = \frac{dq}{dC_A}\frac{dC_A}{dy} = q\frac{dq}{dC_A}$$

Thus:
$$Dq\frac{dq}{dC_A} - kC_A^2 = 0$$

$$q\,dq = (k/D)C_A^2\,dC_A$$

$$(q^2/2) = (k/D)C_A^3/3 + \text{const.}$$

In an infinite system at $y = \infty$, $C_A = 0$ and $dC_A/dy = 0$ and hence the constant $= 0$.

$\therefore$
$$\left(\frac{dC_A}{dy}\right)^2 = \frac{2}{3}\frac{k}{D}C_A^3 \text{ and } \frac{dC_A}{dy} = -\sqrt{\frac{2k}{3D}}C_A^{3/2}$$

noting the negative sign since (dC_A/dy) is negative for all values of C_A.

Thus:
$$-2C_A^{-1/2} = -\sqrt{\frac{2k}{3D}}y + \text{constant}$$

At the free surface, $y = 0$ and $C_A = C_{Ai} = \text{constant}$.

$\therefore$
$$\text{constant} = -2C_{Ai}^{-1/2}$$

and:
$$2\left(C_A^{-1/2} - C_{Ai}^{-1/2}\right) = \sqrt{(2k/3D)}y$$

or:
$$C_A^{-1/2} - C_{Ai}^{-1/2} = \sqrt{(k/6D)}y$$

When $y = y_1$, $C_A = C_{Ai}/2$, and:
$$(C_{Ai}/2)^{-1/2} - C_{Ai}^{-1/2} = \sqrt{(k/6D)}y_1$$

When $y_1 = 10^{-3}$, substituting gives: $C_{Ai}^{-1/2} = 2.42 \times 10^{-3}\sqrt{(k/6D)}$

$\therefore$
$$C_{Ai}^{-1/2} - 2.42 \times 10^{-3}\sqrt{(k/6D)} = \sqrt{(k/6D)}y$$

The mass transfer rate at the interface, where $y = 0$, is:

$$(N_A)_t = -D\left(\frac{dC_A}{dy}\right)_{y=0} = \sqrt{\frac{2kD}{3}}C_{Ai}^{3/2}$$

$$= \sqrt{\frac{2kD}{3}} \times \left(\frac{1}{2.42 \times 10^{-3}\sqrt{(k/6D)}}\right)^3 = 8.47 \times 10^8 D^2/k$$

When $D = 1 \times 10^{-8}$ m^2/s and $N_A = 4 \times 10^{-6}$ kmol/m^{2}s :

$$k = (8.47 \times 10^8) \times (1 \times 10^{-8})^2/4 \times 10^{-6} = \underline{\underline{212 \text{ m}^3/\text{kmol s}}}$$

PROBLEM 10.21

Experiments have been carried out on the mass transfer of acetone between air and a laminar water jet. Assuming that desorption produces random surface renewal with a constant fractional rate of surface renewal, s, but an upper limit on surface age equal to the life of the jet, τ, show that the surface age frequency distribution function, $\phi(t)$, for this case is given by:

$$\phi(t) = s\exp(-st/[1 - \exp(-st)]) \quad \text{for} \quad 0 < t < \tau$$

$$\phi(t) = 0 \quad\quad\quad\quad\quad\quad\quad\quad \text{for} \quad t > \tau.$$

Hence, show that the enhancement, E, for the increase in value of the liquid-phase mass transfer coefficient is:

$$E = [(\pi s\tau)^{1/2} \operatorname{erf}(s\tau)^{1/2}]/\{2[1 - \exp(-s\tau)]\}$$

where E is defined as the ratio of the mass transfer coefficient predicted by conditions described above to the mass transfer coefficient obtained from the penetration theory for a jet with an undisturbed surface. Assume that the interfacial concentration of acetone is practically constant.

Solution

For the penetration theory:

$$\frac{\partial C_A}{\partial t} = D\frac{\partial^2 C_A}{\partial y^2} \quad\quad\quad\quad \text{(equation 10.66)}$$

As shown in Problem 10.19, this equation can be transformed and solved to give:

$$\bar{C}_A = Ae^{\sqrt{(p/D)}y} + Be^{-\sqrt{(p/D)}y}$$

The boundary conditions are:

$$\text{When } y = 0, \quad C_A = C_{Ai}, \quad B = C_{Ai}/p$$

$$\text{and when } y = \infty, \quad C_A = 0 \text{ and } A = 0$$

$$\therefore \quad\quad \bar{C}_A = \frac{C_{Ai}}{p}e^{-\sqrt{(p/D)}y}$$

$$\frac{d\bar{C}_A}{dy} = -C_{Ai}\sqrt{\frac{1}{D}}\sqrt{\frac{1}{p}}e^{-\sqrt{(p/D)}y}$$

From Volume 1, Appendix, Table 12, No 84, the inverse:

$$\frac{dC_A}{dy} = -C_{Ai}\sqrt{\frac{1}{D}}\sqrt{\frac{1}{\pi t}}e^{-y^2/4Dt}$$

At the surface: $\quad (N_A)_t = -D\left(\frac{dC_A}{dy}\right)_{y=0} = C_{Ai}\sqrt{\frac{D}{\pi t}}$ at time t

The average rate over a time τ is:

$$\frac{1}{\tau}C_{Ai}\sqrt{\frac{D}{\pi}}\int_0^\tau \frac{dt}{\sqrt{t}} = 2C_{Ai}\sqrt{\frac{D}{\pi\tau}}$$

In general, $C_{A0} \neq 0$ and $N_A = 2(C_{Ai} - C_{A0})\sqrt{D/\pi\tau}$ for mass transfer without surface renewal.

Random surface renewal is discussed in Section 10.5.2 where it is shown that the age distribution function is:

$$= \text{constant } e^{-st} = k e^{-st}$$

where s is the rate of production of fresh surface per unit total area of surface.

If the maximum age of the surface is τ, then:

$$K \int_0^\tau e^{-st} \, dt = 1$$

$$-\frac{k}{s}[e^{-st}]_0^\tau = 1$$

$$1 - e^{-s\tau} = s/k \text{ and } K = \frac{s}{1 - e^{-s\tau}}$$

$\therefore$ the age distribution function is: $\underline{\underline{\left(\dfrac{s}{1 - e^{-s\tau}}\right) e^{-st}}}$

The mass transfer in time τ is:

$$\int_0^\tau \sqrt{\frac{D}{\pi t}}(C_{Ai} - C_{A0})\frac{s}{1 - e^{-s\tau}}e^{-st} \, dt = \sqrt{\frac{D}{\pi t}}(C_{Ai} - C_{A0})\frac{s}{1 - e^{-s\tau}}\int_0^\tau \frac{e^{-st}}{\sqrt{t}} \, dt$$

The integral is conveniently solved by substituting $st = \beta^2$ and $\sqrt{t} = \beta/\sqrt{s}$ or $s \, dt = 2\beta \, d\beta$ and $dt = 2\beta \, d\beta/s$

Then: $\int_0^\tau \dfrac{\sqrt{s}}{\beta}e^{-\beta}\dfrac{2\beta \, d\beta}{s} = \dfrac{2}{\sqrt{s}}\int_0^\tau e^{-\beta^2} \, d\beta = \sqrt{\dfrac{\pi}{s}}\text{erf}\sqrt{s\tau} = (C_{Ai} - C_{A0})\sqrt{Ds}\dfrac{\text{erf}\sqrt{s\tau}}{1 - e^{-s\tau}}$

The enhancement factor E is given by:

$$E = \frac{(C_{Ai} - C_{A0})\sqrt{Ds}\dfrac{\text{erf}\sqrt{s\tau}}{1 - e^{-s\tau}}}{2\sqrt{\dfrac{D}{\pi\tau}}(C_{Ai} - C_{A0})} = \underline{\underline{\frac{\sqrt{\pi s\tau} \text{ erf }\sqrt{s\tau}}{2(1 - e^{-s\tau})}}}$$

PROBLEM 10.22

Solute gas is diffusing into a stationary liquid, virtually free of solvent, and of sufficient depth for it to be regarded as semi-infinite in extent. In what depth of fluid below the surface will 90% of the material which has been transferred across the interface have accumulated in the first minute? Diffusivity of gas in liquid $= 10^{-9}$ m^2/s.

Solution

As in the previous problem, the basic equation is:

$$\frac{\partial C_A}{\partial t} = D\frac{\partial^2 C_A}{\partial y^2} \qquad \text{(equation 10.66)}$$

which can be solved using the same boundary conditions to give the rate of mass transfer at depth, y, $(N_A)_{y,t}$ as:

$$(N_A)_{y,t} = -D\frac{dC_A}{dy} = C_{Ai}\sqrt{\frac{D}{\pi t}}e^{-y^2/4Dt}$$

At some other value of $y = L$, the amount which has been transferred in time t per unit area is:

$$\int_0^t C_{Ai}\sqrt{\frac{D}{\pi t}}e^{-y^2/4Dt}\,dt$$

This integral can be solved by making the substitution:

$$\beta^2 = y^2/4Dt$$

so that:

$$\beta = y/2\sqrt{Dt}$$

and:

$$t = y^2/4D\beta^2, \quad t^{-1/2} = y/2\beta\sqrt{D}$$

$$dt = (-y^2/2D)\beta^{-3}\,d\beta$$

The amount transferred at depth L is then:

$$= C_{Ai}\sqrt{\frac{D}{\pi}}\frac{y}{\sqrt{D}}\left[\frac{2\sqrt{Dt}}{y}e^{-y^2/4Dt} - \sqrt{\pi}\,\text{erfc}\,\frac{y}{2\sqrt{Dt}}\right]$$

$$= C_{Ai}\left[2\sqrt{\frac{Dt}{\pi}}e^{-y^2/4Dt} - y\,\text{erfc}\,\frac{y}{2\sqrt{Dt}}\right]$$

and:

$$\frac{\text{mass transfer at } L}{\text{mass transfer at } y = 0} = \frac{\left[2\sqrt{\dfrac{Dt}{\pi}}e^{-y^2/4Dt} - y\,\text{erfc}\,\dfrac{y}{2\sqrt{Dt}}\right]}{2\sqrt{\dfrac{Dt}{\pi}}}$$

$$= e^{-y^2/4Dt} - \frac{y}{2\sqrt{Dt}}\sqrt{\pi}\,\text{erfc}\,\frac{y}{2\sqrt{Dt}} = e^{-X^2} - X\sqrt{\pi}\,\text{erfc}\,X$$

where

$$X = y/2\sqrt{Dt}$$

Under the conditions in this problem, this ratio $= 0.1$.

$$\text{erfc}\,X = 1 - \text{erf}\,X$$

so that $\text{erfc}\,X$ can be calculated from Table 13 in the Appendix of Volume 1. Values of X will be assumed and the right hand side evaluated until a value of X is found such that the right hand side $= 0.1$.

X	e^{-X^2}	$\text{erf}\,X$	$\text{erfc}\,X$	$X\sqrt{\pi}\,\text{erfc}\,X$	Right hand side
1	0.368	0.843	0.157	0.278	0.0897
0.9	0.445	0.797	0.203	0.324	0.121
0.97	0.390	0.830	0.170	0.292	0.098
0.96	0.398	0.825	0.175	0.297	0.101

$$\therefore \qquad X = 0.96 = y^2/4Dt$$

$$\therefore \qquad y = (0.96 \times 4 \times 10^{-9} \times 60)^{0.5} = 4.8 \times 10^{-4} \text{ m or } \underline{\underline{0.48 \text{ mm}}}$$

PROBLEM 10.23

A chamber, of volume 1 m^3, contains air at a temperature of 293 K and a pressure of 101.3 kN/m^2, with a partial pressure of water vapour of 0.8 kN/m^2. A bowl of liquid with a free surface of 0.01 m^2 and maintained at a temperature of 303 K is introduced into the chamber. How long will it take for the air to become 90% saturated at 293 K and how much water must be evaporated?

The diffusivity of water vapour in air is 2.4×10^{-5} m^2/s and the mass transfer resistance is equivalent to that of a stagnant gas film of thickness 0.25 mm. Neglect the effects of bulk flow. Saturation vapour pressure of water = 4.3 kN/m^2 at 303 K and 2.3 kN/m^2 at 293 K.

Solution

Moles transferred,

$$n = (DA/L)(C_{As} - C_A)$$

where C_A = concentration (kmol/m^3), C_{As} is the saturation value of C_A at the surface D is the diffusivity and L is the thickness of the stagnant gas film.

If the saturated vapour pressure at the interface is 4.3 kN/m^2 and if at any time the partial pressure in the air is P_A kN/m^2, then the rate of evaporation is given by:

$$\frac{dn}{dt} = \frac{0.01 \times 2.4 \times 10^{-5}}{(0.25/1000)} \left[\frac{1}{22.4} \times \frac{273}{303} \left(\frac{4.3}{101.3} - \frac{P_A}{101.3} \right) \right]$$

$$= 3.81 \times 10^{-7}(4.3 - P_A) \text{ kmol/s}$$

1 m^3 of air at 303 K and 101.3 kN/m^2 is equivalent to $(1/22.4)(273/293) = 0.0416$ kmol

Initial moisture content $= (0.8/101.3) \times 0.0416 = 3.29 \times 10^{-4}$ kmol

Final moisture content $= (0.9 \times 2.3/101.3) \times 0.0416 = 8.50 \times 10^{-4}$ kmol

$$\therefore \text{ Water evaporated} = (8.50 - 3.29) \times 10^{-4} = 5.21 \times 10^{-4} \text{ kmol}$$
$$= 5.21 \times 10^{-4} \times 18$$
$$= 9.38 \times 10^{-3} \text{ kg water}$$

At a pressure P kN/m^2:

$$n = (P_A/101.3)(1/22.4)(273/293) = 4.11 \times 10^{-4}P_A \text{ kmol/m}^3$$

and:
$$dn/dt = 4.11 \times 10^{-4}dP_A/dt$$

$\therefore$
$$4.11 \times 10^{-4}\frac{dP_A}{dt} = 3.81 \times 10^{-7}(4.3 - P_A)$$

$\therefore$
$$\int_{0.8}^{2.3} \frac{dP_A}{(4.3 - P_A)} = 9.27 \times 10^{-4}dt$$

from which $t = \underline{604 \text{ s}}$ (10 min)

PROBLEM 10.24

A large deep bath contains molten steel, the surface of which is in contact with air. The oxygen concentration in the bulk of the molten steel is 0.03% by mass and the rate of transfer of oxygen from the air is sufficiently high to maintain the surface layers saturated at a concentration of 0.16% by weight. The surface of the liquid is disrupted by gas bubbles rising to the surface at a frequency of 120 bubbles per m^2 of surface per second, each bubble disrupts and mixes about 15 cm^2 of the surface layer into the bulk.

On the assumption that the oxygen transfer can be represented by a surface renewal model, obtain the appropriate equation for mass transfer by starting with Fick's second law of diffusion and calculate:

(a) The mass transfer coefficient
(b) The mean mass flux of oxygen at the surface
(c) The corresponding film thickness for a film model, giving the same mass transfer rate.

Diffusivity of oxygen in steel $= 1.2 \times 10^{-8}$ m^2/s. Density of molten steel $= 7100$ kg/m^3.

Solution

If C' is defined as the concentration above a uniform datum value:

$$\frac{\partial C'}{\partial t} = D\frac{\partial^2 C'}{\partial y^2}$$ (equation 10.100)

The boundary conditions are:

$$\begin{array}{llll}
\text{when} & t = 0, & 0 < y < \infty & C' = 0 \\
& t > 0 & y = 0 & C' = C_i' \\
& t > 0 & y = \infty & C' = 0
\end{array}$$

The equation is most conveniently solved using Laplace transforms. The Laplace transform $\bar{C}'$ of C' is:

$$\bar{C}' = \int_0^\infty e^{-pt}C' \, dt$$ (equation 10.101)

Then:
$$\frac{\partial \bar{C}'}{\partial t} = \int_0^\infty e^{-pt}\frac{\partial C'}{\partial t}\,dt \qquad \text{(equation 10.102)}$$

$$= [e^{-pt}C']_0^\infty + p\int_0^\infty e^{-pt}C'\,dt = p\bar{C}' \qquad \text{(equation 10.103)}$$

Since the Laplace transform operation is independent of y,

$$\overline{\frac{\partial^2 C'}{\partial y^2}} = \frac{\partial^2 \bar{C}'}{\partial y^2} \qquad \text{(equation 10.104)}$$

Taking Laplace transforms of both sides of equation 10.100:

$$p\bar{C}' = D\frac{\partial^2 \bar{C}'}{\partial y^2}$$

$$\frac{\partial^2 \bar{C}'}{\partial y^2} - \frac{p}{D}\bar{C}' = 0$$

From which:
$$\bar{C}' = Ae^{\sqrt{(p/D)}y} + Be^{-\sqrt{(p/D)}y} \qquad \text{(equation 10.105)}$$

When $y = \infty$, $\quad \bar{C}' = 0 \quad$ and $\quad A = 0$
When $y = 0$, $\quad \bar{C}' = C_i'/p \quad$ and $\quad B = C_i'/p$

$$\therefore \qquad \bar{C}' = \frac{C_i'}{p}e^{-\sqrt{(p/D)}y}$$

$$\frac{d\bar{C}'}{dy} = -\frac{C_i'}{\sqrt{pD}}e^{-\sqrt{(p/D)}y}$$

Inverting: $\dfrac{\partial C'}{\partial y} = -\dfrac{C_i'}{\sqrt{D}} \times \dfrac{1}{\sqrt{\pi t}}e^{-y^2/4Dt}$ (See Volume 1, Appendix Table 12)

The mass transfer rate at the surface, $(N_A)_t = -D\left(\dfrac{\partial C'}{\partial y}\right)_{y=0} = C_i'\sqrt{\dfrac{D}{\pi t}}$ at time t

The average rate of mass transfer in time t:

$$\frac{1}{t}\int_0^t C_i'\sqrt{\frac{D}{\pi t}}\,dt = 2C_i'\sqrt{\frac{D}{\pi t}}$$

Taking 1 m^2 of surface, the area disrupted by the bubbles per second is:

$$120 \times 15/10000 = 0.18/s$$

$\therefore$ Average surface age duration $= (1/0.18) = 5.55$ s

$C_i' = (0.16 - 0.03)/100 = 0.0013$ kg O$_2$/kg steel $= (0.0013/32) \times 7100$
$$= 0.2885 \text{ kmol/m}^3$$

Then:

(a) The mass transfer coefficient $= 2\sqrt{\dfrac{D}{\pi t}} = 2(1.2 \times 10^{-8}/\pi \times 5.55)^{0.5}$

$$= 5.25 \times 10^{-5} \text{ m/s}$$

(b) The mean rate of transfer, $N_A = 2C_i'(D/\pi t)^{0.5} = 2 \times 0.2885(1.2 \times 10^{-8}/\pi \times 5.55)^{0.5}$

$$= 1.51 \times 10^{-5} \text{ kmol/m}^2 \text{ s}$$

(c) The film thickness L is given by: $N_A = (D/L)\bar{C}_i'$

and: $L = (1.2 \times 10^{-8} \times 0.2885)/(1.51 \times 10^{-5}) = 2.29 \times 10^{-4} \text{ m} = \underline{0.23 \text{ mm}}$

PROBLEM 10.25

Two large reservoirs of gas are connected by a pipe of length $2L$ with a full-bore valve at its mid-point. Initially a gas **A** fills one reservoir and the pipe up to the valve and gas **B** fills the other reservoir and the remainder of the pipe. The valve is opened rapidly and the gases in the pipe mix by molecular diffusion.

Obtain an expression for the concentration of gas **A** in that half of the pipe in which it is increasing, as a function of distance y from the valve and time t after opening. The whole system is at a constant pressure and the ideal gas law is applicable to both gases. It may be assumed that the rate of mixing in the vessels is high so that the gas concentration at the two ends of the pipe do not change.

Solution

The system and nomenclature are shown in Fig. 10c.

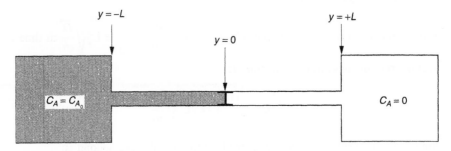

Figure 10c.

When time $t = 0$,
For gas **A**:

$$\frac{\partial C_A}{\partial t} = D\frac{\partial^2 C_A}{\partial y^2}$$

When $t = 0$, $\quad -L < y < 0 \quad\quad C_A = C_{A0}$
When $t = 0$, $\quad 0 < y < L \quad\quad C_A = 0$
When $t > 0$ $\quad\quad y = -L \quad\quad C_A = C_{A0}$
When $t > 0$ $\quad\quad y = +L \quad\quad C_A = 0$

For gas **B**:

$$\frac{\partial C_B}{\partial t} = D\frac{\partial^2 C_B}{\partial y^2}$$

When $t = 0$ $\quad -L < y < 0 \quad\quad C_B = 0$
When $t = 0$ $\quad 0 < y < L \quad\quad C_B = C_{B0}$
When $t > 0$ $\quad\quad y = -L \quad\quad C_B = 0$
When $t > 0$ $\quad\quad y = +L \quad\quad C_B = C_{B0}$

and for all values of y:

$$\frac{\partial C_A}{\partial y} + \frac{\partial C_B}{\partial y} = 0$$

As in previous problems, these equations may be solved by the use of Laplace transforms.
    For $y > 0$:

$$\bar{C}_A = Ae^{\sqrt{(p/D)}y} + Be^{-\sqrt{(p/D)}y}$$

and for $y < 0$:

$$\bar{C}_A = A'e^{\sqrt{(p/D)}y} + B'e^{-\sqrt{(p/D)}y} + C_{A0}/p$$

The boundary conditions may now be used to evaluate the constants thus:

$$A = -\frac{(C_{A0}/p)e^{-2\sqrt{(p/D)}L}}{2(1 - e^{-2\sqrt{(p/D)}L})}$$

$$B = \frac{(C_{A0}/p)}{2(1 - e^{-2\sqrt{(p/D)}L})}$$

$$A' = -B'e^{2\sqrt{(p/D)}L}$$

$$B' = \frac{B(e^{-2\sqrt{(p/D)}L} + 1)}{(e^{2\sqrt{(p/D)}L} + 1)}$$

Substituting these values:

$$C_A = \frac{C_{A0}}{2}\sum_{n=0}^{n=\infty}\left[\text{erfc}\,\frac{2nL + y}{2\sqrt{Dt}} - \text{erfc}\,\frac{2(n+1)L - y}{2\sqrt{Dt}}\right]$$

This relation can be checked as follows:

(a) When $y = 0$: $C_A = \dfrac{C_{A0}}{2}\displaystyle\sum_{0}^{\infty}\left[\text{erfc}\,\dfrac{nL}{\sqrt{Dt}} - \text{erfc}\,\dfrac{(n+1)L}{\sqrt{Dt}}\right] = \dfrac{C_{A0}}{2}$

(b) When $y = L$: $C_A = 0$

PROBLEM 10.26

A pure gas is absorbed into a liquid with which it reacts. The concentration in the liquid is sufficiently low for the mass transfer to be governed by Fick's law and the reaction is first order with respect to the solute gas. It may be assumed that the film theory may be applied to the liquid and that the concentration of solute gas falls from the saturation value to zero across the film. Obtain an expression for the mass transfer rate across the gas-liquid interface in terms of the molecular diffusivity, D, the first-order reaction rate constant k, the film thickness L and the concentration C_{AS} of solute in a saturated solution. The reaction is initially carried out at 293 K. By what factor will the mass transfer rate across the interface change, if the temperature is raised to 313 K? Reaction rate constant at 293 K $= 2.5 \times 10^{-6}$ s^{-1}. Energy of activation for reaction (in Arrhenius equation) $=$ 26430 kJ/kmol. Universal gas constant $\mathbf{R} = 8.314$ kJ/kmol K. Molecular diffusivity $D = 10^{-9}$ m^2/s. Film thickness, $L = 10$ mm. Solubility of gas at 313 K is 80% of solubility at 293 K.

Solution

See Volume 1, Example 10.11

PROBLEM 10.27

Using Maxwell's law of diffusion obtain an expression for the effective diffusivity for a gas $\mathbf{A}$ in a binary mixture of $\mathbf{B}$ and $\mathbf{C}$, in terms of the diffusivities of $\mathbf{A}$ in the two pure components and the molar concentrations of $\mathbf{A}, \mathbf{B}$ and $\mathbf{C}$.

Carbon dioxide is absorbed in water from a 25 per cent mixture in nitrogen. How will its absorption rate compare with that from a mixture containing 35 per cent carbon dioxide, 40 per cent hydrogen and 25 per cent nitrogen? It may be assumed that the gas-film resistance is controlling, that the partial pressure of carbon dioxide at the gas–liquid interface is negligible and that the two-film theory is applicable, with the gas film thickness the same in the two cases. Diffusivity of CO_2 in hydrogen $= 3.5 \times 10^{-5}$ m^2/s; in nitrogen $= 1.6 \times 10^{-5}$ m^2/s.

Solution

Maxwell's Law of Diffusion is discussed in Section 10.3.2 where for a two component gaseous mixture:

$$-dP_A/dy = F_{AB}C_A C_B(u_A - u_B) \qquad \text{(equation 10.77)}$$

For an ideal gas, $\qquad\qquad P_A = C_A \mathbf{R}T \qquad\qquad$ (equation 10.9a)

and from equation 10.78: $\qquad u_A = N'_A/C_A \qquad\qquad$ (equation 10.9b) when **B** is not undergoing mass transfer, or $u_B = 0$, then:

$$-\mathbf{R}T(dC_A/dy) = F_{AB}C_B N'_A$$

$$N'_A = -\frac{\mathbf{R}T}{F_{AB}C_B}\frac{dC_A}{dy} = -\frac{\mathbf{R}T}{F_{AB}C_T}\frac{C_T}{C_B}\frac{dC_A}{dy}$$

By comparison with Stefan's Law:

$$N'_A = -D_{AB}\frac{C_T}{C_B}\frac{dC_A}{dy} \qquad\qquad \text{(equation 10.29)}$$

Then:
$$D_{AB} = \mathbf{R}T/F_{AB}C_T$$

or:
$$F_{AB} = \frac{\mathbf{R}T}{D_{AB}C_T}$$

Applying to **A** in a mixture of **B** and **C**:

$$-dP_A/dy = F_{AB}C_A C_B(u_A - u_B) + F_{AC}C_A C_C(u_A - u_C)$$

For the case where $u_B = u_C = 0$:

$$-\mathbf{R}T\frac{dC_A}{dy} = \frac{\mathbf{R}T}{D_{AB}C_T}C_B N'_A + \frac{\mathbf{R}T}{D_{AC}C_T}C_C N'_A$$

or:
$$N'_A = \frac{-(dC_A/dy)C_T}{(C_B/D_{AB}) + (C_C/D_{AC})} \qquad\qquad \text{(i)}$$

From Stefan's Law:
$$N'_A = -D'\frac{C_T}{C_T - C_A}\frac{dC_A}{dy} = -\frac{(dC_A/dy)C_T}{(C_T - C_A)/D'} \qquad\qquad \text{(ii)}$$

where D' is the effective diffusivity of **A** in the mixture

Comparing equations (i) and (ii):

$$\frac{1}{D'} = \frac{1}{D_{AB}}\frac{C_B}{C_T - C_A} + \frac{1}{D_{AC}}\frac{C_C}{C_T - C_A}$$

For CO_2 in N_2: $\quad N'_A = D\dfrac{C_T}{(C_{N_2})_{lm}}\dfrac{\Delta C_{CO_2}}{L} = D \times \dfrac{1}{(y_{N_2})_{lm}} \times \dfrac{C_T \Delta y_{CO_2}}{L}$ (equation 10.33)

$$D = 1.6 \times 10^{-5} \text{ m}^2/\text{s} \ (y_{N_2})_1 = 1.0, \ (y_{N_2})_2 = 0.75,$$

$\therefore \qquad\qquad (y_{N_2})_{lm} = [(1 - 0.75)/\ln(1/0.75)] = 0.87$

$$\Delta y_{CO_2} = 0.25, \ \Delta C_T \text{ and } L \text{ are unknown}$$

Hence:
$$N'_A = \frac{1.6 \times 10^{-5}}{0.87} \times 0.25\frac{C_T}{L} = 4.6 \times 10^{-6}C_T/L \text{ kmol/m}^2\text{s}$$

For CO_2 in a mixture of H_2 and N_2, the effective diffusivity, derived in the first part of the problem, is used to give D':

$$\frac{1}{D'} = \frac{1}{3.5 \times 10^{-5}} \times \frac{0.4}{1 - 0.35} + \frac{1}{1.6 \times 10^{-5}} \times \frac{0.25}{1 - 0.35} = 2.4 \times 10^{-5} \text{ m}^2/\text{s}$$

$$N_A' = D' \frac{C_T}{(C_{N_2+H_2})_{lm}} \times \frac{\Delta C_{CO_2}}{L} = D' \times \frac{1}{(y_{N_2+H_2})_{lm}} \times \frac{C_T \Delta y_{CO_2}}{L}$$

$$y_{CO_2} = 0.35$$

$$(y_{N_2+H_2})_1 = 1.0, \quad (y_{N_2+H_2})_2 = 0.65$$

and: $$N_A' = 2.4 \times 10^{-5} \times \frac{\ln(1/0.65)}{(1-0.65)} \times 0.35 \frac{\Delta C}{L} = 1.033 \times 10^{-5} C_T/L$$

$\therefore$ The ratio of mass transfer rates $= (1.033 \times 10^{-5}/4.6 \times 10^{-6}) = \underline{\underline{2.25}}$

PROBLEM 10.28

Given that from the penetration theory for mass transfer across an interface, the instantaneous rate of mass transfer is inversely proportional to the square root of the time of exposure, obtain a relationship between exposure time in the Higbie model and surface renewal rate in the Danckwerts model which will give the same average mass transfer rate. The age distribution function and average mass transfer rate from the Danckwerts theory must be derived from first principles.

Solution

Given that the instantaneous mass transfer rate $= Kt^{-1/2}$, then for the Higbie model, the average mass transfer rate for an exposure time t_e is given by:

$$\frac{1}{t_e} \int_0^{t_e} Kt^{-1/2} \, dt = 2Kt_e^{-1/2}$$

For the Danckwerts model, the random surface renewal analysis, presented in Section 10.5.2, shows that the fraction of the surface with an age between t and $t + dt$ is a function of $t = f(t) \, dt$ and that $f(t) = Ke^{-st}$ where s is the rate of production of fresh surface per unit total area.

For a total surface area of unity:

$$\int_0^\infty Ke^{-st} \, dt = 1 = K \left[\frac{e^{-st}}{-s} \right]_0^\infty = K/s$$

$\therefore$ $$K = s \quad \text{and} \quad f(t) = se^{-st} \, dt$$

The rate of mass transfer for unit area is:

$$\int_0^\infty Kt^{-1/2} \times se^{-st} \, dt = Ks \int_0^\infty t^{-1/2} e^{-st} \, dt$$

Substituting $\beta^2 = st$ and $s \, dt = 2\beta \, d\beta$, then:

$$\text{Rate} = Ks \int_0^\infty \frac{\sqrt{s}}{\beta} e^{-\beta^2} \frac{2\beta}{s} \, d\beta = K\sqrt{s} \times 2 \int_0^\infty e^{-\beta^2} \, d\beta = K\sqrt{s} \times 2 \times \sqrt{\pi}/2 = K\sqrt{\pi s}$$

If the rates from each model are equal, then:

$$2Kte^{-1/2} = K\sqrt{\pi s} \quad \text{or} \quad \underline{\underline{st_e = 4/\pi}}$$

PROBLEM 10.29

Ammonia is absorbed in a falling film of water in an absorption apparatus and the film is disrupted and mixed at regular intervals as it flows down the column. The mass transfer rate is calculated from the penetration theory on the assumption that all the relevant conditions apply. It is found from measurements that the mass transfer rate immediately before mixing is only 16 per cent of that calculated from the theory and the difference has been attributed to the existence of a surface film which remains intact and unaffected by the mixing process. If the liquid mixing process takes place every second, what thickness of surface film would account for the discrepancy? Diffusivity of ammonia in water $= 1.76 \times 10^{-9}$ m^2/s.

Solution

For the penetration theory:

$$\frac{\partial C_A}{\partial t} = D\frac{\partial^2 C_A}{\partial y^2} \qquad \text{(equation 10.66)}$$

When $t = 0$, $\quad C_A = 0$

When $t > 0$, $\quad y = 0$, $\quad C_A = C_{Ai} = \text{constant}$

When $t > 0$, $\quad y = \infty$ $\quad C_A = 0$

As shown earlier in problems 10.19 and 10.21, this equation may be transformed and solved to give:

$$\bar{C}_A = Ae^{2\sqrt{(p/D)}y} + Be^{-2\sqrt{(p/D)}y} \qquad \text{(equation 10.105)}$$

When $y = 0$, $\quad \bar{C}_A = C_{Ai}/p$

$$y = \infty, \quad \bar{C}_A = 0$$

and hence: $\qquad A = 0 \quad \text{and} \quad B = C_{Ai}/p$

Hence, $\qquad \bar{C}_A = \dfrac{C_{Ai}}{p}e^{-\sqrt{(p/D)}y}$

and: $\qquad \dfrac{\partial \bar{C}_A}{\partial y} = -C_{Ai}\sqrt{\dfrac{p}{D}}e^{-\sqrt{(p/D)}y}$

At the surface, $(N_A)_t = -D\left(\dfrac{\partial C_A}{\partial y}\right)_{y=0} = \sqrt{\dfrac{D}{\pi t}}C_{Ai}$ in time t (as in Problem 10.21).

For the film, the origin is taken at the interface between the film (whose thickness is L) and the mixed fluid.

Again: $\qquad \bar{C}_A = Ae^{\sqrt{(p/D)}y} + Be^{-\sqrt{(p/D)}y} \qquad \text{(equation 10.105)}$

$$C_{Ax} \text{ at } y = 0 \text{ is now a variable}$$

$$y = \infty, C_A = 0$$

Hence:
$$A = 0 \text{ and } \bar{C}_A = \bar{C}_{Ax}e^{-\sqrt{(p/D)}y}$$

and:
$$\frac{d\bar{C}_A}{dy} = -\sqrt{\frac{p}{D}}\bar{C}_{Ax}e^{-\sqrt{(p/D)}y}$$

and:
$$\left(\frac{d\bar{C}_1}{dy}\right)_{y=0} = -\sqrt{\frac{p}{D}}\bar{C}_{Ax}$$

To maintain mass balance at the film boundary:

$$\frac{D}{L}(C_{Ai} - C_{Ax}) = -D\left(\frac{\partial C_A}{\partial y}\right)_{y=0}$$

Taking Laplace transforms:

$$\frac{D}{L}\left(\frac{C_{Ai}}{p} - \bar{C}_{Ax}\right) = -D\left(\frac{\partial \bar{C}_A}{\partial y}\right)_{y=0} = \sqrt{pD}\,\bar{C}_{Ax}$$

So:
$$\frac{D}{L}\frac{C_{Ai}}{p} = \left(\frac{D}{L} + \sqrt{pD}\right)\bar{C}_{Ax}$$

and:
$$\bar{C}_{Ax} = \frac{\dfrac{D}{L}\dfrac{C_{Ai}}{p}}{\dfrac{D}{L} + \sqrt{pD}}$$

Hence:
$$\frac{d\bar{C}_{Ax}}{dy} = -\sqrt{\frac{p}{D}}\frac{D}{L}\frac{C_{Ai}}{p}\frac{1}{\dfrac{D}{L} + \sqrt{pD}}e^{-\sqrt{(p/D)}y}$$

$$= -C_{Ai}\frac{1}{\sqrt{p}}\frac{1/L}{\dfrac{\sqrt{D}}{L} + \sqrt{p}}e^{-\sqrt{(p/D)}y}$$

Inverting:
$$\frac{\partial C_A}{\partial y} = -\frac{C_{Ai}}{L}e^{y/L}e^{Dt/L^2}\,\text{erfc}\left(\frac{\sqrt{Dt}}{L} + \frac{y}{2\sqrt{Dt}}\right)$$

$$(N_A)_{y=0} = -D\left(\frac{\partial C_A}{\partial y}\right)_{y=0} = C_{Ai}\frac{D}{L}e^{Dt/L^2}\,\text{erfc}\left(\frac{\sqrt{Dt}}{L}\right)$$

$$\frac{C_{Ai}\dfrac{D}{L}e^{Dt/L^2}\,\text{erfc}\left(\dfrac{\sqrt{Dt}}{L}\right)}{\sqrt{\dfrac{D}{\pi t}}C_{Ai}} = 0.16$$

or:
$$\sqrt{\frac{Dt}{L^2}} e^{Dt/L^2} \operatorname{erfc} \sqrt{\frac{Dt}{L^2}} = \sqrt{\frac{1}{\pi}} \times 0.16 = 0.0903$$

writing
$$X = \sqrt{\frac{Dt}{L^2}} \quad \text{then } Xe^{x^2} \operatorname{erfc} X = 0.0903$$

Solving by trial and error: $X = 0.101$

When $t = 1$ s, $D = 1.76 \times 10^{-9}$ m^2/s, and: $L = \underline{\underline{0.42 \text{ mm.}}}$

PROBLEM 10.30

A deep pool of ethanol is suddenly exposed to an atmosphere consisting of pure carbon dioxide and unsteady state mass transfer, governed by Fick's Law, takes place for 100 s. What proportion of the absorbed carbon dioxide will have accumulated in the 1 mm thick layer of ethanol closest to the surface? Diffusivity of carbon dioxide in ethanol = 4×10^{-9} m^2/s.

Solution

See Volume 1, Example 10.6.

PROBLEM 10.31

A soluble gas is absorbed into a liquid with which it undergoes a second-order irreversible reaction. The process reaches a steady-state with the surface concentration of reacting material remaining constant at C_{As} and the depth of penetration of the reactant being small compared with the depth of liquid which can be regarded as infinite in extent. Derive the basic differential equation for the process and from this derive an expression for the concentration and mass transfer rate (moles per unit area and unit time) as a function of depth below the surface. Assume that mass transfer is by molecular diffusion.

If the surface concentration is maintained at 0.04 kmol/m^3, the second-order rate constant k_2 is 9.5×10^3 m^3/kmol s and the liquid phase diffusivity D is 1.8×10^{-9} m^2/s, calculate:

(a) The concentration at a depth of 0.1 mm.
(b) The molar rate of transfer at the surface (kmol/m^{2}s).
(c) The molar rate of transfer at a depth of 0.1 mm.

It may be noted that if:
$$\frac{dC_A}{dy} = q, \quad \text{then: } \frac{d^2C_A}{dy^2} = q\frac{dq}{dC_A}$$

Solution

Considering element of unit area and depth dy, then for a steady state process:

$$\text{RATE IN} - \text{RATE OUT} = \text{REACTION RATE}$$

or:
$$-D\frac{dC_A}{dy} - \left\{-D\frac{dC_A}{dy} + \frac{d}{dy}\left(-D\frac{dC_A}{dy}\right)dy\right\} = (k_2C_A^2)(dy)$$

$$D\frac{d^2C_A}{dy^2} = k_2C_A^2$$

$$\frac{d^2C_A}{dy^2} - \frac{k_2}{D}C_A^2 = 0$$

Putting:
$$q = \frac{dC_A}{dy}, \text{ then: } \frac{d^2C_A}{dy^2} = \frac{dq}{dy} = \frac{dq}{dC_A} \cdot \frac{dC_A}{dy} = q\frac{dq}{dC_A}$$

Substituting:
$$q\frac{dq}{dC_A} - \frac{k_2}{D}C_A^2 = 0$$

$$q\,dq = \frac{k_2}{D}C_A^2$$

Integrating:
$$\frac{q^2}{2} = \frac{k_2}{D}\frac{C_A^3}{3} + K$$

When $y = \infty$, $q = 0$, $C_A = 0$ and: $K = 0$

Thus:
$$\frac{1}{2}\left(\frac{dC_A}{dy}\right)^2 = \frac{k_2}{D}\frac{C_A^3}{3}$$

$$\frac{dC_A}{dy} = \pm\sqrt{\frac{2}{3}\frac{k_2}{D}}C_A^{3/2}$$

As $N_A = -D(dC_A/dy)$ is positive, negative root must apply and:

$$C_A^{-3/2}\,dC_A = -\sqrt{\frac{2}{3}\frac{k_2}{D}}\,dy$$

Integrating:
$$-2C_A^{-1/2} = -\sqrt{\frac{2}{3}\frac{k_2}{D}}y + K$$

When $y = 0$, $C_A = C_{As}$ and: $K = -2C_{As}^{-1/2}$

Thus:
$$C_A^{-1/2} - C_{As}^{-1/2} = \frac{1}{2}\sqrt{\frac{2}{3}\frac{k_2}{D}}y$$

or:
$$C_{As}^{-1/2}\left[\left(\frac{C_A}{C_{As}}\right)^{-1/2} - 1\right] = \sqrt{\frac{1}{6}\frac{k_2}{D}}y$$

$$C_{As}^{-1/2}\left\{\left(\frac{C_{As}}{C_A}\right)^{1/2} - 1\right\} = \sqrt{\frac{1}{6}\frac{k_2}{D}}y.$$

(i) For the conditions given:
$$\sqrt{\frac{1}{6}\frac{k_2}{D}} = \sqrt{\left\{\left(\frac{1}{6}\right)\left(\frac{9.5 \times 10^3}{1.8 \times 10^{-9}}\right)\right\}}$$

$$= 0.938 \times 10^6 \text{ (m/kmol)}^{0.5}$$

At depth of 0.1 mm $= 10^{-4}$ m : $\sqrt{\dfrac{1}{6}\dfrac{k_2}{D}}\,y = 93.8$

and:
$$\left\{\left(\dfrac{C_{As}}{C_A}\right)^{1/2} - 1\right\} = 93.8\,C_{As}^{+1/2} = 18.8$$

$$\left(\dfrac{C_{As}}{C_A}\right)^{1/2} = 19.8$$

$$C_A = \dfrac{0.04}{19.8^2} = \underline{\underline{0.00010\ \text{kmol/m}^3.}}$$

(ii) The molar transfer rate at surface is:

$$N_A - D\dfrac{dC_A}{dy} = -\sqrt{\dfrac{2}{3}\dfrac{k_2}{D}}\,C_{As}^{3/2}(-D).$$

$$= \sqrt{\dfrac{2k_2 D}{3}}\,C_{As}^{3/2}$$

$$= \sqrt{\dfrac{2}{3} \times 9.5 \times 10^3 \times 1.8 \times 10^{-9}}(0.04)^{3/2}$$

$$= 3.38 \times 10^{-3} \times 0.008 = \underline{\underline{2.70 \times 10^{-5}\ \text{kmol/m}^2\text{s.}}}$$

(iii) The molar transfer rate at depth of 0.1 mm is:

$$N_A = \sqrt{\dfrac{2k_2 D}{3}}\,C_A^{3/2}$$

$$= 3.38 \times 10^{-3} \times (0.00010)^{3/2} = \underline{\underline{3.38 \times 10^{-9}\ \text{kmol/m}^2\text{s}}}$$

PROBLEM 10.32

In calculating the mass transfer rate from the *penetration theory*, two models for the age distribution of the surface elements are commonly used — those due to Higbie and to Danckwerts. Explain the difference between the two models and give examples of situations in which each of them would be appropriate.

(a) In the Danckwerts model, it is assumed that elements of the surface have an age distribution ranging from zero to infinity. Obtain the age distribution function for this model and apply it to obtain the average mass transfer coefficient at the surface, given that from the penetration theory the mass transfer coefficient for surface of age t is $\sqrt{[D/(\pi t)]}$, where D is the diffusivity.

(b) If for unit area of surface the surface renewal rate is s, by how much will the mass transfer coefficient be changed if no surface has an age exceeding $2/s$?

(c) If the probability of surface renewal is linearly related to age, as opposed to being constant, obtain the corresponding form of the age distribution function.

It may be noted that:

$$\int_0^\infty e^{-x^2}\, dx = \frac{\sqrt{\pi}}{2}$$

Solution

(a) If the age distribution function be $f(t)$, then the surface in the age group t to $t + dt$ is: $f(t)\, dt$.

Then, the surface of age $(t + dt)$ minus, the surface of age $(t + dt)$ is the surface destroyed in the dt, or:

$$f(t) - f(t + dt) = sf(t)\, dt$$

or:

$$-f'(t + dt)\, dt = sf(t)\, dt$$

As $dt \to 0$,

$$f'(t) + sf(t) = 0 \tag{i}$$

Using the integrating factor $e^{\int s\, dt} = e^{st}$ then:

$$e^{st} f(t) = K \text{ (const)}$$

and:

$$f(t) = Ke^{-st} \tag{ii}$$

$$\text{the total surface} = 1 = K \int_0^\infty e^{-st}\, dt = K \left[\frac{e^{-st}}{-s}\right]_0^\infty = \frac{K}{s}$$

and hence:

$$f(t) = se^{-st}$$

The mass transfer rate into fraction of surface of age t to $t + dt$ (per unit total area of surface) is:

$$\sqrt{\frac{D}{\pi t}} \Delta C_A s e^{-st}\, dt = \sqrt{\frac{D}{\pi}} \Delta C_A e^{-st} t^{-1/2}\, dt$$

The mass transfer rate per unit area,

$$N_A = \sqrt{\frac{D}{\pi}} \Delta C_A \int_0^\infty t^{-1/2} e^{-st}\, dt$$

Putting $st = \beta^2$: $s\, dt = 2\beta\, d\beta$

and:

$$N_A = \sqrt{\frac{D}{\pi}} \Delta C_A \int_0^\infty \frac{s^{1/2}}{\beta} e^{-\beta^2} 2\beta\, d\beta$$

$$= 2\sqrt{\frac{sD}{\pi}} \Delta C_A \int_0^\infty e^{-\beta^2}\, d\beta$$

$$= 2\sqrt{\frac{sD}{\pi}} \Delta C_A \frac{\sqrt{\pi}}{2} = \sqrt{Ds} \Delta C_A \tag{iii}$$

Thus: the mass transfer coefficient $= \dfrac{N_A}{\Delta C_A} = \sqrt{Ds}$

(b) For an age range of surface from 0 to $2/s$, application of equation (ii) gives:

$$1 = K \int_0^{2/s} e^{-st} \, dt = K \left[\frac{e^{-st}}{-s} \right]_0^{2/s} = \frac{K}{s}(1 - e^{-2})$$

or:

$$K = \frac{s}{1 - e^{-2}}$$

and:

$$f(t) = \frac{s}{1 - e^{-2}} e^{-st}$$

Thus N_A from equation (ii) is multiplied by factor $1/(1 - e^{-2})$ or:

$$N_A = \frac{1}{1 - e^{-2}} \sqrt{Ds} \Delta C_A = 1.16 \text{ times the value in equation 3.}$$

The mass transfer rate per unit total area of surface is:

$$N_A = \int_0^{2/s} \left(\sqrt{\frac{D}{\pi t}} \Delta C_A \right) (1.16 s e^{-st} \, dt) = \sqrt{\frac{D}{\pi}} 1.16 s \Delta C_A \int_0^{2/s} t^{-1/2} e^{-st} \, dt$$

Putting:

$$st = \beta^2$$

then:

$$t^{-1/2} = \frac{\sqrt{s}}{\beta}$$

and:

$$s \, dt = 2\beta \, d\beta$$

Integrating:

$$\int_0^{\sqrt{2}} \frac{\sqrt{s}}{\beta} e^{-\beta^2} \frac{2\beta}{s} \, d\beta = \frac{2}{\sqrt{s}} \int_0^{\sqrt{2}} e^{-\beta^2} \, d\beta$$

$$= \frac{2}{\sqrt{s}} \frac{\sqrt{\pi}}{2} \operatorname{erf} \sqrt{2}$$

$$= \sqrt{\frac{\pi}{s}} \operatorname{erf} \sqrt{2}$$

$$N_A = \sqrt{\frac{D}{\pi}} 1.16 s \sqrt{\frac{\pi}{s}} \Delta C_A \operatorname{erf} \sqrt{2} = 1.16 \times 0.954 \sqrt{Ds} \Delta C_A$$

$$= 1.107 \sqrt{Ds} \Delta C_A$$

The mass transfer coefficient, $h_D = 1.107 \sqrt{Ds}$, an increase of 10.7%

(c) For probability of surface renewal being linearly related to age, $s = kt$ (where k is a constant)

Equation (i) becomes:

$$f'(t) + ktf(t) = 0.$$

The integrating factor is: $e^{\int kt \, dt} = e^{kt^2/2}$

$$e^{kt^2/2} f(t) = K$$

and:

$$f(t) = K e^{-kt^2/2}$$

The total surface is: $1 = K \int_0^\infty e^{-kt^2/2} \, dt = K\sqrt{\dfrac{2}{k}} \int_0^\infty e^{-kt^2/2} \, d(\sqrt{k}/2)t$

Putting $\sqrt{\dfrac{k}{2}}t = X$:

$1 = K\sqrt{\dfrac{2}{k}} \int_0^\infty e^{-X^2} \, dX = K\sqrt{\dfrac{2}{k}}\dfrac{\sqrt{\pi}}{2} = K\sqrt{\dfrac{\pi}{2k}}$,

then: $\qquad\qquad\qquad\qquad K = \sqrt{\dfrac{2k}{\pi}}$

and the age distribution function is: $\underline{\underline{\sqrt{\dfrac{2k}{\pi}}e^{-kt^2/2}}}$

PROBLEM 10.33

Explain the basis of the *penetration theory* for mass transfer across a phase boundary. What are the assumptions in the theory which lead to the result that the mass transfer rate is inversely proportional to the square root of the time for which a surface element has been expressed? (Do *not* present a solution of the differential equation.) Obtain the age distribution function for the surface:

(a) On the basis of the Danckwerts' assumption that the probability of surface renewal is independent of its age.
(b) On the assumption that the probability of surface renewal increases linearly with the age of the surface.

Using the Danckwerts surface renewal model, estimate:

(c) At what age of a surface element is the mass transfer rate equal to the mean value for the whole surface for a surface renewal rate (s) of $0.01 \text{ m}^2/\text{m}^2\text{s}$?
(d) For what proportion of the total mass transfer is surface of an age exceeding 10 seconds responsible?

Solution

(a) *Danckwerts age distribution function*
    Dividing the total unit surface into elements each of duration dt, then:

If the fraction of surface in age band t to $t + dt$ is $f(t) \, dt$, then:

the fraction of surface in age band $t - dt$ to t will be $f(t - dt) \, dt$.

The surface not going from $t - dt/t$ to $t/t + dt = f(t - dt)\,dt - f(t)\,dt$
$$= -f'(t - dt)\,dt\,dt$$

This will be surface destroyed in time dt
$$= (\text{destruction rate} \times \text{area of surface} \times \text{time interval})$$

$$= s[f(t - dt)dt]dt$$

Thus: $-f'(t - dt)dt\,dt = s[f(t - dt)dt]dt$

As $dt \to 0$ then: $f'(t) + sf(t) = 0$

Using the Danckwerts model: $s = \text{const.}$
$$e^{st}f'(t) + se^{st}f(t) = 0$$

$$e^{st}f(t) = K$$

$$f(t) = Ke^{-st}$$

There is no upper age limit to surface

Thus total surface $= 1 = K \displaystyle\int_0^\infty e^{-st}dt = K\left[\dfrac{e^{-st}}{-s}\right]_0^\infty = K/s$ or: $K = s$

$$f(t) = se^{-st}.$$

(b) $s = at$ where a is a constant.

Thus:
$$e^{at^2/2}f'(t) + ate^{at^2/2}f(t) = 0$$

$$e^{at^2/2}f(t) = K'$$

and:
$$f(t) = K'e^{-at^2/2}$$

For unit total surface: $1 = K' \displaystyle\int_0^\infty e^{-at^2/2}\,dt = K'\sqrt{\dfrac{2}{a}}\int_0^\infty e^{-at^2/2}\,d\left(\sqrt{\dfrac{a}{2}}t\right)$

$$= K'\sqrt{\dfrac{2}{a}}\dfrac{\sqrt{\pi}}{2}$$

and: $K' = \sqrt{\dfrac{2a}{\pi}}$ and $f(t) = \sqrt{\dfrac{2a}{\pi}}e^{-at^2/2}$

(c) *regarding surface renewal as random*

For unit total surface, mass transfer (mol/area time) $= \displaystyle\int_0^\infty kt^{-1/2}(se^{-st})dt\,\Delta C_A$

$$= \Delta C_A ks \int_0^\infty t^{-1/2}e^{-st}dt.$$

where ΔC_A is the concentration driving force in moles per unit volume
Putting $st = x^2$, then: $s\,dt = 2x\,dx$

The mass transfer rate $= ks\Delta C_A \displaystyle\int_0^\infty x^{-1}\sqrt{s}\,e^{-x^2}2x\,dx.\dfrac{1}{s}$

$$= \Delta C_A 2k\sqrt{s}\int_0^\infty e^{-x^2}\,dx = \Delta C_A 2k\sqrt{s}\dfrac{\sqrt{\pi}}{2} = \Delta C_A k\sqrt{\pi s}$$

The mass transfer rate at time t is: $\Delta C_A k t^{-1/2}$

Thus the age of surface at which rate $=$ average is given by or:

$$kt^{-1/2} = k\sqrt{\pi s}$$

$$t = \frac{1}{\pi s} = \frac{100}{\pi} = \underline{\underline{31.8 \text{ s}}}$$

(d) *Surface of age less than 10 seconds*

The mass transfer taking place into surface of age up to 10 s is given by the same expression as for the whole surface but with upper limit of 10 s instead of infinity

or: $$\Delta C_A 2k\sqrt{s} \int_0^{k10^{-t}} e^{-x^2} dx$$

when $t = 10$ s: $x = \sqrt{st} = \sqrt{0.01 \times 10} = 0.316$

The mass transfer into surface up to 10 s age is then:

$$\Delta C_A 2k\sqrt{s} \int_0^{0.316} e^{-x^2} dx = \Delta C_A 2k\sqrt{s} \frac{\sqrt{\pi}}{2} \text{ erf } 0.316 = \Delta C_A k\sqrt{\pi s} \times 0.345$$

Thus a fraction: $\underline{\underline{0.345}}$ is contributed by surface of age $0 - 10$ s and:

a fraction: $\underline{\underline{0.655}}$ by surface of age 10 s to infinity.

PROBLEM 10.34

At a particular location in a distillation column, where the temperature is 350 K and the pressure 500 m Hg, the mol fraction of the more volatile component in the vapour is 0.7 at the interface with the liquid and 0.5 in the bulk of the vapour. The molar latent heat of the more volatile component is 1.5 times that of the less volatile. Calculate the mass transfer rates (kmol m^{-2}s^{-1}) of the two components. The resistance to mass transfer in the vapour may be considered to lie in a stagnant film of thickness 0.5 mm at the interface. The diffusivity in the vapour mixture is 2×10^{-5} m^{2}s^{-1}.

Calculate the mol fractions and concentration gradients of the two components at the mid-point of the film. Assume that the ideal gas law is applicable and that the Universal Gas Constant $\mathbf{R} = 8314$ J/kmol K.

Solution

In this case:

$$T = 350 \text{ K},$$

$$P = 500 \text{ mm Hg} = \left(\frac{500}{760} \times 101{,}300\right) = 0.666 \times 10^5 \text{ N/m}^2$$

$$D = 2 \times 10^{-5} \text{ m}^2/\text{s}$$

and: $$C_T = \frac{P}{RT} = \left(\frac{0.666 \times 10^5}{8314 \times 350}\right) = 0.0229 \text{ kmol/m}^3$$

If: $\qquad$ $\mathbf{A} = MVC \qquad \mathbf{B} = LVC$, then:

$$\lambda_A = 1.5\lambda_B$$

$$N'_A\lambda_A = -N'_B\lambda_B$$

and: $\qquad N'_B = -N'_A, \dfrac{\lambda_A}{\lambda_B} = -1.5N'_A$

$$N'_A = -DC_T\frac{dx_A}{dy} + u_F C_A \qquad \text{(from equations 10.46a and 10.19)}$$

$$= -DC_T\frac{dx_A}{dy} + \frac{N'_A + N'_B}{C_T}C_A$$

$$= -DC_T\frac{dx_A}{dy} + (N'_A - 1.5N'_A)x_A$$

$$N'_A(1 + 0.5x_A) = -DC_T\frac{dx_A}{dy} \qquad\qquad\qquad \text{(i)}$$

$$N'_A \int_0^{5\times 10^{-4}} dy = -DC_T \int_{0.7}^{0.5} \frac{dx_A}{1 + 0.5x_A}$$

$$N'_A \times 5 \times 10^{-4} = -DC_T 2\Big[\ln(1 + 0.5x_A)\Big]_{0.7}^{0.5}$$

$$N'_A = -2DC_T \ln\frac{1.25}{1.35} \times \frac{1}{5 \times 10^{-4}}$$

$$= 1.41 \times 10^{-4} \text{ kmol/m}^2\text{s}$$

and: $\qquad\qquad N'_B = (-)2.11 \times 10^{-4} \text{ kmol/m}^2\text{s}$

At the mid-point: $y = 2.5 \times 10^{-4}$ m

$$N'_A \times 2.5 \times 10^{-4} = -2 \times 10^{-5} \times 0.0229 \times 2\Big[\ln(1 + 0.5x_A)\Big]_{0.7}^{x_A}$$

$$\ln\frac{1 + 0.5x_A}{1.35} = -\frac{1.41 \times 10^{-4} \times 2.5 \times 10^{-4}}{2 \times 10^{-5} \times 0.0229 \times 2}$$

$$\ln\frac{1.35}{1 + 0.5x_A} = 0.0384$$

and: $\qquad\qquad\qquad x_A = \underline{0.598}$

The concentration gradient is given by equation (i) or:

$$\frac{dx_A}{dy} = \frac{N'_A(1 + 0.5x_A)}{-DC_T}$$

When x_A is 0.598, then:

$$\frac{dx_A}{dy} = \frac{1.41 \times 10^{-4}(1 + 0.5 \times 0.598)}{-2 \times 10^{-5} \times 0.0229}$$

$$= 400 \text{ m}^{-1} = 0.4 \text{ mm}^{-1}$$

and:
$$\frac{dC_A}{dy} = C_T \frac{dx_A}{dy} = \underline{\underline{9.16 \text{ kmol/m}^4}}.$$

PROBLEM 10.35

For the diffusion of carbon dioxide at atmospheric pressure and a temperature of 293 K, at what time will the concentration of solute 1 mm below the surface reach 1 per cent of the value at the surface? At that time, what will the mass transfer rate (kmol m^{-2}s^{-1}) be:

(a) At the free surface?
(b) At the depth of 1 mm?

The diffusivity of carbon dioxide in water may be taken as 1.5×10^{-9} m^{2}s^{-1}. In the literature, Henry's law constant K for carbon dioxide at 293 K is given as 1.08×10^6 where $K = P/X$, P being the partial pressure of carbon dioxide (mm Hg) and X the corresponding mol fraction in the water.

Solution

$$\frac{\partial C_A}{\partial t} = D \frac{\partial^2 C_A}{\partial y^2}$$

where C_A is concentration of solvent undergoing mass transfer.

The boundary conditions are:

$$\begin{array}{llll} y = 0 \text{ (interface)} & C_A = C_{As} \text{ (solution value)} & t > 0 \\ y = \infty & C_A = 0 \\ t = 0 & C_A = 0 & 0 < y < \infty \end{array}$$

Taking Laplace transforms then:

$$\frac{\overline{\partial C_A}}{\partial t} = \int_0^\infty \left(\frac{\partial C_A}{\partial t} \right) e^{-pt} dt$$

$$= [C_A e^{-pt}]_0^\infty - \int_0^\infty (-pe^{pt}) C_A \, dt = 0 + p\bar{C}_A$$

$$\frac{\overline{\partial^2 C_A}}{\partial y^2} = \frac{\partial^2 \bar{C}_A}{\partial y^2}$$

Thus:
$$p\bar{C}_A = D \frac{\partial^2 \bar{C}_A}{\partial y^2}$$

$$\frac{\partial^2 \bar{C}_A}{\partial y^2} - \frac{p}{D} \bar{C}_A = 0$$

$$\bar{C}_A = A e^{\sqrt{p/D} y} + B e^{-\sqrt{p/D} y}$$

For $t > 0$; when $y = \infty$ $C_A = 0$, $\bar{C}_A = 0$ $\therefore A = 0$.

when $y = 0$ $C_A = C_{As}$ $\bar{C}_{As} = \int_0^\infty C_{As} e^{-pt} \, dt = C_{As} \left[\frac{e^{-pt}}{-p} \right]_0^\infty = \frac{C_{As}}{p}$.

Thus: $\dfrac{C_{As}}{p} = B.1$

and: $\bar{C}_A = \dfrac{C_{As}}{p}\mathrm{e}^{-\sqrt{p/D}y}.$

Inverting: $\dfrac{C_A}{C_{As}} = \mathrm{erfc}\,\dfrac{y}{2\sqrt{Dt}}$ (See Table in Volume 1, Appendix)

Differentiating with respect to y:

$$\frac{1}{C_{As}}\frac{\partial C_A}{\partial y} = \frac{\partial}{\partial y}\left\{\frac{2}{\sqrt{\pi}}\int_{y/2\sqrt{Dt}}^{\infty}\mathrm{e}^{-y^2/4Dt}\,\mathrm{d}\left(\frac{y}{2\sqrt{Dt}}\right)\right.$$

$$= \frac{2}{\sqrt{\pi}}\cdot\frac{1}{2\sqrt{Dt}}(-\mathrm{e}^{-y^2/4Dt}) = -\frac{1}{\sqrt{\pi Dt}}\mathrm{e}^{-y^2/4Dt}$$

The mass transfer rate at t, y, $N = -D\left\{\dfrac{-1}{\sqrt{\pi Dt}}\mathrm{e}^{-y^2/4Dt}\right\}C_{As}$

when: $t > 0$, then: $\left(-D\dfrac{\partial C_A}{\partial y}\right)_{y=y} = C_{As}\sqrt{\dfrac{D}{\pi Dt}}\mathrm{e}^{-y^2/4Dt}$ \hfill (i)

At $t > 0$ and $y = 0$, then: $N_0 = C_{As}\sqrt{\dfrac{D}{\pi t}}$ \hfill (ii)

For a concentrated 1% of surface value at $y = 1$ mm, $C_A/C_{As} = 0.01$ and:

$$0.01 = \mathrm{erfc}\left\{\frac{10^{-3}}{2\sqrt{1.5\times10^{-9}t}}\right\}$$

Writing $\quad\mathrm{erf}\,x = 1 - \mathrm{erfc}\,x$, then:

$0.99 = \mathrm{erf}(12.91t^{-1/2})$

From tables $\quad 1.82 = 12.91t^{-1/2}$

$t = \underline{\underline{50.3\text{ s}}}$

The mass transfer rate at the interface at $t = 50.3$ s is given by equation (ii) as:

$$N_0 = C_{As}\sqrt{\frac{D}{\pi t}} = C_{As}\sqrt{\frac{1.5\times10^{-9}}{\pi\times50.3}}$$

$$= 3.08\times10^{-6}C_{As}\ \text{kmol/m}^2\ \text{s}$$

The mass transfer rate at $y = 1$ mm. and $t = 50.3$ s is given by equation (i) as:

$$N = N_0\mathrm{e}^{-y^2/4Dt} = 3.08\times10^{-6}\mathrm{e}^{-10^{-6}/(4\times1.5\times10^{-9}\times50.3)}C_{As}$$

$$= 1.121\times10^{-7}C_{As}\ \text{where }C_{As}\text{ is in kmol/m}^3.$$

Henry's law constant, $K = 1.08 \times 10^6$, where: $K = P/X$.

Also: $P = 760$ mm Hg

$X =$ Mol fraction in liquid

$X = 760/(1.08 \times 10^6) = 7.037 \times 10^{-4}$ kmol CO_2 kmol solution ($\approx$ per kmol water)

Water molar density $= (1000/18)$ kmol/m^3

and:
$$C_{As} = 7.037 \times 10^{-4} \left(\frac{1000}{18} \right) = 0.0391 \text{ kmol/m}^3$$

When $y = 0$, then: $\quad N_{A0} = \underline{\underline{1.204 \times 10^{-7}}}$ kmol/m^2 s.

When $y = 1$ mm, then: $\quad N_A = \underline{\underline{4.38 \times 10^{-9}}}$ kmol/m^2 s.

PROBLEM 10.36

Experiments are carried out at atmospheric pressure on the absorption into water of ammonia from a mixture of hydrogen and nitrogen, both of which may be taken as insoluble in the water. For a constant mole fraction of 0.05 of ammonia, it is found that the absorption rate is 25 per cent higher when the molar ratio of hydrogen to nitrogen is changed from $1:1$ to $4:1$. Is this result consistent with the assumption of a steady-state gas-film controlled process and, if not, what suggestions have you to make to account for the discrepancy?

Neglect the partial pressure attributable to ammonia in the bulk solution.

Diffusivity of ammonia in hydrogen $= 52 \times 10^{-6}$ m^2/s

Diffusivity of ammonia in nitrogen $= 23 \times 10^{-6}$ m^2/s

Solution

Using Maxwell's Law for mass transfer of **A** in **B** then:

$$\frac{-dP_A}{dy} = FC_A C_B(u_A - u_B) \qquad \text{(equation 10.77)}$$

or:
$$-\mathbf{R}T \cdot \frac{dC_A}{dy} = FC_A C_B(N'_A/C_A - N'_B/C_B)$$

For an absorption process $N'_B = 0$ and:

$$-\frac{dC_A}{dy} = N'_A \frac{FC_B}{\mathbf{R}T} \qquad \text{(i)}$$

From Stefan's Law: $-\dfrac{dC_A}{dy} = N'_A \dfrac{1}{D} \cdot \dfrac{C_B}{C_T}$ (equation 10.30)

Thus: $\qquad D = \dfrac{RT}{FC_T} \quad$ or: $\quad F = \dfrac{RT}{DC_T}$ (ii)

For the three-component system, equation 1 may be written:

$$-\dfrac{dC_A}{dy} = \dfrac{N'_A}{RT}(F_{AB}C_B + F_{AC}C_C)$$

From (2): $\qquad F_{AB} = \dfrac{RT}{D_{AB}C_T} \text{ and } F_{AC} = \dfrac{RT}{D_{AC}C_T}.$

Substituting: $\qquad -\dfrac{dC_A}{dy} = N'_A\left(\dfrac{C_B}{D_{AB}} + \dfrac{C_C}{D_{AC}}\right)\dfrac{1}{C_T}$ (iii)

Stefan's law for 3-components system may be written as:

$$N'_A = -D' \cdot \dfrac{C_T}{C_T - C_A}\dfrac{dC_A}{dy} \text{ where } D' \text{ is the effective diffusivity}$$

or: $\qquad -\dfrac{dC_A}{dy} = N'_A\dfrac{1}{D'} \cdot \dfrac{C_T - C_A}{C_T}$ (iv)

Comparing equations (iii) and (iv):

$$\dfrac{1}{D'} = \dfrac{1}{D_{AB}} \cdot \dfrac{C_B}{C_T - C_A} + \dfrac{1}{D_{AC}} \cdot \dfrac{C_C}{C_T - C_A}$$

$$= \dfrac{x'_B}{D_{AB}} + \dfrac{x'_C}{D_{AC}}$$

where x'_B and x'_C are mole fractions of B, C in the "stationary" gas. Taking A as NH$_3$, B as H$_2$ and C as N$_2$, then:

Case 1

$$x'_B = 0.5 \quad x'_C = 0.5$$

$$\dfrac{1}{D'} = \left(\dfrac{0.5}{52 \times 10^{-6}}\right) + \left(\dfrac{0.5}{23 \times 10^{-6}}\right) = 0.03135 \times 10^6 \text{ s/m}^2$$

and: $\qquad D' = 31.9 \times 10^{-6} \text{ m}^2/\text{s}$

$$x_{BM} = \dfrac{1 - 0.95}{\ln[(1/0.95)]} = \dfrac{0.05}{\ln(0.95)^{-1}} = 0.975, \text{ and } \dfrac{x_T}{x_{Bm}} = 1.026$$

Mass transfer rate, $\qquad N'_A = \dfrac{D'}{L}\Delta C_A\dfrac{x_T}{x_{Bm}} = \dfrac{1}{L}\Delta C_A\dfrac{x_T}{x_{Bm}}D'$

$$= \dfrac{1}{L}\Delta C_A\dfrac{x_T}{x_{Bm}} \cdot 31.9$$

Case 2

$$x'_B = 0.8 \quad x'_C = 0.2$$

$$\dfrac{1}{D'} = \left(\dfrac{0.8}{52 \times 10^{-6}}\right) + \left(\dfrac{0.2}{23 \times 10^{-6}}\right) = 0.0241 \times 10^6 \text{ s/m}^2$$

and: $\qquad D' = 41.5 \times 10^{-6} \text{ m}^2/\text{s}$

x_T/x_{Bm}, L and ΔC_A are as in case 1.

$$\therefore \qquad N_A' = \frac{D'}{L}\Delta C_A \frac{x_T}{x_{Bm}} = \frac{1}{L}\Delta C_A \frac{x_T}{x_{Bm}} \cdot 41.5$$

$$\frac{(N_A')_2}{(N_A')_1} = \frac{41.5}{31.9} = \underline{1.30} \text{ or } 1.3 \text{ times greater in the second case.}$$

The observed ratio of two rates is only 1.25. This may be explained by:

1. Steady-state film conditions do not exist and there is some periodic partial disruption of the film.
   Penetration model $\rightarrow N_A' \propto D^{0.5}c/f. \propto D$ for film model.
   In this problem observed result would be accounted for by:

$$N_A' \propto D'^m.$$

   where $(1.3)^m = 1.25$ or $m = 0.85$.

2. The assumption of a gas-film controlled process may not be valid. If there is a liquid-film resistance, the effect of increasing the gas-film diffusivity will be less than predicted for a gas-film controlled process.
3. The value of the film thickness L is not the same because of different hydrodynamic conditions (second mixture having a lower viscosity).
   In this case, the film thickness would be expected to be reduced giving rise to the reverse effect so this is not a plausible explanation.
4. Experimental inaccuracies!

PROBLEM 10.37

Using a steady-state film model, obtain an expression for the mass transfer rate across a laminar film of thickness L in the vapour phase for the more volatile component in a binary distillation process:

(a) where the molar latent heats of two components are equal.
(b) where the molar latent heat of the less volatile component (LVC) is f times that of the more volatile component (MVC).

For the case where the ratio of the molar latent heats f is 1.5. what is the ratio of the mass transfer rate in case (b) to that in case (a) when the mole fraction of the MVC falls from 0.75 to 0.65 across the laminar film?

Solution

Case (a): With equal molar latent heats, equimolecular counter diffusion takes place and there is no bulk flow.

Writing Fick's Law for the MVC gives:

$$N_A = -D\frac{dC_A}{dy} \qquad\qquad \text{(equation 10.4)}$$

Integrating for the steady state, then:

$$N_A = D\frac{C_{A_1} - C_{A_2}}{y_2 - y_1} = DC_T\frac{(x_{A_1} - x_{A_2})}{L} \qquad \text{(i)}$$

Case (b): The net heat effect at the interface must be zero and hence:

$$N'_A \lambda_A + N'_B \lambda_B = 0$$

and:

$$N'_B = -\frac{\lambda_A}{\lambda_B}N'_A = -\frac{1}{f}N'_A$$

In this case:

Total flux of **A** = Diffusional flux + Bulk-flow

or:

$$N'_A = -D\frac{dC_A}{dy} + u_F C_A = -DC_T\frac{dx_A}{dy} + x_A\left(N'_A - \frac{1}{f}N'_A\right)$$

Thus: $N'_A\left[1 - x_A\left(1 - \frac{1}{f}\right)\right] = -DC_T\frac{dx_A}{dy}$

$$N'_A\int_0^L dy = -DC_T\int_{C_{A_1}}^{C_{A_2}}\frac{1}{1 - x_A\left(1 - \frac{1}{f}\right)}dx_A$$

$$\therefore \qquad N'_A L = -\frac{DC_T}{\left(1 - \frac{1}{f}\right)}\left[\ln\left[1 - x_A\left(1 - \frac{1}{f}\right)\right]\right]_{x_{A_1}}^{x_{A_2}}$$

$$= \frac{DC_T}{\left(1 - \frac{1}{f}\right)L}\ln\frac{1 - x_{A_2}\left(1 - \frac{1}{f}\right)}{1 - x_{A_1}\left(1 - \frac{1}{f}\right)} \qquad \text{(ii)}$$

From equations (i) and (ii):

$$R = \frac{(N'_A)\text{ case b}}{(N_A)\text{ case a}} = \frac{1}{1 - \frac{1}{f}}\frac{1}{x_{A_1} - x_{A_2}}\ln\frac{1 - x_{A_2}\left(1 - \frac{1}{f}\right)}{1 - x_{A_1}\left(1 - \frac{1}{f}\right)}$$

Substituting $f = 1.5$ then:

$$1 - \frac{1}{f} = \frac{1}{3}$$

$$x_{A_1} = 0.75 \qquad x_{A_2} = 0.65$$

$$R = \frac{1}{\frac{1}{3}}\cdot\frac{1}{0.75 - 0.65}\ln\frac{1 - 0.65 \times \frac{1}{3}}{1 - 0.75 \times \frac{1}{3}} = \underline{\underline{1.303}}$$

For **B**: $\qquad R = (1.303/1.5) = \underline{\underline{0.869}}.$

PROBLEM 10.38

Based on the assumptions involved in the penetration theory of mass transfer across a phase boundary, the concentration C_A of a solute A at a depth y below the interface at a time t after the formation of the interface is given by:

$$\frac{C_A}{C_{Ai}} = \text{erfc}\left[\frac{y}{2\sqrt{(Dt)}}\right]$$

where C_{Ai} is the interface concentration, assumed constant and D is the molecular diffusivity of the solute in the solvent. The solvent initially contains no dissolved solute. Obtain an expression for the molar rate of transfer of A per unit area at time t and depth y, and at the free surface ($y = 0$).

In a liquid-liquid extraction unit, spherical drops of solvent of uniform size are continuously fed to a continuous phase of lower density which is flowing vertically upwards, and hence countercurrently with respect to the droplets. The resistance to mass transfer may be regarded as lying wholly within the drops and the penetration theory may be applied. The upward velocity of the liquid, which may be taken as uniform over the cross-section of the vessel, is one-half of the terminal falling velocity of the droplets in the still liquid.

Occasionally, two droplets coalesce forming a single drop of twice the volume. What is the ratio of the mass transfer rate (kmol/s) at a coalesced drop to that at a single droplet when each has fallen the same distance, that is to the bottom of the column?

The fluid resistance force acting on the droplet should be taken as that given by Stokes' law, that is $3\pi\mu du$ where μ is the viscosity of the continuous phase, d the drop diameter and u its velocity relative to the continuous phase.

It may be noted that:

$$\text{erfc}(x) = \frac{2}{\sqrt{\pi}} \int_x^\infty e^{x^2} \, dx.$$

Solution

$$\frac{C_A}{C_{Ai}} = \text{erfc} \frac{y}{2\sqrt{Dt}} = \frac{2}{\sqrt{\pi}} \int_{y/(2\sqrt{Dt})}^\infty e^{-y^2/4Dt} \, d\left(\frac{y}{2\sqrt{Dt}}\right)$$

Differentiating with respect to y at constant t gives:

$$\frac{1}{C_{Ai}} \frac{\partial C_A}{\partial y} = \frac{2}{\sqrt{\pi}} \frac{\partial}{\partial y} \left\{ \frac{1}{2\sqrt{Dt}} \int_{y/(2\sqrt{Dt})}^\infty e^{-y^2/4Dt} \, dt \right\}$$

$$= \frac{1}{\sqrt{\pi Dt}} (-e^{-y^2/4Dt})$$

Thus: $(N_A)_t = -D\dfrac{\partial C_A}{\partial y} = C_{Ai}(-D)\sqrt{\dfrac{1}{\pi Dt}}(-e^{-y^2/4Dt}) = C_{Ai}\sqrt{\dfrac{D}{\pi t}} e^{-y^2/4Dt}$

At the interface, when $y = 0$: $N_A = C_{Ai}\sqrt{\dfrac{D}{\pi t}} = bt^{-1/2}$

From Stokes' Law, the terminal falling velocity of the droplet is given by:

$$3\pi\mu d u_0 = \frac{\pi}{6}d^3(\rho_s - \rho)g$$

or:

$$u_0 = \frac{d^2 g}{18\mu}(\rho_s - \rho) = Kd^2$$

Thus, the time taken for the droplet to travel the depth H of the rising liquid is:

$$\frac{H}{\frac{1}{2}Kd^2}$$

Since the liquid is rising at a velocity of $\frac{1}{2}Kd^2$ and the relative velocity is $Kd^2 - \frac{1}{2}Kd^2 = \frac{1}{2}Kd^2$, the mass transfer rate (kmol/m²s) to droplet at end of travel is:

$$b\sqrt{\frac{K}{2H}}d$$

The mass transfer rate to the drop is:

$$b\sqrt{\frac{K}{2H}}d(\pi d)^2 = \pi b\sqrt{\frac{K}{2H}}d^3 = \pi b\sqrt{\frac{K}{H} - \frac{1}{\sqrt{2}}}d^3$$

For *coalesced* drops, the new diameter is: $2^{1/3}d$

The terminal falling velocity is: $K2^{2/3}d^2$

Its velocity relative to the liquid is: $K2^{2/3}d^2 - \frac{1}{2}Kd^2 = Kd^2(2^{2/3} - \frac{1}{2})$

Thus:

$$\text{Time of fall of drop} = \frac{H}{Kd^2(2^{2/3} - \frac{1}{2})}$$

$$\text{Mass transfer rate at end of travel} = bd\sqrt{\frac{K}{H}}d\sqrt{(2^{2/3} - \frac{1}{2})} \text{ kmol/m}^2\text{s}$$

$$\text{Mass transfer rate to drop} = bd\sqrt{\frac{K}{H}}d\sqrt{(2^{2/3} - \frac{1}{2})}(\pi 2^{1/3}d)^2 \text{ kmol/s}$$

$$= \pi b\sqrt{\frac{K}{H}}\sqrt{(2^{2/3} - \frac{1}{2})}2^{2/3}d^3 \text{ kmol/s}$$

The ratio of the mass transfer rate for the coalesced drop to the mass transfer rate for the single droplet is then:

$$= \frac{\sqrt{(2^{2/3} - \frac{1}{2})2^{2/3}}}{1/\sqrt{2}} = \underline{\underline{2.34}}$$

PROBLEM 10.39

In a drop extractor, a dense organic solvent is introduced in the form of spherical droplets of diameter d and extracts a solute from an aqueous stream which flows upwards at a velocity equal to half the terminal falling velocity u_0 of the droplets. On increasing the flowrate of the aqueous stream by 50 per cent, whilst maintaining the solvent rate constant, it is found that the average concentration of solute in the outlet stream of organic phase is decreased by 10 per cent. By how much would the effective droplet size have had to change to account for this reduction in concentration? Assume that the penetration theory is applicable with the mass transfer coefficient inversely proportional to the square root of the contact time between the phases and that the continuous phase resistance is small compared with that within the droplets. The drag force F acting on the falling droplets may be calculated from Stokes' Law, $F = 3\pi\mu d u_0$, where μ is the viscosity of the aqueous phase. Clearly state any assumptions made in your calculation.

Solution

For droplets in the Stokes' law region, the terminal falling velocity is given by:
$$(\pi/6)d^3(\rho_s - \rho)g = 3\pi\mu d u_0$$

or:
$$u_0 = \frac{d^2 g}{18\mu}(\rho_s - \rho) = kd^2$$

The mass transfer rate to the droplet is $Kt_e^{-1/2}$ moles per unit area per unit time

$$\propto Kt_e^{1/2} \text{ moles per unit area in time } t_e$$

$$\propto Kt_e^{1/2}d^2 \text{ moles per drop during time of rise}$$

The concentration of solute in drop: $\propto Kt_e^{1/2}d^2/d^3 \propto Kt_e^{1/2}d^{-1}$.

Initial case: $u_0 = kd^2$

$$\text{Rising velocity of liquid} = \frac{kd^2}{2}$$

$$\text{Velocity of liquid relative to container} = kd^2 - \frac{kd^2}{2} = \frac{kd^2}{2}$$

$$\text{Time of exposure in height } H = \frac{H}{kd^2/2} = t_e.$$

Thus the concentration in the drop $\propto K\sqrt{\dfrac{2H}{kd^2}}d^{-1} \propto K\sqrt{\dfrac{2H}{k}}d^{-2} = C_1$

Second case: New drop diameter $= d'$

$$\text{Rising velocity of liquid} = 1.5 \times \frac{kd^2}{2} = \frac{3}{4}kd^2$$

$$\text{Rising velocity of drop relative to liquid} = kd'^2$$

$$\text{Velocity relative to container} = kd'^2 - \tfrac{3}{4}kd^2$$

$$\text{Time of exposure} = \frac{H}{k(d'^2 - \tfrac{3}{4}d^2)}$$

Concentration of solute in drop $= K t_e^{1/2} (d')^{-1}$

$\therefore$ $$C_2 = K \left\{ \frac{H}{k(d'2 - \frac{3}{4}d^2)} \right\}^{1/2} (d')^{-1}$$

Given that: $C_2/C_1 = 0.9$, then:

$$\frac{K \left\{ \dfrac{H}{k(d'^2 - \frac{3}{4}d^2)} \right\}^{1/2} (d'^{-1})}{K \sqrt{\dfrac{2H}{k}} d^{-2}} = 0.9$$

Squaring gives: $$\frac{d'^{-2}}{d'^2 - \frac{3}{4}d^2} = 1.62 d^{-4}$$

Writing $R = \dfrac{d'}{d}$, then: $$\frac{1}{R^4 - \frac{3}{4}R^2} = 1.62$$

$$R^4 - \tfrac{3}{4}R^2 - 0.6173 = 0$$

$$R = \underline{\underline{1.11 \text{ or } 11.1 \text{ per cent increase}}}$$

PROBLEM 10.40

According to the penetration theory for mass transfer across an interface, the ratio of the concentration C_A at a depth y and time t to the surface concentration C_{As} at the liquid is initially free of solute, is given by

$$\frac{C_A}{C_{As}} = \text{erfc} \frac{y}{2\sqrt{Dt}}$$

where D is the diffusivity. Obtain a relation for the instantaneous rate of mass transfer at time t both at the surface $(y = 0)$ and at a depth y.

What proportion of the total solute transferred into the liquid in the first 90 s of exposure will be retained in a 1 mm layer of liquid at the surface, and what proportion will be retained in the next 0.5 mm? The diffusivity is 2×10^{-9} m^2/s.

Solution

For a rectangular particle:

Thiele Modulus $= \phi = \lambda L$

Thus: $$\lambda = \sqrt{\frac{k}{D}} = \sqrt{\frac{5 \times 10^{-4}}{2 \times 10^{-9}}} = 500 \text{ m}^{-1}$$

$$L = \frac{8}{2} = 4 \text{ mm or } 0.004 \text{ m}$$

$$\phi = (500 \times 0.004) = 2$$

Thus the effectiveness factor, $\eta = \dfrac{\tanh \phi}{\phi} = \dfrac{\tanh 2}{2} = \dfrac{0.96}{2} = \underline{\underline{0.48}}$ (equation 10.202)

For a spherical particle:

$$\text{Thiele Modulus} = \phi = \lambda R$$

$$R = \frac{\text{Volume}}{\text{Surface}} = \frac{\frac{4}{3}\pi r_o^3}{4\pi r_o^2} = \frac{r_o}{3}$$

$$R = \frac{5 \times 10^{-3}}{3} \text{m}$$

$$\phi = \left(\underset{(\lambda)}{500} \times \underset{(R)}{\frac{5 \times 10^{-3}}{3}} \right) = 0.833$$

Thus the effectiveness factor, $\eta = \dfrac{1}{\phi} \coth 3\phi - \dfrac{1}{3\phi^2}$ (equation 10.215)

$$= \frac{1}{0.833} \coth 2.5 - \frac{1}{3 \times 0.833^2}$$

$$= (1.217 - 0.480) = \underline{\underline{0.736}}$$

PROBLEM 10.41

Obtain an expression for the effective diffusivity of component **A** in a gaseous mixture of **A**, **B** and **C** in terms of the binary diffusion coefficients D_{AB} for **A** in **B**, and D_{AC} for **A** in **C**.

The gas-phase mass transfer coefficient for the absorption of ammonia into water from a mixture of composition NH_3 20%, N_2 73%. H_2 7% is found experimentally to be 0.030 m/s. What would you expect the transfer coefficient to be for a mixture of composition NH_3 5%, N_2 60%, H_2 35%? All compositions are given on a molar basis. The total pressure and temperature are the same in both cases. The transfer coefficients are based on a steady-state film model and the effective film thickness may be assumed constant. Neglect the solubility of N_2 and H_2 in water.

Diffusivity of NH_3 in $N_2 = 23 \times 10^{-6}$ m²/s.

Diffusivity of NH_3 in $H_2 = 52 \times 10^{-6}$ m²/s.

Solution

For case 1:

The effective diffusivity D' is given by: $\dfrac{1}{D'} = \left(\dfrac{73/80}{23 \times 10^{-6}} \right) + \left(\dfrac{7/80}{52 \times 10^{-6}} \right)$

                                                         (from equation 10.90)

or: $\qquad\qquad D' = 24.2 \times 10^{-6} \text{ m}^2/\text{s}$

The mass transfer coefficient is: $\dfrac{D'}{L} \left(\dfrac{1}{\log \text{ mean 1 and } 0.8} \right)$

$$= \frac{24.2 \times 10^{-6}}{L} \left(\frac{0.2}{\ln \frac{1}{0.8}} \right)^{-1} = \frac{1}{L} \times 27 \times 10^{-6} = \underline{\underline{0.030 \text{ m/s}}}$$

and hence: $\quad L = 0.90 \times 10^{-3}$ m

Case 2:

The effective diffusivity D' is given by: $\dfrac{1}{D'} = \left(\dfrac{60/95}{23 \times 10^{-6}} \right) + \left(\dfrac{35/95}{52 \times 10^{-6}} \right)$

or: $\qquad\qquad D' = 28.9 \times 10^{-6} \text{ m}^2/\text{s}$

The mean transfer coefficient is: $\dfrac{D'}{L} \cdot \left(\dfrac{1}{\log \text{ mean 1 and } 0.95} \right)$

$$= \left(\frac{28.9 \times 10^{-6}}{0.9 \times 10^{-3}} \right) \left(\frac{0.05}{\ln \frac{1}{0.95}} \right)^{-1} = \underline{\underline{0.033 \text{ m/s}}}.$$

PROBLEM 10.42

State the assumptions made in the penetration theory for the absorption of a pure gas into a liquid. The surface of an initially solute-free liquid is suddenly exposed to a soluble gas and the liquid is sufficiently deep for no solute to have time to reach the far boundary of the liquid. Starting with Fick's second law of diffusion, obtain an expression for (i) the concentration, and (ii) the mass transfer rate at a time t and a depth y below the surface.

After 50 s, at what depth y will the concentration have reached one tenth the value at the surface? What is the mass transfer rate (i) at the surface, and (ii) at the depth y, if the surface concentration has a constant value of 0.1 kmol/m^3?

Solution

$$\frac{C_A}{C_{AS}} = \text{erfc} \, \frac{y}{2\sqrt{Dt}} \qquad\qquad \text{(equation 10.108)}$$

Differentiating with respect to y:

$$\frac{1}{C_{AS}} \frac{\partial C_A}{\partial y} = \frac{\partial}{dy} \left\{ \frac{2}{\sqrt{\pi}} \int_{y/2\sqrt{Dt}}^{\infty} e^{-y^2/4Dt} \, d \left(\frac{y}{2\sqrt{Dt}} \right) \right.$$

$$= -\frac{1}{\sqrt{\pi Dt}} e^{-y^2/4Dt}$$

Thus: $\qquad (N_A)_{y,t} = -D \left\{ -\dfrac{1}{\sqrt{\pi Dt}} e^{-y^2/4Dt} \right\} C_{AS} = \left(\sqrt{\dfrac{D}{\pi t}} e^{-y^2/4Dt} \right) C_{AS}$

When $t = 50$ s and $C_A/C_{AS} = 0.1$, then:

$$0.1 = \text{erfc}\ \frac{y}{2\sqrt{10^{-9} \times 50}}$$

or:

$$0.9 = \text{erf}\ \frac{y}{2\sqrt{Dt}}$$

From Table 13 in the Appendix of Volume 1, the quantity whose error fraction $= 0.9$ is:

$$\frac{y}{2\sqrt{Dt}} = 1.16$$

At the surface: $(N_A)_{y=0,t} = \sqrt{\dfrac{D}{\pi t}} C_{AS}$

$$= \sqrt{\frac{10^{-9}}{\pi \times 50}} \times 0.1$$

$$= 0.252 \times 10^{-6} \text{ kmol/m}^2 \text{ s.}$$

At a depth y: $N_A = 0.252 \times 10^{-6} \times e^{-(1.16)^2} = \underline{\underline{0.0656 \times 10^{-6} \text{ kmol/m}^2\text{s.}}}$

PROBLEM 10.43

In a drop extractor, liquid droplets of approximately uniform size and spherical shape are formed at a series of nozzles and rise countercurrently through the continuous phase which is flowing downwards at a velocity equal to one half of the terminal rising velocity of the droplets. The flowrates of both phases are then increased by 25 per cent. Because of the greater shear rate at the nozzles, the mean diameter of the droplets is, however, only 90 per cent of the original value. By what factor will the overall mass transfer rate change?

It may be assumed that the penetration model may be used to represent the mass transfer process. The depth of penetration is small compared with the radius of the droplets and the effects of surface curvature may be neglected. From the penetration theory, the concentration C_A at a depth y below the surface at time t is given by:

$$\frac{C_A}{C_{AS}} = \text{erfc}\left[\frac{y}{2\sqrt{(Dt)}}\right] \quad \text{where erfc } X = \frac{2}{\sqrt{\pi}}\int_X^\infty e^{-x^2}\, dx$$

where C_{AS} is the surface concentration for the drops (assumed constant) and D is the diffusivity in the dispersed (droplet) phase. The droplets may be assumed to rise at their terminal velocities and the drag force F on the droplet may be calculated from Stokes' Law, $F = 3\pi\mu\, du$.

Solution

Case 1: For a volumetric flowrate Q_1, the numbers of drops per unit time is:

$$\frac{Q_1}{(\frac{1}{6})\pi d^3} = \frac{6Q_1}{\pi d_1^3}$$

The rising velocity is given by a force balance:

$$3\pi\mu d_1 u = \tfrac{1}{6}\pi d_1^3(\rho_1 - \rho_2)g$$

or:
$$u_1 = \frac{d_1^2 g}{18\mu}(\rho_1 - \rho_2) = Kd_1^2 \text{ relative to continuous phase}$$

The downward liquid velocity is $\tfrac{1}{2}Kd_1^2$

The upward droplet velocity relative to container is:

$$Kd_1^2 - \tfrac{1}{2}Kd_1^2 = \tfrac{1}{2}Kd_1^2$$

and the time of contact during rise through height H is:

$$t_c = \frac{H}{\tfrac{1}{2}Kd_1^2}. \tag{i}$$

The mass transfer rate is: $-D(\partial C_A/\partial y)$.

Thus:
$$\frac{1}{C_{AS}}\frac{\partial C_A}{\partial y} = \frac{\partial}{\partial y}\left\{\text{erfc}\frac{y}{2\sqrt{Dt}}\right\} = \frac{\partial}{\partial y}\left\{\frac{2}{\sqrt{\pi}}\int_{y/2\sqrt{Dt}}^{\infty} e^{-y^2/4Dt}\,d\left(\frac{y}{2\sqrt{Dt}}\right)\right\}$$

$$= \frac{\partial}{\partial y}\cdot\frac{1}{2\sqrt{Dt}} - \frac{2}{\sqrt{\pi}}\int_y^{\infty} e^{-y^2/4Dt}\,dt.$$

or:
$$\frac{\partial C_A}{\partial y} = -\frac{C_{AS}}{\sqrt{\pi Dt}}e^{-y^2/4Dt}\quad \left(\frac{\partial C_A}{\partial y}\right)_{y=0} = -\frac{C_{AS}}{\sqrt{\pi Dt}}$$

The mass transfer rate at the surface is: (moles/area $\times$ time).

$$-D\left(-\frac{C_{AS}}{\sqrt{\pi Dt}}\right) = \sqrt{\frac{D}{\pi t}}C_{AS}$$

The mass transfer in time t_{e1} is:

$$\sqrt{\frac{D}{\pi}}C_{AS}\int_0^{t_{e1}} t^{-1/2}\,dt = 2\sqrt{\frac{D}{\pi}}t_{e1}^{1/2}C_{AS} = Kt_{e1}^{1/2}$$

Substituting from equation (i):

$$\text{Mass transfer in moles per unit area of drop} \propto \frac{\sqrt{2H}}{\sqrt{Kd_1}}$$

The mass transfer per drop is proportional to:

$$\sqrt{2}\sqrt{\frac{H}{K}}d_1^{-1}d_1^2 \propto \sqrt{2}\sqrt{\frac{H}{K}}d_1$$

The mass transfer per unit time = Mass transfer per drop $\times$ drops/time

or:
$$\text{proportional to: } \sqrt{2}\sqrt{\frac{H}{K}}d_1 \times \frac{6Q_1}{\pi d_1^3} \propto 8.48\sqrt{\frac{H}{K}\frac{Q_1}{\pi d_1^2}}$$

Case 2: diameter $= 0.9d = d_2$, $Q_2 = 1.25Q_1$

The number of drops per unit time is:

$$\frac{6Q_2}{\pi d_2^3} = \frac{7.5Q_1}{\pi(0.9d_1)^3} = \frac{10.29Q_1}{\pi d_1^3}$$

Rising velocity $= Kd_1^2 = K(0.9d_1)^2 = 0.81Kd_1^2$

Downward liquid velocity $= \frac{5}{4} \times (\frac{1}{2}Kd_1^2) = 0.625\ Kd_1^2$ relative to continuous phase

Rising velocity relative to container $= 0.81\ Kd_1^2 - 0.625\ Kd_1^2 = 0.185Kd_1^2$

Contact time, $t_{e_2} = \dfrac{H}{0.185Kd_1^2}$

Mass transfer in time t_{e_2} per unit area $\propto kt_{e_2}^{1/2} \propto \sqrt{\dfrac{H}{0.185K}}\dfrac{1}{d_1} \propto 2.325\sqrt{\dfrac{H}{K}}d_1^{-1}$

Mass transfer per drop $\qquad\propto 2.325\sqrt{\dfrac{H}{K}}d_1^{-1} \times (0.9d_1)^2 \propto 1.883\sqrt{\dfrac{H}{K}}d_1$

Mass transfer per unit time $\qquad\propto 1.883\sqrt{\dfrac{H}{K}}d_1 \times \dfrac{10.29Q_1}{\pi d_1^3}$

$$\propto 19.37\sqrt{\dfrac{H}{K}}\dfrac{Q_1}{\pi d_1^2}$$

Thus, the factor by which mass transfer rate is increased is: $(19.37/8.48) = \underline{\underline{2.28}}$

PROBLEM 10.44

According to Maxwell's law, the partial pressure gradient in a gas which is diffusing in a two-component mixture is proportional to the product of the molar concentrations of the two components multiplied by its mass transfer velocity relative to that of the second component. Show how this relationship can be adapted to apply to the absorption of a soluble gas from a multicomponent mixture in which the other gases are insoluble, and obtain an effective diffusivity for the multicomponent system in terms of the binary diffusion coefficients.

Carbon dioxide is absorbed in alkaline water from a mixture consisting of 30% CO_2 and 70% N_2, and the mass transfer rate is 0.1 kmol/s. The concentration of CO_2 in the gas in contact with the water is effectively zero. The gas is then mixed with an equal molar quantity of a second gas stream of molar composition 20% CO_2, 50%, N_2 and 30% H_2. What will be the new mass transfer rate, if the surface area, temperature and pressure remain unchanged? It may be assumed that a steady-state film model is applicable and that the film thickness is unchanged.

Diffusivity of CO_2 in $N_2 = 16 \times 10^{-6}$ m^2/s.

Diffusivity of CO_2 in $H_2 = 35 \times 10^{-6}$ m^2/s.

Solution

For a binary system, Maxwell's Law gives:

$$-\frac{dC_A}{dy} = FC_AC_B(u_A - u_B) \qquad \text{(equation 10.77)}$$

For an ideal gas mixture:

$$-\mathbf{R}T\frac{dC_A}{dy} = FC_AC_B\left(\frac{N'_A}{C_A} - \frac{N'_B}{C_B}\right)$$

For a gas absorption process: $N'_B = 0$ if **B** is insoluble

or:

$$-\frac{dC_A}{dy} = \frac{FC_B}{\mathbf{R}T}N'_A \qquad \text{(i)}$$

from Stefan's law (equation 10.30):

$$N'_A = -D\frac{C_T}{C_B}\frac{dC_A}{dy}$$

or:

$$-\frac{dC_A}{dy} = N'_A\frac{1}{D}\frac{C_B}{C_T} \qquad \text{(ii)}$$

$$= N'_A\frac{1}{D}\frac{C_T - C_A}{C_T} \qquad \text{(iii)}$$

Comparing equations (i) and (ii):

$$\frac{F}{\mathbf{R}T} = \frac{1}{DC_T}$$

or:

$$F = \frac{\mathbf{R}T}{DC_T} \text{ or } D = \frac{\mathbf{R}T}{FC_T}$$

Applying Maxwell's law to a multicomponent system gives:

$$-\frac{dC_A}{dy} = F_{AB}C_AC_B(u_A - u_B) + F_{AC}C_AC_C(u_A - u_C) + \cdots$$

For **B, C** ... insoluble $N'_B, N'_C \ldots = 0$

Writing:

$$F_{AB} = \frac{\mathbf{R}T}{D_{AB}C_T}, \text{ and } F_{AC} = \frac{\mathbf{R}T}{D_{AC}C_T}$$

then:

$$-\mathbf{R}T\frac{dC_A}{dy} = \frac{\mathbf{R}T}{D_{AC}C_T}N'_AC_B + \frac{\mathbf{R}T}{D_{AC}C_T}N'_AC_C + \cdots.$$

$$\therefore \qquad -\frac{dC_A}{dy} = \frac{N'_A}{C_T}\left(\frac{C_B}{D_{AB}} + \frac{C_C}{D_{AC}} + \cdots\right)$$

From equation (iii), using an effective diffusivity D' for a multicomponent system gives:

$$-\frac{dC_A}{dy} = N'_A\frac{1}{D'}\frac{C_T - C_A}{C_T} \qquad \text{(iv)}$$

Comparing equations (iii) and (iv), then:

$$\frac{1}{D'} = \frac{1}{C_T - C_A} \left\{ \frac{C_B}{D_{AB}} + \frac{C_C}{D_{AC}} + \dots \right\}$$

$$= \frac{x'_B}{D_{AB}} + \frac{x'_C}{D_{AC}} + \dots$$

where $x'_B, x'_C \dots = \dfrac{C_B}{C_T - C_A} > \dfrac{C_C}{C_T - C_A} \dots$

= mole fraction of **B, C** ... in mixture of **B, C**

For absorption of CO_2 from mixture with N_2, the concentration driving force is:

$$\Delta C_A = C_T(0.3 - 0) = 0.3C_T$$

where C_T is the total molar concentration

The mass transfer rate for an area $A = AN'_A = \Delta C_A \dfrac{D}{L} \cdot \dfrac{C_T}{C_{Bm}} \cdot A$

where: C_{Bm} is the log mean of $1 \times C_T$ and $0.7C_T$

or: $C_{Bm} = \dfrac{C_T - 0.7C_T}{\ln \dfrac{C_T}{0.7C_T}} = 0.841C_T$ and: $\dfrac{C_T}{C_{Bm}} = 1.189$

Mass transfer rate, $AN'_A = 0.3C_T \dfrac{16 \times 10^{-6}}{L} 1.189 \, A = 5.71 \times 10^{-6} \dfrac{C_T A}{L} = 0.10$ kmol/s

or: $\dfrac{C_T A}{L} = 0.0175 \times 10^6$ kmol/m^2.

For absorption of CO_2 from mixed stream, stream composition must be calculated.

100 moles stream 1 $\rightarrow$ 30 moles CO_2 70 moles N_2

100 moles stream 2 $\rightarrow$ 20 moles CO_2 50 moles N_2 30 moles H_2

200 moles mixture $\rightarrow$ 50 moles CO_2 120 moles N_2 30 moles H_2

100 moles mixture $\rightarrow$ 25 moles CO_2 60 moles N_2 15 moles H_2

$$x'_{N_2} = \frac{60}{75} = 0.8 \quad x'_{H_2} = \frac{15}{75} = 0.2$$

Diffusivity of CO_2 in mixture is given by:

$$\frac{1}{D'} = \frac{0.8}{16 \times 10^{-6}} + \frac{0.2}{35 \times 10^{-6}}$$

$$\frac{10^{-6}}{D'} = \frac{1}{20} + \frac{1}{175} = 0.0557 \text{ s/m}^2$$

$$D' = 17.9 \times 10^{-6} \text{ m}^2/\text{s}$$

Thus:
$$\Delta C_A = C_T(0.25 - 0) = 0.25C_T$$

$$C_{Bm} = \frac{1 - 0.75}{\ln \frac{1}{0.75}} C_T = 0.869C_T \frac{C_T}{C_{Bm}} = 1.150$$

and the mass transfer rate:

$$AN_{A'} = 0.25C_T \frac{17.9 \times 10^{-6}}{L} 1.150A = 5.15 \times 10^{-6} \frac{AC_T}{L} \text{ kmol/s} = \underline{\underline{0.090 \text{ kmol/s}}}$$

PROBLEM 10.45

What is the penetration theory for mass transfer across a phase boundary? Give details of the underlying assumptions.

From the penetration theory, the mass transfer rate per unit area N_A is given in terms of the concentration difference ΔC_A between the interface and the bulk fluid, the molecular diffusivity D and the age t of the surface element by:

$$N_A = \sqrt{\frac{D}{\pi t}} \Delta C_A \quad \text{kmol/m}^2\text{s (in SI units)}$$

What is the mean rate of transfer if all elements of the surface are exposed for the same time t_e before being remixed with the bulk?

Danckwerts assumed a random surface renewal process in which the probability of surface renewal is independent of its age. If s is the fraction of the total surface renewed per unit time, obtain the age distribution function for the surface and show that the mean mass transfer rate N_A over the whole surface is:

$$N_A = \sqrt{Ds}\Delta C_A \quad \text{(kmol/m}^2\text{s, in SI units)}$$

In a particular application, it is found that the older surface is renewed more rapidly than the recently formed surface, and that after a time s^{-1}, the surface renewal rate doubles, that is it increases from s to $2s$. Obtain the new age distribution function.

Solution

Assuming the age spread of the surface ranges for $t = 0$, to $t = \infty$, consider the mass transfer per unit area in each age group is t to $t + dt$ and so on.

Then the mass transfer to surface in age group t to $t + dt$ is:

$$= \sqrt{\frac{D}{\pi}} \Delta C_A t^{-1/2} \, dt$$

Thus the total mass transfer per unit area is:

$$\sqrt{\frac{D}{\pi}} \Delta C_A \int_0^{t_e} t^{-1/2} \, dt = \sqrt{\frac{D}{\pi}} \Delta C_A \left[\frac{t^{1/2}}{\frac{1}{2}} \right]_0^{t_e} = 2\sqrt{\frac{D}{\pi}} \Delta C_A t_e^{1/2}$$

The average mass transfer rate is:

$$\frac{1}{t_e}\left\{2\sqrt{\frac{D}{\pi}}\Delta C_A t_e^{1/2}\right\} = 2\sqrt{\frac{D}{\pi t_e}}\Delta C_A$$

In the steady state if $f(t)$ is the age distraction function of the surface, then:

surface in age group t to $t + dt = f(t)\,dt$

and: surface in age group $t - dt$ to $t = f(t - dt)\,dt$

Surface of age $t - dt$ to t which is destroyed is that not entering the next age group

$$f(t - dt)\,dt - f(t)\,dt = s\{f(t - dt)\,dt\}\,dt$$

As $dt \rightarrow 0$, then: $-f'(t)\,dt\,dt = sf(t)\,dt\,dt$

$$e^{st}f'(t) + e^{st}sf(t) = 0$$

Integrating gives: $e^{st}f(t) = K$

then: $f(t) = Ke^{-st}$ (i)

As the total surface $= 1$, $K\int_0^\infty e^{-st}\,dt = K\left[\dfrac{e^{-st}}{-s}\right]_0^\infty = \dfrac{K}{s}$ $\therefore K = s$

and: $f(t) = se^{-st}$ (ii)

The mass transfer rate into the fraction of surface in the age group $t - t\,dt$ is:

$$\sqrt{\frac{D}{\pi t}}\Delta C_A se^{-st}\,dt$$

The mass transfer rate into the surface over the age span $t = 0$ to $t = \infty$ is:

$$= \sqrt{\frac{D}{\pi}}\Delta C_A s \int_0^\infty t^{-1/2}e^{-st}\,dt = \sqrt{\frac{D}{\pi}}\Delta C_A$$

Putting $= st = \beta^2$, then: $t^{-1/2} = \dfrac{\sqrt{s}}{\beta}$

and: $s\,dt = 2\beta\,d\beta$.

$$I = \int_0^\infty \frac{\sqrt{s}}{\beta}e^{-\beta^2}\frac{2\beta\,d\beta}{s} = \frac{2}{\sqrt{s}}\int_0^\infty e^{-\beta^2}\,d\beta = \frac{2}{\sqrt{s}}\cdot\frac{\sqrt{\pi}}{2} = \sqrt{\frac{\pi}{s}}$$

Thus the mass transfer rate per unit area for the surface as a whole is:

$$\sqrt{\frac{D}{\pi}}\Delta C_A s\sqrt{\frac{\pi}{s}} = \sqrt{Ds}\Delta C_A$$

With the new age distribution function:

$$\left.\begin{array}{l}0 < t < \dfrac{1}{s}, \quad \text{surface renewal rate/area} = s \quad f(t) = Ke^{-st} \\[2mm] \dfrac{1}{s} < t < \infty, \quad \text{surface renewal rate/area} = 2s \quad f(t) = K'e^{-2st}\end{array}\right\} \text{ from equation (ii)}$$

Fraction of surface of age 0 to $\dfrac{1}{s} = K \displaystyle\int_0^{1/s} e^{-st} \, dt = K \left[\dfrac{e^{-st}}{-s}\right]_0^{1/s} = \dfrac{K}{s}(1 - e^{-1})$

Fraction of surface of age $\dfrac{1}{s}$ to $\infty = K' \displaystyle\int_{1/s}^{\infty} e^{-2st} \, dt = K' \left[\dfrac{e^{-2st}}{-2s}\right]_{1/s}^{\infty} = \dfrac{K'}{2s}e^{-2}$

The total surface is unity or: $\quad \dfrac{K}{s}(1 - e^{-1}) + \dfrac{K'}{2s}e^{-2}$

At $t = (1/s)$, both age distribution functions must apply, and:

$$Ke^{-1} = K'e^{-2} \text{ or: } K' = K \cdot e$$

Thus, for the total surface: $1 = \dfrac{K}{s}(1 - e^{-1}) + \dfrac{Ke}{2s}e^{-2}$

$$= \dfrac{K}{s}\left(1 - e^{-1} + \tfrac{1}{2}e^{-1}\right) = \dfrac{K}{s}\left(1 - \tfrac{1}{2}e^{-1}\right)$$

Thus: $\qquad K = s(1 - \tfrac{1}{2}e^{-1})^{-1} \quad K' = se(1 - \tfrac{1}{2}e^{-1})^{-1}$

Thus $0 < t < \dfrac{1}{s} \qquad f(t) = s\left(1 - \tfrac{1}{2}e^{-1}\right)^{-1} e^{-st}$

$\dfrac{1}{s} < t < \infty \qquad f(t) = se\left(1 - \tfrac{1}{2}e^{-1}\right)^{-1} e^{-2st} = \underline{\underline{s\left(1 - \tfrac{1}{2}e^{-1}\right)^{-1} e^{1-2st}}}$

PROBLEM 10.46

Derive the partial differential equation for unsteady-state unidirectional diffusion accompanied by an nth-order chemical reaction (rate constant k):

$$\frac{\partial C_A}{\partial t} = D\frac{\partial^2 C_A}{\partial y^2} - kC_A^n$$

where C_A is the molar concentration of reactant at position y at time t.

Explain why, when applying the equation to reaction in a porous catalyst particle, it is necessary to replace the molecular diffusivity D by an effective diffusivity D_e.

Solve the above equation for a first-order reaction under steady-state conditions, and obtain an expression for the mass transfer rate per unit area at the surface of a catalyst particle which is in the form of a thin platelet of thickness $2L$.

Explain what is meant by the effectiveness factor η for a catalyst particle and show that it is equal to $(1/\phi)\tanh\phi$ for the platelet referred to previously where ϕ is the Thiele modulus $L\sqrt{(k/D_e)}$.

For the case where there is a mass transfer resistance in the fluid external to the particle (mass transfer coefficient h_D), express the mass transfer rate in terms of the bulk concentration C_{Ao} rather than the concentration C_{AS} at the surface of the particle.

For a bed of catalyst particles in the form of flat platelets it is found that the mass transfer rate is increased by a factor of 1.2 if the velocity of the external fluid is doubled.

The mass transfer coefficient h_D is proportional to the velocity raised to the power of 0.6. What is the value of h_D at the original velocity?

$$k = 1.6 \times 10^{-3} \text{ s}^{-1}, \quad D_e = 10^{-8} \text{ m}^2/\text{s}$$

catalyst pellet thickness $(2L) = 10$ mm.

Solution

(i) The partial differential equation for unsteady-state diffusion accompanied by chemical reaction is derived in Volume 1 as equation 10.170

(ii) The molecular diffusivity D must be replaced by an effective diffusivity D_e because of the complex internal structure of the catalyst particle which consists of a multiplicity of interconnected pores, and the molecules must take a tortuous path. The effective distance the molecules must travel is consequently increases. Furthermore, because the pores are very small, their dimensions may be less than the mean free path of the molecules and Knudsen diffusion effects may arise

(iii) Equation 10.170 is solved in Volume 1 to give equation 10.199 for a catalyst particle in the form of a flat platelet

(iv) The effectiveness factor is the ratio of the actual rate of reaction to that which would be achieved in the absence of a mass-transfer resistance. For a platelet, it is evaluated in terms of the Thiele modulus as equation 10.202

(v) For the case, where there is an external mass transfer resistance, the reaction rate is expressed in terms of the bulk concentration as equation 10.222:

$$R_v = \frac{k C_{Ao}}{(1/\eta) + (kL/h_D)}$$

For $k = 1.6 \times 10^{-3}$ s^{-1}, $\quad D_e = 10^{-8}$ m^2/s, $\quad L = 5 \times 10^{-3}$ m:

$$\phi = L\sqrt{\frac{k}{D_e}} = 5 \times 10^{-3}\sqrt{\frac{1.6 \times 10^{-3}}{10^{-3}}} = 2$$

$$\eta = \frac{1}{\phi}\tanh\phi = \frac{1}{2}\tanh 2 = \frac{0.96}{2} = 0.48$$

$$\therefore \quad \frac{1}{\eta k L} = \frac{1}{0.48 \times 1.6 \times 10^{-3} \times 5 \times 10^{-3}} = \frac{1}{3.84 \times 10^{-6}} = 0.260 \times 10^6.$$

If the original value of mass transfer coefficient is h_D, the new value at twice original velocity $= h_D(2)^{0.6} = 1.516\, h_D$. Given that the overall rate is increased by a factor of 1.2:

$$\left(\frac{1}{0.260 \times 10^6 + \dfrac{1}{1.516 h_D}}\right) \Bigg/ \left(\frac{1}{0.260 \times 10^6 + \dfrac{1}{h_D}}\right) = 1.2$$

$$\frac{\left(0.260 \times 10^6 + \dfrac{1}{h_D}\right)}{\left(0.260 \times 10^6 + 0.66\dfrac{1}{h_D}\right)} = 1.2$$

$$0.260 \times 10^6 + \frac{1}{h_D} = 0.312 \times 10^6 + 0.792 \frac{1}{h_D}$$

$$\frac{1}{h_D} = 0.25 \times 10^6 \quad \text{and:} \quad \underline{\underline{h_D = 4.0 \times 10^{-6} \text{ m/s}}}.$$

PROBLEM 10.47

Explain the basic concepts underlying the two-film theory for mass transfer across a phase boundary and obtain an expression for film thickness.

Water evaporates from an open bowl at 349 K at the rate of 4.11×10^3 kg/m²s. What is the effective gas-film thickness?

The water is replaced by ethanol at 343 K. What will be its rate of evaporation in kg/m²s if the film thickness is unchanged?

At the surface of the ethanol, what proportion of the total mass transfer will then be attributable to bulk flow?

Data.

> Vapour pressure of water at 349 K = 34 mm Hg
>
> Vapour pressure of ethanol at 343 K = 544 mm Hg
>
> Neglect the partial pressure of vapour in the surrounding atmosphere
>
> Diffusivity of water vapour in air = 26×10^{-6} m²/s
>
> Diffusivity of ethanol in air = 12×10^{-6} m²/s
>
> Density of mercury = 13,600 kg/m³
>
> Universal gas constant $\mathbf{R}$ = 8314 J/kmol K

Solution

For evaporation for a free surface, Stefan's law is applicable or:

$$N'_A = -D \frac{C_T}{C_B} \frac{dC_T}{dy} = +D \frac{C_T}{C_B} \frac{dC_B}{dy} \qquad \text{(from equation 10.30)}$$

Integration gives:
$$N'_A = \frac{D}{y_2 - y_1} C_T \ln \frac{C_{B_2}}{C_{B_1}}$$

$$= \frac{D}{y_2 - y_1} \frac{C_T}{C_{Bm}} (C_{A1} - C_{A2})$$

For water: Vapour pressure, P_{A1} = 301 mm Hg at 349 K

At 349 K:
$$C_{A1} = \frac{P_{A1}}{\mathbf{R}T} = \frac{(0.301 \times 13,600 \times 9.81)}{(8314 \times 349)}$$

$$= 0.0138 \text{ kmol/m}^3$$

$$C_T = \frac{P}{RT} = \frac{(0.760 \times 13{,}600 \times 9.81)}{(8314 \times 349)}$$

$$= 0.0350 \text{ kmol/m}^3$$

For air: $$C_{B1} = C_T - C_{A1} = (0.0350 - 0.0138)$$

$$= 0.0212 \text{ kmol/m}^3$$

and: $$C_{B2} = C_T = 0.0350 \text{ kmol/m}^3.$$

The evaporation rate of water is: $N_A' = \dfrac{26 \times 10^{-6}}{y_2 - y_1} \times 0.0350 \ln\left(\dfrac{0.0350}{0.0212}\right)$

$$= \frac{0.456 \times 10^{-6}}{y_2 - y_1} \text{ kmol/m}^2\text{s}$$

But: $N_A' = 4.11 \times 10^{-3}$ kg/m^{2}s $= 0.228 \times 10^{-3}$ kmol/m^{2}s.

giving a film thickness of: $(y_2 - y_1)\, 2 \times 10^{-3}$ m = **2 mm**

For ethanol: Vapour pressure $= 541$ mm Hg at 343 K.

At 343 K: $$C_{A1} = \frac{P_{A1}}{RT} = \frac{(0.541 \times 13{,}600 \times 9.81)}{(8314 \times 343)}$$

$$= 0.0253 \text{ kmol/m}^3$$

$$C_T = \frac{P}{RT} = \frac{(0.760 \times 13{,}600 \times 9.81)}{(8314 \times 343)}$$

$$= 0.0356 \text{ kmol/m}^3$$

$$C_{B1} = (C_T - C_{A1}) = 0.0103 \text{ kmol/m}^3$$

and: $$C_{B2} = C_T = 0.0356 \text{ kmol/m}^3$$

The evaporation rate of ethanol, $N_A' = \left(\dfrac{12 \times 10^{-6}}{0.002}\right) \times 0.0356 \times \ln\left(\dfrac{0.0356}{0.0103}\right)$

$$= 0.265 \times 10^{-3} \text{ kmol/m}^2\text{s}$$

$$= \underline{\underline{12.2 \times 10^{-3} \text{ kg/m}^2\text{s}}}$$

The total flux at any location = diffusional flux + bulk flow

$$\frac{\text{bulk flow}}{\text{total flux}} = 1 - \frac{\text{diffusional flux}}{\text{total flux}} = 1 - \frac{C_B}{C_T}$$

At the ethanol surface, the proportion of flux due to bulk flow is:

$$1 - \frac{0.0103}{0.0356} = \underline{\underline{0.71}}$$

SECTION 11

The Boundary Layer

PROBLEM 11.1

Calculate the thickness of the boundary layer at a distance of 75 mm from the leading edge of a plane surface over which water is flowing at a rate of 3 m/s. Assume that the flow in the boundary layer is streamline and that the velocity u of the fluid at a distance y from the surface can be represented by the relation $u = a + by + cy^2 + dy^3$, where the coefficients a, b, c, and d are independent of y. The viscosity of water is 1 mN s/m^2.

Solution

At a distance y from the surface: $u = a + by + cy^2 + dy^3$.

When $y = 0$, $u = 0$, and hence $a = 0$.

The shear stress within the fluid: $R_0 = -\mu(\partial u/\partial y)_{y=0}$ and since $(\partial u/\partial y)$ is constant for small values of y, $(\partial^2 u/\partial y^2)_{y=0} = 0$.

At the edge of the boundary layer, $y = \delta$ and $u = u_s$, the main stream velocity.

$$\partial u/\partial y = 0 \text{ and } u = by + cy^2 + dy^3$$

$\therefore$ $$\partial u/\partial y = b + 2cy + 3dy^2 \text{ and } \partial^2 u/\partial y^2 = 2c + 6dy$$

When $y = 0$, $\partial^2 u/\partial y^2 = 0$, and hence $c = 0$.

When $y = \delta$, $u = b\delta + d\delta^3 = u_s$

and: $$\partial u/\partial y = b + 3d\delta^2 = 0$$

$\therefore$ $$b = -3d\delta^2$$

$\therefore$ $$d = -u_s/2\delta^3 \text{ and } b = 3u_s/2\delta$$

The velocity profile is given by, $u = (3u_s y/2\delta) - (u_s/2)(y/\delta)^3$

or: $$u/u_s = 1.5(y/\delta) - 0.5(y/\delta)^3 \quad\quad \text{(equation 11.12)}$$

The integral in the momentum equation 11.9 is now evaluated, and substituting from equations 11.14 and 11.15 into equation 11.9:

$$(\delta/x) = 4.64\,Re_x^{-0.5}$$

$$Re_x = (0.075 \times 3 \times 1000/1 \times 10^{-3}) = 225{,}000$$

$$\delta/x = (4.64 \times 225{,}000^{-5}) = 0.00978$$

and: $$\delta = (0.00978 \times 0.075) = 0.000734 \text{ m or } \underline{\underline{0.734 \text{ mm}}}$$

PROBLEM 11.2

Water flows at a velocity of 1 m/s over a plane surface 0.6 m wide and 1 m long. Calculate the total drag force acting on the surface if the transition from streamline to turbulent flow in the boundary layer occurs when the Reynolds group $Re_x = 10^5$.

Solution

See Volume 1, Example 11.1

PROBLEM 11.3

Calculate the thickness of the boundary layer at a distance of 150 mm from the leading edge of a surface over which oil, of viscosity 50 mN s/m^2 and density 990 kg/m^3, flows with a velocity of 0.3 m/s. What is the displacement thickness of the boundary layer?

Solution

See Volume 1, Example 11.2

PROBLEM 11.4

Calculate the thickness of the laminar sub-layer when benzene flows through a pipe 50 mm diameter at 0.003 m^3/s. What is the velocity of the benzene at the edge of the laminar sub-layer? Assume fully developed flow exists within the pipe.

Solution

See Volume 1, Example 11.3

PROBLEM 11.5

Air is flowing at a velocity of 5 m/s over a plane surface. Derive an expression for the thickness of the laminar sub-layer and calculate its value at a distance of 1 m from the leading edge of the surface.

Assume that within the boundary layer outside the laminar sub-layer, the velocity of flow is proportional to the one-seventh power of the distance from the surface and that the shear stress R at the surface is given by:

$$(R/\rho u_s^2) = 0.03(u_s \rho x/\mu)^{-0.2}$$

where ρ is the density of the fluid (1.3 kg/m^3 for air), μ is the viscosity of the fluid (17×10^{-6} N s/m^2 for air), u_s is the stream velocity (m/s), and x is the distance from the leading edge (m).

Solution

The shear stress in the fluid at the surface: $R = -\mu u_x/y$

From the equation given: $\quad R = 0.03\rho u_s^2(\mu/u_s\rho x)^{0.2}$

$\therefore \quad\quad\quad\quad u_x = (0.03\rho u_s^2 y/\mu)(\mu/u_s\rho x)^{0.2}$

If the velocity at the edge of the laminar sub-layer is u_b, $u_x = u_b$ when $y = \delta_b$.

$\therefore \quad\quad\quad\quad u_b = (0.03\rho u_s^2 \delta_b/\mu)(\mu/u_s\delta\rho)^{0.2}$

and: $\quad\quad\quad (\delta_b/\delta) = 33.3(u_b/u_s)(\mu/u_s\delta\rho)^{0.8}$

From equation 11.24, the velocity distribution is given by:

$$(\delta_b/\delta)^{1/7} = (u_b/u_s)$$

and hence: $\quad\quad (u_b/u_s)^7 = 33.3(u_b/u_s)(\mu/u_s\delta\rho)^{0.8}$

or: $\quad\quad\quad (u_b/u_s) = 1.65(\mu/u_s\delta\rho)^{0.115}$

$$= 1.65\,Re_\delta^{-0.115}$$

Substituting $0.376x^{0.8}(\mu/u_s\rho)^{0.2}$ for δ from equation 11.29:

$$(u_b/u_s) = 1.65[0.376u_s\rho x^{0.8}\mu^{0.2}/(\mu u_s^{0.2}\rho^{0.2})]^{-0.115}$$

$$= (1.65/0.376^{0.115})(u_s^{0.8}x^{0.8}\rho^{0.8}/\mu^{0.8})^{-0.115}$$

$$= 1.85\,Re_x^{-0.09}$$

Now: $\quad\quad\quad (\delta_b/\delta) = (u_b/u_s)^7 = 74.2\,Re_x^{0.63}$

From equation 11.31: $\quad\quad (\delta/x) = 0.376\,Re_x^{-0.2}$

$\therefore \quad\quad\quad (\delta_b/x) = (74.2 \times 0.376)/(Re_x^{0.63}\,Re_x^{0.2})$

$$= 27.9\,Re_x^{-0.83}$$

In this case: $\quad\quad Re_x = (1 \times 5 \times 1.3/17 \times 10^{-6})$

$$= 3.82 \times 10^5$$

$$\delta_b = 1.0 \times 27.9(3.82 \times 10^5)^{-0.83}$$

$$= 6.50 \times 10^{-4} \text{ m or } \underline{0.65 \text{ mm}}$$

PROBLEM 11.6

Obtain the momentum equation for an element of the boundary layer. If the velocity profile in the laminar region can be represented approximately by a sine function, calculate the boundary layer thickness in terms of distance from the leading edge of the surface.

Solution

The derivation of the momentum equation for an element of the boundary layer is presented in detail in Section 11.2 and the final expression is:

$$-R_o = (\rho\partial/\partial x) \int_0^l (u_s - u_x)u_x \, dy$$

A sine function may be developed as follows. When $y = 0$, $u_x = 0$ and when $y = \delta$, $u_x = u_s$.

Thus:
$$u_x = u_s \sin(ay)$$

and when $y = \delta$, $\sin ay = \pi/2$ or $a\delta = \pi/2$ and $a = \pi/2\delta$.

$$\therefore \qquad u_x = u_s \sin(\pi y/2d)$$

and over the range $0 < y < \delta$,

$$(u_x/u_s) = \sin[(\pi/2)(y/\delta)]$$

The integral in the momentum equation may now be evaluated for the laminar boundary layer considering the ranges $0 < y < \delta$ and $\delta < y < l$ separately.

$$\therefore \int_0^l (u_s - u_x)u_x \, dy = \int_0^\delta u_s^2 \{1 - \sin[(\pi/2)(y/\delta)]\}\{\sin[(\pi/2)(y/\delta)]\} \, dy$$

$$+ \int_\delta^l (u_s - u_s)u_s \, dy$$

$$= u_s^2 \int_0^\delta [\sin(\pi y/2\delta) - \sin^2(\pi y/2\delta)] \, dy$$

$$= u_s^2[-[\cos(\pi y/2\delta)]/(\pi/2\delta) - y/2 + \sin(\pi y/\delta)/(2\pi/\delta)]_0^\delta$$

$$= u_s^2 \delta[(2/\pi) - (1/2)]$$

$$R_0 = -\mu(\partial u_x/\partial y)_{y=0} = -\mu u_s \pi/2\delta$$

and substituting in the momentum equation:

$$\rho\partial \left[u_s^2 \delta \left(\frac{2}{\pi} - \frac{1}{2}\right)\right] \Big/ \partial x = \mu u_s \pi/2\delta$$

$$\therefore \qquad \delta \, d\delta = \mu\pi^2 \, dx/\rho u_s(4 - \pi)$$

$$\therefore \qquad \delta^2/2 = [\pi^2/(4 - \pi)](\mu x/\rho u_s)$$

and:
$$\delta = 4.80(\mu x/\rho u_s)^{0.5}$$

$$\therefore \qquad (\delta/x) = 4.80(\mu/x\rho u_s)^{0.5} = 4.80 \, Re_x^{-0.5}$$

PROBLEM 11.7

Explain the concepts of "momentum thickness" and "displacement thickness" for the boundary layer formed during flow over a plane surface. Develop a similar concept to displacement thickness in relation to heat flux across the surface for laminar flow and heat transfer by thermal conduction, for the case where the surface has a constant temperature and the thermal boundary layer is always thinner than the velocity boundary layer. Obtain an expression for this 'thermal thickness' in terms of the thicknesses of the velocity and temperature boundary layers.

Similar forms of cubic equations may be used to express velocity and temperature variations with distance from the surface.

For a Prandtl number, Pr, less than unity, the ratio of the temperature to the velocity boundary layer thickness is equal to $Pr^{-1/3}$. Work out the 'thermal thickness' in terms of the thickness of the velocity boundary layer for a value of $Pr = 0.7$.

Solution

Consideration is given to the streamline portion of the boundary layer in Section 11.3 where, assuming:

$$u_x = u_o + ay + by^2 + cy^3 \qquad \text{(equation 11.10)}$$

it is shown that the equation for the velocity profile is:

$$(u_x/u_s) = 1.5(y/\delta) - 0.5(y/\delta)^3 \qquad \text{(equation 11.12)}$$

The equivalent equation for the thermal boundary layer will be:

$$(\theta/\theta_s) = 1.5(y/\delta_t) - 0.5(y/\delta_t)^3$$

where δ_t is the thickness of the thermal boundary layer.

The heat flow is given by:

$$Q = \int_0^l C_p \rho u_x T \, dy$$

$$= u_s T_s C_p \rho \int_0^l [1.5(y/\delta_t) - 0.5(y/\delta_t)^3][1.5(y/\delta) - 0.5(y/\delta)^3] \, dy$$

This is made up of two components:
the heat flow through the thermal boundary layer:

$$= u_s T_s C_p \rho \int_0^{(\delta_t/\delta)} \{(2.25y^2/\delta \cdot \delta_t) - 0.75y^4/(\delta^3 \delta_t) - 0.75y^4/(\delta \cdot \delta_t^3)$$

$$+ (0.25y^6/\delta^2\delta_t^5)\} \, d(y/\delta)$$

and the heat flow through the velocity boundary layer between $y = \delta_t$ and $y = \delta$:

$$= u_s T_s C_p \rho \delta \int_{(\delta_t/\delta)}^1 [1.5y/\delta - 0.5(y/\delta)^3] \, d(y/\delta)$$

Thus, putting $\sigma = (\delta_t/\delta)$, the heat flow becomes:

$$Q = u_s T_s C_p \rho \delta \left[\int_0^x (2.25(y/\delta)^2/\sigma - 0.75(y/\delta)^4(1/\sigma + 1/\sigma^3)) + 0.25(y/\delta)^6/\sigma^3) \, d(y/\delta) \right.$$

$$\left. + \int_x^l 1.5(y/\delta) - 0.5(y/\delta)^3 \, d(y/\delta) \right]$$

$$= u_s T_s C_p \rho \delta [(0.75\sigma^3/\sigma - 0.15\sigma^5(1/\sigma + 1/\sigma^3) + 0.036\sigma^7/\sigma^3)$$

$$+ (0.75(1 - \sigma^2) - 0.125(1 - \sigma^4))]$$

$$= u_s T_s C_p \rho \delta (0.625 - 0.15\sigma^2 + 0.0107\sigma^4)$$

The heat flow from δ to δ_t^* in the absence of boundary layers $= (\delta - \delta_t^*) u_s T_s \rho C_p$.

$\therefore$ $\qquad\qquad\qquad (\delta - \delta_t^*) = \delta(0.625 - 0.15\sigma^2 + 0.0107\sigma^4)$

and: $\qquad\qquad\qquad (\delta_t^*/\delta) = 0.375 + 0.15\sigma^2 - 0.0107\sigma^4$

When $\sigma < 1$, then $\sigma = Pr^{-0.33}$
and neglecting the σ^4 term, an approximate value is:

$$(\delta_t^*/\delta) = 0.375 + 0.15 Pr^{-0.67}$$

When $Pr = 0.7$, $Pr^{0.67} = 0.788$

and: $\qquad\qquad\qquad (\delta_t^*/\delta) = (0.375 + 0.15/0.788) = \underline{\underline{0.185}}$

(Since this is much less than 1, neglecting the σ^4 term is justified.)

PROBLEM 11.8

Explain why it is necessary to use concepts, such as the displacement thickness and the momentum thickness, for a boundary layer in order to obtain a boundary layer thickness which is largely independent of the approximation used for the velocity profile in the neighbourhood of the surface.

It is found that the velocity u at a distance y from the surface can be expressed as a simple power function ($u \propto y^n$) for the turbulent boundary layer at a plane surface. What is the value of n if the ratio of the momentum thickness to the displacement thickness is 1.78?

Solution

The first part of this problem is discussed in Section 11.1. If the displacement and the momentum thicknesses are δ^* and δ_m respectively, then:

$$\text{the momentum flux} = \int_0^\delta u_y \, dy \rho u_y = \int_{\delta_m}^\delta u_s \, dy \rho u_s$$

and: $\qquad\qquad\qquad \text{the mass flux} = \int_0^\delta u_y \, dy \rho = \int_{\delta*}^\delta u_s \, dy \rho$

$$\therefore \qquad \int_0^\delta u_y^2 \, \mathrm{d}y = u_s^2(\delta - \delta_m)$$

and:
$$\int_0^\delta u_y \, \mathrm{d}y = u_s(\delta - \delta^*)$$

$$\int_0^1 (u_y/u_s)^2 \, \mathrm{d}(y/\delta) = 1 - (\delta_m - \delta)$$

and:
$$\int_0^1 (u_y/u_s) \, \mathrm{d}(y/\delta) = 1 - (\delta^*/\delta)$$

$$\therefore \qquad (\delta_m/\delta^*) = \left[1 - \int_0^1 (u_y/u_s)^2 \, \mathrm{d}(y/\delta)\right] \bigg/ \left[1 - \int_0^1 (u_y/u_s) \, \mathrm{d}(y/\delta)\right]$$

If $(u_y/u_s) = (y/\delta)^n$, then:

$$\int_0^1 (u_y/u_s) \, \mathrm{d}(y/\delta) = \int_0^1 (y/\delta)^n \, \mathrm{d}(y/\delta) = 1/(n+1)$$

and:
$$\int_0^1 (u_y/u_s) \, \mathrm{d}(y/\delta) = \int_0^1 (y/\delta)^{2n} \, \mathrm{d}(y/\delta) = 1/(2n+1)$$

$$\therefore \qquad (\delta_m/\delta^*) = [1 - 1/(2n+1)]/[1 - 1/(n+1)] = 2(n+1)/(2n+1)$$

When $(\delta_m/\delta^*) = 1.78$:

$$1.78 = 2n + 2/(2n+1) = 3.56n + 1.78 = 2n + 2$$

and:
$$n = (0.22/1.56) = 0.141 \text{ or approximately } \underline{\underline{1/7}}.$$

PROBLEM 11.9

Derive the momentum equation for the flow of a fluid over a plane surface for conditions where the pressure gradient along the surface is negligible. By assuming a sine function for the variation of velocity with distance from the surface (within the boundary layer) for streamline flow, obtain an expression for the boundary layer thickness as a function of distance from the leading edge of the surface.

Solution

Using the nomenclature of Fig. 11.5, the argument presented in Section 11.2 results in the expression known as the momentum equation, given in equation 11.9, which may be expressed as:

$$\rho \partial/\partial x \int_0^l u_x(u_s - u_x) \, \mathrm{d}y = -R_0 \qquad (i)$$

If the velocity within the boundary layer may be represented by a sine function:

$$(u_x/u_s) = \sin[(\pi/2)(y/\delta)] \tag{ii}$$

the integral: $$\int_0^l u_x(u_s - u_x)\, dy = u_s^2 \delta \int_0^1 (u_x/u_s)(1 - (u_x/u_s))\, d(y/\delta)$$

and, substituting from equation (ii):

$$\int_0^l u_x(u_s - u_x)\, dy = u_s^2 \delta \int_0^1 [\sin(\pi/2)(y/\delta) - \sin^2(\pi/2)(y/\delta)]\, d(y/\delta)$$

$$= u_s^2 \delta \left\{ \left[-(2/\pi)\cos[(\pi/2)(y/\delta)] \right]_0^1 - 0.5 \int_0^1 (1 - \cos \pi y/\delta)\, d(y/\delta) \right\}$$

$$= u_s^2 \delta \{ (2/\pi - 0.5 + \left[1/\pi \right] \sin(\pi y/\delta)]_0^1 \}$$

$$= u_s^2 \delta (2/\pi - 0.5) = 0.1366 u_s^2 \delta$$

From equation (ii): $$u_x = u_s \sin[(\pi/2)(y/\delta)]$$

∴ $$du_x/dy = u_s(\pi/2\delta)\cos[(\pi/2)(y/\delta)]$$

and when $y = 0$:

$$(du_x/dy)_{y=0} = (\pi/2)(u_s/\delta)$$

But: $$R_0 = -\mu(du_x/dy)_{y=0} = -(\pi/2)(\mu u_s/\delta)$$

Therefore, substituting in equation (i):

$$\rho \partial/\partial x (0.1366 u_s^2 \delta) = (\pi/2)(\mu u_s/\delta)$$

$$\delta\, d\delta = (\mu/\rho u_s)(\pi/0.2732)\, dx$$

$$\delta^2/x = (\mu x/\rho u_s)(\pi/0.2732)$$

$$\delta^2/x^2 = (\mu/\rho u_s x)(\pi/0.1366)$$

and: $$\underline{\underline{(\delta/x) = 4.80\, Re_x^{-0.5}}}$$

PROBLEM 11.10

Derive the momentum equation for the flow of a viscous fluid over a small plane surface.

Show that the velocity profile in the neighbourhood of the surface can be expressed as a sine function which satisfies the boundary conditions at the surface and at the outer edge of the boundary layer.

Obtain the boundary layer thickness and its displacement thickness as a function of the distance from the leading edge of the surface, when the velocity profile is expressed as a sine function.

Solution

The derivation of the momentum equation is given in Section 11.2 to give:

$$\rho\frac{\partial}{\partial x}\int_0^l u_x(u_s - u_x)\,dy = -R_0 = \mu(du_x/dy)_{y=0} \qquad \text{(equation 11.9) (i)}$$

If the velocity profile is a sine function, then:

$$(u_x/u_s) = \sin[(\pi/2)(y/\delta)]$$

Differentiating:

$$(1/u_s)\partial u_x/\partial y = (\pi/2\delta)\cos[(\pi/2)(y/\delta)]$$

When $y = \delta$, $u_x = u_s$ and $(\partial u_x/\partial y) = 0$
Differentiating again:

$$(1/u_s)\partial^2 u_x/\partial y^2 = (-\pi^2/4\delta^2)\sin[(\pi/2)(y/\delta)]$$

When $y = 0$, $u_x = 0$ and $\partial^2 u_x/\partial y^2 = 0$
Substituting, noting that $u_x = u_s$ when $y > \delta$:

$$\int_0^l u_x(u_x - u_s)\,dy = u_s^2 \int_0^\delta \sin(y/\delta)(\pi/2)[1 - \sin(y/\delta)(\pi/2)]\,dy$$

$$= u_s^2\delta\int_0^1 [\sin(y/\delta)(\pi/2) - 0.5(1 - \cos(y\pi/\delta))]\,d(y/\delta)$$

$$= u_s^2\delta\left([-(2/\pi\cos(y/\delta)(\pi/2))]_0^1 - 0.5\,[y/\delta]_0^1 + 0.5\,[(1/\pi)\sin(y\pi/\delta)]_0^1\right)$$

$$= u_s^2\delta((2/\pi) - 0.5 + 0) = 0.1366u_s^2\delta \qquad \text{(as in Problem 11.8)}$$

Substituting in equation (i):

$$\rho\partial(0.1366u_s^2\delta)/\partial x = \mu(\pi/2)u_s/\delta$$

$$\therefore \qquad\qquad \delta d\delta = ((\pi/2)/0.1366)(\mu/\rho u_s)\,dx$$

If $\delta = 0$ when $x = 0$:

$$\delta^2/x = 11.5(\mu x/\rho u_s^2)$$

and: $\qquad (\delta/x)^2 = 11.5(\mu/\rho u_s x)$ and $\underline{\underline{(\delta/x) = 3.39\,Re_x^{-0.5}}}$

The displacement thickness, δ^* is given by:

$$u_s(\delta - \delta^*) = \int_0^\delta u_s\sin[(y/\delta)(\pi/2)]\,dy = u_s(2\delta/\pi)\left[-\cos[(y/\delta)(\pi/2)]\right]_0^1$$

$\therefore \qquad\qquad \delta - \delta^* = 2\delta/\pi = 0.637\delta$

and: $\qquad\qquad \delta^* = 0.363\delta$

$\therefore \qquad\qquad (\delta^*/x) = (0.363 \times 3.39)\,Re_x^{-0.5} = \underline{\underline{1.23\,Re_x^{-0.5}}}$

PROBLEM 11.11

Derive the momentum equation for the flow of a fluid over a plane surface for conditions where the pressure gradient along the surface is negligible. By assuming a sine function for the variation of velocity with distance from the surface (within the boundary layer) for streamline flow, obtain an expression for the boundary layer thickness as a function of distance from the leading edge of the surface.

Solution

The total mass flowrate through plane 1–2 is: $\int_0^l \rho u_x \, dy$.

The total momentum flux through plane 1–2 is: $\int_0^l \rho u_x^2 \, dy$

Change in mass flowrate from 1–2 to 3–4 is: $\dfrac{\partial}{\partial x}\left\{\int_0^l \rho u_x \, dy\right\} dx$

Change in momentum flux from 1–2 to 3–4 is: $\dfrac{\partial}{\partial x}\left\{\int_0^l \rho u_x^2 \, dy\right\} dx$.

Change in momentum flux is attributable, in the absence of pressure gradient, to:

(a) Momentum of fluid entering through 2–4

    Since all this fluid has velocity u_s, the momentum flux is:

$$\left\{\frac{\partial}{\partial x}\left[\int_0^l \rho u_x \, dy\right] dx\right\} u_s$$

(b) Force due to shear stress at surface $= R_0 \, dx$.

Thus a momentum balance gives:

$$\frac{\partial}{\partial x}\left\{\int_0^l \rho u_x^2 \, dy\right\} dx = \frac{\partial}{\partial x}\left\{\int_0^l \rho u_x \, dy\right\} dx.u_s + R_0 \, dx.$$

$$\therefore \qquad \rho\frac{\partial}{\partial x}\left\{\int_0^l u_x(u_s - u_x) \, dy\right\} = -R_0$$

Representing velocity within boundary layer by a sine function, then:

$$\frac{u_x}{u_s} = \sin\left(\frac{\pi}{2}\frac{y}{\delta}\right)$$

Thus:

$$\int_0^l u_x(u_s - u_x) \, dy = u_s^2\delta \int_0^1 \frac{u_x}{u_s}\left(1 - \frac{u_x}{u_s}\right) d\left(\frac{y}{\delta}\right)$$

$$= u_s^2\delta \int_0^1 \left(\sin\frac{\pi}{2}\cdot\frac{y}{\delta} - \sin^2\frac{\pi}{2}\frac{y}{\delta}\right) d\frac{y}{\delta}$$

$$= u_s^2 \delta \left\{ \left[-\frac{2}{\pi} \cos \frac{\pi}{2} \frac{y}{\delta} \right]_0^l - \frac{1}{2} \int_0^1 \left(1 - \cos \frac{\pi y}{\delta} \right) d \left(\frac{y}{\delta} \right) \right\}$$

$$= u_s^2 \delta \left\{ \frac{2}{\pi} - \frac{1}{2} + \left[\frac{1}{\pi} \sin \pi \frac{y}{\delta} \right]_0^l \right\}$$

$$= u_s^2 \delta \left\{ \frac{2}{\pi} - \frac{1}{2} \right\} = 0.1366 u_s^2 \delta.$$

Differentiating:
$$\frac{du_x}{dy} = u_s \cdot \left(\frac{\pi}{2\delta} \cos \frac{\pi}{2} \frac{y}{\delta} \right)$$

$$\left(\frac{du_x}{dy} \right)_{y=0} = \left(\frac{\pi}{2} \right) \cdot \left(\frac{u_s}{\delta} \right)$$

and:
$$R_0 = -\mu \left(\frac{du_x}{dy} \right)_{y=0} = -\frac{\pi}{2} \frac{\mu u_s}{\delta}.$$

Thus:
$$\rho \frac{\partial}{\partial x} (0.1366 u_s^2 \delta) = \frac{\pi}{2} \frac{\mu u_s}{\delta}$$

$$\therefore \qquad \delta \, d\delta = \left(\frac{\mu}{\rho u_s} \right) \left(\frac{\pi}{0.2732} \right) dx.$$

$$\therefore \qquad \frac{\delta^2}{2} = \left(\frac{\mu x}{\rho u_s} \right) \left(\frac{\pi}{0.2732} \right)$$

$$\therefore \qquad \frac{\delta^2}{x^2} = \left(\frac{\mu}{\rho u_s x} \right) \left(\frac{\pi}{0.1366} \right)$$

and:
$$\frac{\delta}{x} = 4.796 \, Re_x^{-1/2}$$

PROBLEM 11.12

Derive the momentum equation for the flow of a viscous fluid over a small plane surface. Show that the velocity profile in the neighbourhood of the surface may be expressed as a sine function which satisfies the boundary conditions at the surface and at the outer edge of the boundary layer.

Obtain the boundary layer thickness and its displacement thickness as a function of the distance from the leading edge of the surface, when the velocity profile is expressed as a sine function.

Solution

The momentum flux across 1–2 is: $\int_0^L \rho u_x^2 \, dy$

The change from 1–2 to 3–4 is: $\rho \frac{\partial}{\partial x} \left\{ \int_0^L u_x^2 \, dy \right\} dx$

The mass flux across $1-2$ is: $\int_0^L \rho u_x \, dy$.

The change from $1-2$ to $3-4$ is: $\rho \dfrac{\partial}{\partial x} \left\{ \int_0^L u_x \, dy \right\} dx$.

The rate of momentum entering through $2-4$ is:

$$\rho \frac{\partial}{\partial x} \left\{ \int_0^L u_x u_s \, dy \right\} dx,$$

Assuming $u_s \neq f(x)$, then:

$$\frac{\partial P}{\partial x} = 0.$$

A momentum balance gives:

$$\rho \frac{\partial}{\partial x} \left\{ \int_0^L u_x^2 \, dy \right\} dx = \rho \frac{\partial}{\partial x} \left\{ \int_0^L u_x u_s \, dy \right\} dx + R_0 \, dx$$

or: $\quad \rho \dfrac{\partial}{\partial x} \left\{ \int_0^L u_x (u_s - u_x) \, dy \right\} = -R_0 = \mu \left(\dfrac{\partial u_x}{\partial y} \right)_{y=0} \quad$ for a Newtonian fluid

The sine function is:

$$u_x = K \sin ky, \quad \text{so} \quad \frac{\partial u_x}{\partial y} = Kk \cos ky \quad \text{and} \quad \frac{\partial^2 u_x}{\partial y^2} \equiv Kk^2 \sin ky$$

is satisfied for all finite values of K and k.

The boundary conditions are:

$$y = 0 \quad u_x = 0$$

$$y = 0 \quad \frac{\partial^2 u_x}{\partial y^2} = 0$$

$$y = \delta \quad u_x = u_s \quad K = u_s$$

$$y = \delta \quad \frac{\partial u_x}{\partial y} = 0$$

Thus: $\qquad\qquad\qquad\qquad k = \dfrac{\pi}{2} \text{ and } k = \dfrac{\pi}{2\delta}.$

Hence: $\qquad\qquad u_x = u_s \sin \dfrac{\pi}{2} \dfrac{y}{\delta}, \quad \dfrac{\partial u_x}{\partial y} = \dfrac{\pi u_s}{2\delta} \cos \dfrac{\pi}{2} \dfrac{y}{\delta}$

and: $\qquad\qquad \left\{ \int_y^L \right\} = 0$

Thus: $\qquad\qquad \rho u_s^2 \dfrac{\partial}{\partial x} \left\{ \int_0^\delta \sin \dfrac{\pi}{2} \dfrac{y}{\delta} \left(1 - \sin \dfrac{\pi}{2} \dfrac{y}{\delta} \right) dy \right\} = \dfrac{\mu \pi u_s}{2\delta}$

$$\frac{\partial}{\partial x} \delta \left\{ \int_0^1 \left[\sin \frac{\pi}{2} \frac{y}{\delta} - \sin^2 \frac{\pi}{2} \frac{y}{\delta} \right] d\frac{y}{\delta} \right\} = \frac{\mu \pi}{2\delta \rho u_s}$$

$$\frac{\partial}{\partial x}\delta\left\{\int_0^1\left[\sin\frac{\pi}{2}\frac{y}{\delta}-\frac{1}{2}+\frac{1}{2}\cos\pi\frac{y}{\delta}\right]d\frac{y}{\delta}=\frac{\mu\pi}{2\delta\rho u_s}\right.$$

$$\frac{\partial}{\partial x}\delta\left[\frac{2}{\pi}\cos\frac{\pi}{2}\frac{y}{\delta}-\frac{1}{2}\frac{y}{\delta}+\frac{1}{2\pi}\sin\pi\frac{y}{\delta}\right]_0^1=\frac{\mu\pi}{2\delta\rho u_s}$$

$$\frac{\partial}{\partial x}\delta\left[\frac{2}{\pi}-\frac{1}{2}\right]=\frac{\mu\pi}{2\delta\rho u_s}$$

$$\left(\frac{4}{\pi}-1\right)\frac{\delta^2}{2}=\frac{\mu\pi x}{\rho u_s}\quad\text{assuming }\delta=0\text{ at }x=0$$

$$\frac{\delta^2}{x^2}=\left(\frac{\mu}{\rho u_s x}\right)\left(\frac{2\pi}{(4/\pi-1)}\right)$$

$\therefore$
$$\frac{\delta}{x}=\sqrt{23.1}\,Re_x^{-1/2}=4.80\,Re_x^{-1/2}$$

$$(\delta-\delta^*)u_s=\int_0^\delta u_s\sin\left(\frac{\pi}{2}\frac{y}{\delta}\right)dy=\left[u_s\left(-\frac{2\delta}{\pi}\right)\cos\frac{\pi}{2}\frac{y}{\delta}\right]_0^\delta=u_s\frac{2\delta}{\pi}$$

$$\delta^*=\delta\left(1-\frac{2}{\pi}\right)$$

and :
$$\frac{\delta^*}{\delta}=\underline{\underline{0.363.}}$$

SECTION 12

Momentum, Heat and Mass Transfer

PROBLEM 12.1

If the temperature rise per metre length along a pipe carrying air at 12.2 m/s is 66 deg K, what will be the corresponding pressure drop for a pipe temperature of 420 K and an air temperature of 310 K? The density of air at 310 K is 1.14 kg/m^3.

Solution

For a pipe of diameter d, the mass flow $= (u\rho\pi d^2)/4$
and the rate of heat transfer, $q = (u\rho\pi d^2/4)C_p\Delta T$ where ΔT is the temperature rise.

Also $q = hA(T_w - T_m) = h\pi dl(T_w - T_m)$ W
where T_w and T_m are the mean wall and fluid temperatures.

Thus, $$(h/C_p\rho u) = d\Delta T/4l(T_w - T_m)$$

From equation 12.102: $$R/\rho u^2 = d\Delta T/4(T_w - T_m)l$$

Substituting in equation 3.18:
$$-\Delta P = 4d\Delta T(l/d)\rho u^2/4(T_w - T_m) = (\Delta T\rho u^2 l)/(T_w - T_m)$$
$$= (66 \times 1.14 \times 12.2^2 \times 1.0)/(420 - 310)$$
$$= \underline{\underline{101.8 \ (N/m^2)/m}}$$

PROBLEM 12.2

It is required to warm a quantity of air from 289 K to 313 K by passing it through a number of parallel metal tubes of inner diameter 50 mm maintained at 373 K. The pressure drop must not exceed 250 N/m^2. How long should the individual tubes be?

The density of air at 301 K is 1.19 kg/m^3 and the coefficients of heat transfer by convection from the tube to air are 45, 62, and 77 W/m^2 K for velocities of 20, 24, and 30 m/s at 301 K respectively.

Solution

From equations 12.102 and 3.18:
$$-\Delta P = 4(h/C_p\rho u)(l/d)\rho u^2 = 4hlu/C_p d$$

$\therefore$ $$250 = 4(hlu/C_p \times 0.050) \text{ or } h = 3.125C_p/lu \text{ W/m}^2 \text{ K} \qquad \text{(i)}$$

The heat transferred to the air $= u(\pi d^2/4)\rho C_p (T_2 - T_1)$

$$= u(\pi \times 0.050^2/4)1.19 C_p (313 - 289) = 0.056 C_p u \text{ W}$$

This is equal to: $h\pi dl(T_w - T_m) = h\pi \times 0.050 l(373 - 301) = 11.3 hl$ W

$\therefore \qquad\qquad\qquad\qquad 11.31 hl = 0.056 C_p u \text{ or } h = 0.0050 C_p u/l \qquad\qquad\qquad$ (ii)

From equation (i): $\qquad (C_p/l) = 0.32 hu$

and substituting in equation (ii): $h = (0.005 \times 0.32 hu^2)$ and $u = 25$ m/s

For this velocity, interpolation of the given data gives a value of $h = 64$ W/m^2 K.

$\therefore \qquad\qquad\qquad (C_p/l) = (0.32 \times 64 \times 25) = 512$ J/kg K m

For air: $\qquad\qquad\qquad C_p = 1000$ J/kg K

and hence: $\qquad\qquad\qquad l = (1000/512) = \underline{\underline{1.95 \text{ m}}}$

PROBLEM 12.3

Air at 330 K, flowing at 10 m/s, enters a pipe of inner diameter 25 mm, maintained at 415 K. The drop of static pressure along the pipe is 80 N/m^2 per metre length. Using the Reynolds analogy between heat transfer and friction, estimate the temperature of the air 0.6 m along the pipe.

Solution

See Volume 1, Example 12.2.

PROBLEM 12.4

Air flows at 12 m/s through a pipe of inside diameter 25 mm. The rate of heat transfer by convection between the pipe and the air is 60 W/m^2 K. Neglecting the effects of temperature variation, estimate the pressure drop per metre length of pipe.

Solution

From equations 3.18 and 12.96, $-\Delta P = 4(h/C_p \rho u)(l/d)\rho u^2$

Taking $C_p = 1000$ J/kg K and $l = 1$ m, then:

$$-\Delta P = 4(60/1000\rho \times 12)(1.0/0.025)\rho \times 12^2 = \underline{\underline{115.2 \text{ N/m}^2 \text{ per metre}}}$$

PROBLEM 12.5

Air at 320 K and atmospheric pressure is flowing through a smooth pipe of 50 mm internal diameter and the pressure drop over a 4 m length is found to be 1.5 kN/m². Using the Reynolds analogy, by how much would the air temperature be expected to fall over the first metre of pipe length if the wall temperature there is kept constant at 295 K?

Viscosity of air $= 0.018$ mN s/m². Specific heat capacity of air $= 1.05$ kJ/kg K.

Solution

(Essentially, this is the same as Problem 9.40 though, here, an alternative solution is presented.)

From equations 3.18 and 12.102: $-\Delta P = 4(h/C_p\rho u)(l/d)\rho u^2$.

For a length of 4 m: $1500 = 4(h/C_p\rho u)(4.0/0.050)\rho u^2$

$$\therefore \qquad\qquad (hu/C_p) = 4.69 \text{ kg/ms}^2 \qquad\qquad\qquad\text{(i)}$$

The rate of heat transfer $= h\pi dl(T_m - T_w)$, which for a length of 1 m is:

$$(h\pi \times 0.050 \times 1.0)(0.5(320 + T_2) - 295) = 0.157h(0.5T_2 - 135)$$

The heat lost by the air $= u(\pi d^2/4)\rho C_p(T_1 - T_2)$

$$= u(\pi \times 0.050^2/4)\rho C_p(320 - T_2) = 0.00196u\rho C_p(320 - T_2)$$

$$\therefore \qquad 80.1(h/C_p\rho u) = (320 - T_2)/(0.5T_2 - 135)$$

Substituting from equation (i):

$$80.1(4.69/\rho u^2) = (320 - T_2)/(0.5T_2 - 135) \qquad\qquad\text{(ii)}$$

From equation 12.139:

$$(h/C_p\rho u) = 0.032(du\rho/\mu)^{-0.25} = 0.032(\mu/du\rho)^{0.25}$$

At 320 K and 101.3 kN/m², $\rho = (28.9/22.4)(273/320) = 1.10$ kg/m³

$$\therefore \qquad\qquad (h/C_p\rho u) = (hu/C_p)/(\rho u^2) = 4.69/(1.10u^2)$$

$$\therefore \qquad 4.69/(1.10u^2) = 0.032[0.018 \times 10^{-3}/(0.050 \times 1.10u)]^{0.25} \text{ and } u = 51.4 \text{ m/s}$$

Substituting, in equation (ii):

$$(80.1 \times 4.69)/(1.10 \times 51.4^2) = (320 - T_2)/(0.5T_2 - 135) \text{ and } T_2 = 316 \text{ K}$$

The temperature drop over the first metre is therefore 4 deg K which agrees with the solution to Problem 9.40.

(It may be noted that, in those problems, an arithmetic mean temperature difference is used rather than a logarithmic value for ease of solution. This is probably justified in view of the small temperature changes involved and also the approximate nature of the Reynolds analogy.)

PROBLEM 12.6

Obtain an expression for the simple Reynolds analogy between heat transfer and friction. Indicate the assumptions which are made in the derivation and the conditions under which you would expect the relation to be applicable.

The Reynolds number of a gas flowing at 2.5 kg/m^{2}s through a smooth pipe is 20,000. If the specific heat of the gas at constant pressure is 1.67 kJ/kg K, what will the heat transfer coefficient be?

Solution

The derivation of the simple Reynolds analogy and its application is presented in detail in Section 12.8.

For a Reynolds number of 2.0×10^4, from Fig. 3.7, $(R/\rho u^2) = 0.0032$ for a smooth pipe.

From equation 12.102: $(h/C_p \rho u) = 0.0032$

$$\rho u = 2.5 \text{ kg/m}^2\text{s}$$

and hence: $h = (0.0032 \times 1670 \times 2.5) = \underline{\underline{13.4 \text{ W/m}^2 \text{ K}}}$

PROBLEM 12.7

Explain Prandtl's concept of a 'mixing length'. What parallels may be drawn between the mixing length and the mean free path of the molecules in a gas?

The ratio of the mixing length to the distance from the pipe wall has a constant value of 0.4 for the turbulent flow of a fluid in a pipe. What is the value of the pipe friction factor if the ratio of the mean velocity to the axial velocity is 0.8?

Solution

Transfer by molecular diffusion is discussed in Section 12.2 and the concept of the mixing length in Section 12.3.2. By analogy with kinetic theory, the eddy kinematic viscosity, E, is given by:

$$E \propto \lambda_E u_E \qquad \text{(equation 12.18)}$$

where λ_E is the mixing length and u_E is some measure of the linear velocity of the fluid in the eddies.

As shown in equation 12.21: $u_E \propto \lambda_E |du_x/dy|$

Combining this with the previous equation: $E \propto \lambda_E(\lambda_E|du_x/dy|)$

Putting the proportionality constant equal to unity, $E = \lambda_E^2 |du_x/dy|$ (equation 12.23)

In the absence of momentum transfer by molecular movement, the shear stress is given by:

$$R_y = -E\mathrm{d}(\rho u_x)/\mathrm{d}y = -\rho\lambda_E^2|du_x/dy|du_x/dy. \qquad \text{(equation 12.20)}$$

Since near a surface du_x/dy is positive and assuming R_y is approximately constant at a value at the pipe wall, that is $R_y = R_o = -R$, then:

$$R = \rho\lambda_E^2(du_x/dy)^2$$

or: $\sqrt{(R/\rho)} = \lambda_E(du_x/dy)$ (equation 12.26)

Here, $\sqrt{(R/\rho)}$, the shearing stress or friction velocity is usually denoted by u^*.

Since from equation 12.35, $\lambda_E = 0.4y$, then:

$$u^* = 0.4y\,du_x/dy$$

Rearranging: $du_x/u^* = dy/0.4y$

and integrating: $(u_x/u^*) = 2.5\ln y + \text{const.}$ (i)

At $y = r$:

$$(u_{max}/u^*) = 2.5\ln r + \text{const.}$$

or: $\text{const.} = (u_{max}/u^*) - 2.5\ln r$

Substituting in equation (i):

$$(u_{max}/u^*) - 2.5\ln r = (u_x/u^*) - 2.5\ln y$$

and: $(u_{max} - u_x)/u^* = 2.5\ln(r/y)$ (ii)

The mean velocity: $u = \int_0^r (2\pi(r - y)\,dyu_x)/\pi r^2$

and dividing by r: $u = 2\int_0^1 (1 - y/r)\,d(y/r)u_x$

Substituting for u_x from equation (ii):

$$u = 2\int_0^1 (1 - y/r)\,d(y/r)(u_{max} + 2.5u^*\ln(y/r))$$

$$= 2\left([u_{max} + 2.5\ln(y/r)][(y/r) - 0.5(y/r)^2]\Big|_0^1\right.$$

$$\left. -u^*\int_0^1 2.5(r/y)[(y/r) - 0.5(y/r)^2]d(y/r)\right)$$

$$= 2\left(u_{max}(0.5) - 2.5u^*\left[(y/r) - 0.25(y/r)^2\right]_0^1\right) = u_{max} - 3.75u^*$$

When $(u/u_{max}) = 0.8$, $u = (u/0.8) - 3.75u\sqrt{(R/\rho u^2)}$

∴ $\sqrt{(R/\rho u^2)} = 0.0667$ and $\underline{\underline{(R/\rho u^2) = 0.00444}}$

PROBLEM 12.8

The velocity profile in the neighbourhood of a surface for a Newtonian fluid may be expressed in terms of a dimensionless velocity u^+ and a dimensionless distance y^+ from the surface. Obtain the relation between u^+ and y^+ in the laminar sub-layer. Outside the laminar sub-layer, the relation is:

$$u^+ = 2.5 \ln y^+ + 5.5$$

At what value of y^+ does the transition from the laminar sub-layer to the turbulent zone occur?

In the "Universal Velocity Profile", the laminar sub-layer extends to values of $y^+ = 5$ and the turbulent zone starts at $y^+ = 30$ and the range $5 < y^+ < 30$, the buffer layer, is covered by a second linear relation between u^+ and $\ln y^+$. What is the *maximum* difference between the values of u^+, in the range $5 < y^+ < 30$, using the two methods of representation of the velocity profile?

Definitions: $u^+ = (u_x/u^*)$, $y^+ = (yu^*\rho)/\mu$ and $u^{*2} = R/\rho$ where u_x is the velocity at a distance y from the surface, R is the wall shear stress and ρ and μ are the density and viscosity of the fluid respectively.

Solution

As discussed in Section 12.4.2, if the velocity gradient du_x/dy approaches a constant value near the surface, (d^2u_x/dy^2) approaches zero and $R = \mu u_x/y$.

$\therefore$
$$u^{*2} = \mu u_x/(\rho y)$$

and, as given in equation 12.39:

$$(u_x/u^*) = yu^*\rho/\mu = y^+$$

Hence: $$u^+ = y^+$$ (equation 12.40)

Since $u^+ = 2.5 \ln y^+ + 5.5$ and $y^+ = 2.5 \ln y^+ + 5.5$, then solving by trial and error, the transition from the laminar sub-layer to the turbulent zone occurs when:

$$y^+ = \underline{11.6} \text{ and } u^+ = 2.5 \ln 11.6 + 5.5 = \underline{11.62}$$

For the buffer layer, $u^+ = a \ln y^+ + a'$ (equation 12.41)

When $y^+ = 5$, $u^+ = 5$ and when $y^+ = 30$, $u^+ = 2.5 \ln 30 + 5.5 = 14$

$\therefore$ $$5 = a \ln 5 + a'$$

$\therefore$ $$a' = 5 - a \ln 5 = 5 - 1.609a$$

and: $$14 = a \ln 30 + a' = a \ln 30 + 5 - 1.609a$$

$\therefore$ $$9 = 3.401a - 1.609a \text{ and } a = 5.02$$

and: $$a' = 5 - (1.609 \times 5.02) = -3.08$$

The difference between the two values of u^+ is a maximum when $y^+ = 11.6$.

From the two-layer theory: $u^+ = 11.6$

From the buffer-layer theory:

$$u^+ = 5.02 \ln 11.6 - 3.08 = 9.2$$

and hence, the maximum difference is $(11.6 - 9.2) = \underline{\underline{2.4}}$

PROBLEM 12.9

Calculate the rise in temperature of water flowing at 4 m/s through a smooth 25 mm diameter pipe 6 m long. The water enters at 300 K and the temperature of the wall of the tube can be taken as approximately constant at 330 K. Use:

(a) The simple Reynolds analogy,
(b) The Taylor-Prandtl modification,
(c) The buffer layer equation,
(d) $Nu = 0.023 \, Re^{0.8} \, Pr^{0.33}$.

Comment on the differences in the results so obtained.

Solution

See Volume 1, Example 12.3.

PROBLEM 12.10

Calculate the rise in temperature of a stream of air, entering at 290 K and flowing at 4 m/s through the tube maintained at 350 K; other conditions remaining the same as detailed in Problem 12.9.

Solution

See Volume 1, Example 12.4.

PROBLEM 12.11

Air flows through a smooth circular duct of internal diameter 0.25 m at an average velocity of 15 m/s. Calculate the fluid velocity at points 50 mm and 5 mm from the wall. What will be the thickness of the laminar sub-layer if this extends to $u^+ = y^+ = 5$? The density of air may be taken as 1.12 kg/m^3 and the viscosity of air as 0.02 mN s/m^2.

Solution

See Volume 1, Example 12.1.

PROBLEM 12.12

Obtain the Taylor–Prandtl modification of the Reynolds analogy for momentum and heat transfer, and give the corresponding relation for mass transfer (no bulk flow).

An air stream at approximately atmospheric temperature and pressure, and containing a low concentration of carbon disulphide vapour, is flowing at 38 m/s through a series of 50 mm diameter tubes. The inside of the tubes is covered with a thin film of liquid and both heat and mass transfer are taking place between the gas stream and the liquid film. The film heat transfer coefficient is found to be 100 W/m² K. Using a pipe friction chart and assuming the tubes to behave as smooth surfaces, calculate:

(a) the film mass transfer coefficient, and
(b) the gas velocity at the interface between the laminar sub-layer and the turbulent zone of the gas. Specific heat of air $= 1.0$ kJ/kg K. Viscosity of air $= 0.02$ mN s/m². Diffusivity of carbon disulphide vapour in air $= 1.1 \times 10^{-5}$ m²/s. Thermal conductivity of air $= 0.024$ W/m K.

Solution

The Taylor–Prandtl modification of the Reynolds analogy for heat transfer and mass transfer is discussed in Section 12.8.3 and the relevant equations are:

For heat transfer: $(R/\rho u^2) = (h/C_p \rho u_s)(1 + \alpha(Pr - 1)) = \phi$ (equation 12.119)

or: $(h/C_p \rho u_s) = \phi/(1 + \alpha(Pr - 1))$ (i)

For mass transfer: $(R/\rho u^2) = (h_D/u_s)(1 + \alpha(Sc - 1)) = \phi$ (equation 12.120)

or: $(h_D/u_s) = \phi/(1 + \alpha(Sc - 1))$ (ii)

Taking the molecular mass of air as 29 kg/kmol and atmospheric temperature as 293 K, the density, $\rho = (29/22.4)(273/293) = 1.206$ kg/m³.

∴ $Re = du\rho/\mu = (50 \times 10^{-3} \times 38 \times 1.206)/(0.02 \times 10^{-3}) = 114{,}570$

and from Fig. 3.7: $\phi = (R/\rho u^2) = 0.0021$.

From Table 1.3, the Prandtl number, $Pr = C_p \mu/k$
$$= (1.0 \times 10^3 \times 0.02 \times 10^{-3})/0.024 = 0.833$$

From Table 1.3, the Schmidt number, $Sc = \mu/\rho D$
$$= (0.02 \times 10^{-3})/(1.206 \times 1.1 \times 10^{-5})$$
$$= 1.508$$

Substituting in equation (i):

$$(100/(1.0 \times 10^3 \times 1.206 \times 38)) = 0.0021/(1 + \alpha(0.833 - 1))$$

∴ $0.00218 = 0.0021/(1 - 0.167\alpha)$ and $\underline{\underline{\alpha = 0.22}}$

Substituting in equation (ii):

$$(h_D/38) = 0.0021/(1 + 0.22(1.508 - 1)) = 0.00189 \text{ and } h_D = \underline{\underline{0.072 \text{ m/s}}}$$

The gas velocity at the interface of the laminar sub-layer and the turbulent zone, u_b may also be estimated from:

$$(u_b/u) = 2.32\,Re^{-0.125}$$ (equation 12.60)

or: $u_b = (38 \times 2.32)(114{,}570)^{-0.125} = \underline{\underline{20.6 \text{ m/s}}}$

PROBLEM 12.13

Obtain the Taylor–Prandtl modification of the Reynolds' analogy between momentum and heat transfer and give the corresponding analogy for mass transfer. For a particular system a mass transfer coefficient of 8.71×10^{-6} m/s and a heat transfer coefficient of 2730 W/m^{2}K were measured for similar flow conditions. Calculate the ratio of the velocity in the fluid where the laminar sub-layer terminates, to the stream velocity. Molecular diffusivity $= 1.5 \times 10^{-9}$ m^2/s. Viscosity $= 1$ mN s/m^2. Density $= 1000$ kg/m^3. Thermal conductivity $= 0.48$ W/m K. Specific heat capacity $= 4.0$ kJ/kg K.

Solution

The Taylor–Prandtl modification to heat and mass transfer is discussed in Section 12.8.3 resulting in the modified Lewis relation:

$$h_D = (h/C_p\rho)[1 + \alpha(Pr - 1)]/[1 + \alpha(Sc - 1)] \qquad \text{(equation 12.121)}$$

In this case: $Pr = C_p\mu/k = (4 \times 10^3 \times 1 \times 10^{-3})/0.48 = 8.33$

and: $Sc = \mu/\rho D = (1 \times 10^{-3})/(1000 \times 1.5 \times 10^{-9}) = 667$

Substituting:

$$(8.71 \times 10^{-6}) = [2730/(4000 \times 1000)][1 + \alpha(8.33 - 1)]/[1 + \alpha(667 - 1)]$$

$$0.01276 = (1 + 7.33\alpha)/(1 + 666\alpha) \text{ and } \underline{\alpha = 0.844}$$

PROBLEM 12.14

Heat and mass transfer are taking place simultaneously to a surface under conditions where the Reynolds analogy between momentum, heat and mass transfer may be applied. The mass transfer is of a single component at a high concentration in a binary mixture, the other component of which undergoes no net transfer. Using the Reynolds analogy, obtain a relation between the coefficients for heat transfer and for mass transfer.

Solution

The solution to this problem is presented in Sections 12.8.1 and 12.8.2 and the relation between the coefficients for heat transfer and mass transfer is:

$$h = (C_{Bw}/C_T)C_p\rho h_D \qquad \text{(equation 12.112)}$$

PROBLEM 12.15

Derive the Taylor–Prandtl modification of the Reynolds analogy between momentum and heat transfer.

In a shell and tube type condenser, water flows through the tubes which are 10 m long and 40 mm diameter. The pressure drop across the tubes is 5.6 kN/m^2 and the effects of entry and exit losses may be neglected. The tube walls are smooth and flow may be taken as fully developed. The ratio of the velocity at the edge of the laminar sub-layer to the mean velocity of flow may be taken as $2\,Re^{-0.125}$, where Re is the Reynolds number in the pipeline.

If the tube walls are at an approximately constant temperature of 393 K and the inlet temperature of the water is 293 K, estimate the outlet temperature. Physical properties of water: density $= 1000$ kg/m^3, viscosity $= 1$ mNs/m^2, thermal conductivity $= 0.6$ W/m K, specific heat capacity $= 4.2$ kJ/kg K.

Solution

The Taylor–Prandtl modification of the Reynolds analogy to heat transfer, discussed in Section 12.8.3, leads to:

$$St = (h/C_p\rho u) = (R/\rho u^2)/[1 + \alpha(Pr - 1)] \qquad \text{(equation 12.117) (i)}$$

From equation 3.23:

$$(R/\rho u^2)\,Re^2 = -\Delta P_f d^3 \rho/(4l\mu^2)$$

$$= (5600 \times (40/1000)^3 \times 1000)/(4 \times 10 \times (1 \times 10^{-3})^2) = 8{,}960{,}000$$

From Fig. 3.8: $Re = 62{,}000$ and $R/\rho u^2 = 8{,}960{,}000/(62{,}000)^2 = 0.0023$

The ratio of the velocity at the edge of the laminar sub-layer to the mean velocity of flow is:

$$\alpha = 2\,Re^{-0.125} = 2/(62000)^{0.125} = 0.5035$$

The Prandtl group, $Pr = C_p\mu/k = (4200 \times 1 \times 10^{-3})/0.6 = 7.0$

and from equation (i), the Stanton group, $St = (R/\rho u^2)/[1 + \alpha(Pr - 1)]$

$$= 0.0023/[1 + 0.5035(7.0 - 1)] = 0.000572$$

$$\therefore \qquad h = 0.000572 C_p\rho u = 0.000572 C_p\,Re\,\mu/d$$

$$= (0.000572 \times 4200 \times 62000 \times 1 \times 10^{-3})/(40/1000) = 3724 \text{ W/m}^2 \text{ K.}$$

The area for heat transfer per unit length of pipe $= (40\pi/1000) = 0.04\pi = 0.126$ m^2/m and making a heat balance over unit length of pipe dl:

$$\therefore \qquad hA\,\mathrm{d}l\,(T_w - T) = \rho u C_p A_c \mathrm{d}T = (Re\,\mu/d)C_p A_c \mathrm{d}T$$

$$\therefore \qquad 3724 \times 0.126\,\mathrm{d}l\,(393 - T) = (62000 \times 1 \times 10^{-3}/(40/1000))$$

$$\times 4200 \times (\pi/4)(40/1000)^2\,\mathrm{d}T$$

$$\therefore \qquad 0.0572\,\mathrm{d}l = \mathrm{d}T/(393 - T)$$

Integrating:

$$0.0572 \int_0^{10} \mathrm{d}l = \int_{293}^{T_0} \mathrm{d}T/(393 - T)$$

$$(0.0572 \times 10) = \ln[(393 - 293)/(393 - T_0)]$$

$$\therefore \qquad 1.772 = 100/(393 - T_0) \text{ and } T_0 = \underline{\underline{336.6 \text{ K}}}$$

PROBLEM 12.16

Explain the importance of the universal velocity profile and derive the relation between the dimensionless derivative of velocity u^+, and the dimensionless derivative of distance from the surface y^+, using the concept of Prandtl's mixing length λ_E.

It may be assumed that the fully turbulent portion of the boundary layer starts at $y^+ = 30$, that the ratio of the mixing length λ_E to distance y from the surface, $\lambda_E/y = 0.4$, and that for a smooth surface $u^+ = 14$ at $y^+ = 30$.

If the laminar sub-layer extends from $y^+ = 0$ to $y^+ = 5$, obtain the equation for the relation between u^+ and y^+ in the buffer zone, and show that the ratio of the eddy viscosity to the molecular viscosity increases linearly from 0 to 5 through this buffer zone.

Solution

The importance of the universal velocity profile is discussed in Section 12.4. From equation 12.18, for isotropic turbulence, the eddy kinematic viscosity, $E \propto \lambda_E u_E$ where λ_E is the mixing length and u_E is some measure of the linear velocity of the fluid in the eddies. The momentum transfer rate per unit area in a direction perpendicular to the surface at position y is then:

$$R_y = -E\,\mathrm{d}(\rho u_x)/\,\mathrm{d}y$$

and for constant density,
$$R_y = -E\rho\,\mathrm{d}u_x/\,\mathrm{d}y \qquad \text{(equation 12.20)}$$

or:
$$-R_y/\rho = E\,\mathrm{d}u_x/\,\mathrm{d}y$$

where $\sqrt{(R_y/\rho)}$, the friction velocity, may be denoted by u^* and then $u^{*2} = E\,\mathrm{d}u_x/\mathrm{d}y$

Assuming $E = \lambda_E u_E$, that is a proportionality constant of unity, and $u_E = \lambda_E |\,\mathrm{d}u_x/\,\mathrm{d}y|$, then:

$$u^{*2} = \lambda_E^2 (\mathrm{d}u_x/\mathrm{d}y)|(\mathrm{d}u_x/\mathrm{d}y)|$$

and hence near the surface where $(\mathrm{d}u_x/\mathrm{d}y)$ is positive:

$$u^* = \lambda_E (\mathrm{d}u_x/\mathrm{d}y)$$

Assuming $\lambda_E = Ky$:
$$u^*\mathrm{d}y/y = K\mathrm{d}u_x \qquad \text{(equation 12.28)}$$

Integrating: $\quad u_x/u^* = (1/K)\ln y + B$ where B is a constant

or: $\quad u_x/u^* = (1/K)\ln(yu^*\rho/\mu) + B' \qquad \text{(equation 12.29)}$

Since $(u^*\rho/\mu)$ is constant, B' is also constant and, writing the dimensionless velocity term, (u_x/u^*) as u^+ and the dimensionless derivative of $y(yu^*\rho/\mu)$ as y^+, then:

$$u^+ = (1/K)\ln y^+ + B' \qquad \text{(equation 12.30)}$$

Given that $K = 0.4$, then: $\quad u^+ = 2.5\ln y^+ + B'$

Given that for a smooth surface, $u^+ = 14$ at $y^+ = 30$, then:

$$B' = 14 - 2.5 \ln 30 = 5.5$$

and: $$u^+ = 2.5 \ln y^+ + 5.5 \qquad \text{(equation 12.37)}$$

For molecular transfer in the laminar sub-layer near the wall, from Section 12.4.2:

$$R_y = -\mu u_x/y$$

or: $$(-R_y/\rho) = u^{*2} = \mu u_x/(y\rho)$$

$\therefore$ $$\qquad (u_x/u^*) = y\rho u^*/\mu \text{ and } u^+ = y^+ \qquad \text{(equation 12.40)}$$

If the buffer zone stretches from $y^+ = 5$ to $y^+ = 30$ at which u^+ is 5 and 14 respectively, then in equation 12.41:

$$u^+ = a \ln y^+ + a'$$

or: $$5 = a \ln 5 + a' \text{ and } 14 = a \ln 30 + a'$$

and: $$u^+ = 5.0 \ln y^+ - 3.05 \qquad \text{(equation 12.42)}$$

From equation 12.46, the velocity gradient, $du^+/dy^+ = 5/y^+$

From equation 12.61: $R_y = -(\mu + E\rho)du_x/dy$

and substituting $u^* = \sqrt{(R_y/\rho)}$:

$$(du_x/dy) = u^{*2}/(E + \mu/\rho) \qquad \text{(equation 12.62)}$$

$\therefore$ $$(du^+/dy^+) = (\mu/\rho)(1/[E + \mu/\rho]) = 5/y^+$$

and: $$E/(\mu/\rho) = (y^+/5) - 1 \qquad \text{(equation 12.63)}$$

Hence as y^+ goes from 5 to 30, the ratio of the eddy kinematic viscosity to the kinematic viscosity goes from 0 to 5.

PROBLEM 12.17

Derive the Taylor–Prandtl modification of the Reynolds analogy between heat and momentum transfer and express it in a form in which it is applicable to pipe flow.

If the relationship between the Nusselt number Nu, Reynolds number Re and Prandtl number Pr is:

$$Nu = 0.023 \, Re^{0.8} \, Pr^{0.33}$$

calculate the ratio of the velocity at the edge of the laminar sub-layer to the velocity at the pipe axis for water ($Pr = 10$) flowing at a Reynolds number of 10,000 in a smooth pipe. Use the pipe friction chart.

Solution

The derivation of the Taylor–Prandtl modification of the Reynolds analogy between heat and momentum transfer is presented in Section 12.8.3 and the result is summarised as:

$$St = h/(C_p\rho u_s) = (R/\rho u_s^2)/[1 + \alpha(Pr - 1)] \quad \text{(equation 12.117)}$$

or: $(R/\rho u_s^2) = [h/(C_p\rho u_s)][1 - \alpha(1 - Pr)]$

For turbulent pipe flow, u_s is approximately equal to $(u_{mean}/0.82)$ and:

$$0.82(R/\rho u^2) = [h/(C_p\rho u)][1 - \alpha(1 - Pr)]$$

When $Re = 10,000$, then from Fig. 3.1, $(R/\rho u^2) = 0.0038$, and for $Pr = 10$:

$$(0.82 \times 0.0038) = St[1 - \alpha(1 - 10)]$$

or: $St(1 + 9\alpha) = 0.0031$ (i)

But: $Nu = 0.023\,Re^{0.8}\,Pr^{0.33}$

and: $St = Nu/(Re \cdot Pr) = 0.023\,Re^{-0.2}\,Pr^{-0.67}$

$$= 0.023(10,000)^{-0.2}(10)^{-0.67} = 0.000777$$

Hence, substituting in equation (i):

$$0.000777(1 + 9\alpha) = 0.0031 \text{ and } \alpha = (u_b/u_s) = \underline{0.33}$$

PROBLEM 12.18

Obtain a dimensionless relation for the velocity profile in the neighbourhood of a surface for the turbulent flow of a liquid, using Prandtl's concept of a "Mixing Length" (Universal Velocity Profile). Neglect the existence of the buffer layer and assume that, outside the laminar sub-layer, eddy transport mechanisms dominate. Assume that in the turbulent fluid the mixing length λ_E is equal to 0.4 times the distance y from the surface and that the dimensionless velocity u^+ is equal to 5.5 when the dimensionless distance y^+ is unity.

Show that, if the Blasius relation is used for the shear stress R at the surface, the thickness of the laminar sub-layer δ_b is approximately 1.07 times that calculated on the assumption that the velocity profile in the turbulent fluid is given by Prandtl's one seventh power law.

Blasius Equation:

$$\frac{R}{\rho u_s^2} = 0.0228 \left(\frac{u_s \delta \rho}{\mu}\right)^{-0.25}$$

where ρ, μ are the density and viscosity of the fluid, u_s is the stream velocity, and δ is the total boundary layer thickness.

Solution

The Universal Velocity Profile is discussed in detail in Section 12.4, and in the region where eddy transport dominates ($y^+ > 30$) and making all the stated assumptions:

$$u^+ = 2.5 \ln y^+ + 5.5 \qquad\qquad \text{(equation 12.37)}$$

If, in the laminar sub-layer (from equation 12.40), $u^+ = y^+$ then:

$$y^+ = 2.5 \ln y^+ + 5.5$$

and, solving by trial and error: $y^+ = 11.6 = u_s^* \delta_b \rho / \mu$ $\qquad$ (from equation 12.44)

Since $u^* = \sqrt{(R/\rho)}$, then: $\delta_b = 11.6\mu/(\rho u^*) = 11.6\mu/\sqrt{(R\rho)}$ $\qquad\qquad$ (i)

But, from the Blasuis equation: $R/\rho u^2 = 0.0228(u_s \delta \rho/\mu)^{-0.25}$

and substituting for R in equation (i), $\delta_b = (11.6\mu/\sqrt{\rho})(1/u_s\sqrt{\rho})(1/0.0228)^{-0.5}$ $(u_s \delta \rho/\mu)^{0.125}$

and: $\qquad\qquad\qquad\qquad (\delta_b/\delta) = 76.8(u_s \delta \rho/\mu)^{-0.875}$

Using Prandtl's one seventh power law, $(u_b/u_s) = (\delta_b/\delta)^{1/7} = (\delta_b/\delta)^{0.143}$

But: $\qquad\qquad\qquad R = \mu u_b/\delta_b = 0.0228\rho u_s^2 (u_s \delta \rho/\mu)^{-0.25}$

∴ $\qquad\qquad (\mu/\delta_b)u_s(\delta_b/\delta)^{0.143} = 0.0228\rho u_s^2 (u_s \delta \rho/\mu)^{-0.25}$

and: $\qquad\qquad\qquad\qquad (\delta_b/\delta) = 82.38(u_s \delta \rho/\mu)^{-0.875}$

The ratio of the values obtained using the two approaches to the problem is:

$$(82.38/76.8) = \underline{\underline{1.073}}$$

PROBLEM 12.19

Obtain the Taylor–Prandtl modification of the Reynolds analogy between momentum transfer and mass transfer (equimolecular counterdiffusion) for the turbulent flow of a fluid over a surface. Write down the corresponding analogy for heat transfer. State clearly the assumptions which are made. For turbulent flow over a surface, the film heat transfer coefficient for the fluid is found to be 4 kW/m^2 K. What would the corresponding value of the mass transfer coefficient be, given the following physical properties? Diffusivity $D = 5 \times 10^{-9}$ m^2/s. Thermal conductivity, $k = 0.6$ W/m K. Specific heat capacity $C_p = 4$ kJ/kg K. Density, $\rho = 1000$ kg/m^3. Viscosity, $\mu = 1$ mNs/m^2.

Assume that the ratio of the velocity at the edge of the laminar sub-layer to the stream velocity is (a) 0.2, (b) 0.6.

Comment on the difference in the two results.

Solution

The discussion of the Reynolds analogy is presented in Section 12.8 and consideration of mass transfer in Sections 12.8.2 and 12.8.3 leads to the modified Lewis relation which may be written as:

$$h_D[1 - \alpha(1 - Sc)] = (h/C_p\rho)[1 - \alpha(1 - Pr)] \quad \text{(equation 12.121) (i)}$$

The Schmidt group, $Sc = \mu/\rho D = (1 \times 10^{-3})/(1000 \times 5 \times 10^{-9}) = 200$

The Prandtl group, $Pr = C_p\mu/k = (4 \times 10^3 \times 1 \times 10^{-3})/0.6 = 6.67$

and: $(h/C_p\rho) = 4000/(4 \times 10^3 \times 1000) = 0.001$

Thus, in equation (i):

$$h_D[1 - \alpha(1 - 200)] = 0.001[1 - \alpha(1 - 6.67)]$$

and: $h_D = 0.001(1 + 5.67\alpha)/(1 + 199\alpha)$ m/s   (ii)

When $\alpha = 0.2$, then from equation (ii),

$$h_D = 0.001(1 + 1.134)/(1 + 39.8) = \underline{\underline{5.2 \times 10^{-5}}} \text{ m/s}$$

When $\alpha = 0.6$, then from equation (ii),

$$h_D = 0.001(1 + 3.402)/(1 + 119.4) = \underline{\underline{3.6 \times 10^{-5}}} \text{ m/s}$$

It is worth noting that even with a very large variation in α (threefold in fact) the change in the mass transfer coefficient is less than 50%.

PROBLEM 12.20

By using the simple Reynolds analogy, obtain the relation between the heat transfer coefficient and the mass transfer coefficient for the gas phase for the absorption of a soluble component from a mixture of gases. If the heat transfer coefficient is 100 W/m² K, what will the mass transfer coefficient be for a gas of specific heat capacity 1.5 kJ/kg K and density 1.5 kg/m³? The concentration of the gas is sufficiently low for bulk flow effects to be negligible.

Solution

From Section 12.8.1, the heat transfer coefficient is given by:

$$(R/\rho u^2) = h/(C_p\rho u_s) \qquad \text{(equation 12.102)}$$

and the mass transfer coefficient by: $(R/\rho u^2) = h_D/u_s$ \qquad (equation 12.103)

Hence: $\underline{\underline{h_D = h/(C_p\rho)}}$ \qquad (equation 12.105)

In this case: $h_D = 100/(1.5 \times 10^3 \times 1.5) = \underline{\underline{0.044 \text{ m/s}}}$

PROBLEM 12.21

The velocity profile in the neighbourhood of a surface for a Newtonian fluid may be expressed in terms of a dimensionless velocity u^+ and a dimensionless distance y^+ from the surface. Obtain the relation between u^+ and y^+ in the laminar sub-layer. Outside the laminar sub-layer, the relation takes the form:

$$u^+ = 2.5 \ln y^+ + 5.5$$

At what value of y^+ does the transition from the laminar sub-layer to the turbulent zone occur?

In the *Universal Velocity Profile*, the laminar sub-layer extends to values of $y^+ = 5$, the turbulent zone starts at $y^+ = 30$ and the range $5 < y^+ < 30$, the buffer layer, is covered by a second linear relation between u^+ and $\ln y^+$. What is the *maximum* difference between the values of u^+, in the range $5 < y^+ < 30$, using the two methods of representation of the velocity profile?

Definitions:

$$u^+ = \frac{u_x}{u^*}$$

$$y^+ = \frac{y u^* \rho}{\mu}$$

$$u^{*2} = R/\rho$$

where u_x is velocity at distance y from surface

R is wall shear stress

ρ, μ are the density and viscosity of the fluid respectively.

Solution

Laminar sub-layer

If the velocity gradient approaches constant value in the laminar sub-layer, then $d^2 u_x/dy^2 \to 0$ and:

$$R = \mu \frac{u_x}{y}$$

$$u^{*2} = \frac{\mu u_x}{\rho y}$$

and: $\qquad \dfrac{u_x}{u^*} = \dfrac{u^* y \rho}{\mu}$ or: $u^+ = y^+$, by definition

The point of interaction of $u^+ = y^+$

and: $\qquad\qquad\qquad\qquad u^+ = 2.5 \ln y^+ + 5.5$

is given by: $\qquad\qquad\qquad\qquad y^+ = 2.5 \ln y^+ + 5.5$

Evaluating: y^+:

y^+	RHS
10	11.26
15	12.27
20	13.00
25	13.5
8	10.7
7	10.4
12	11.71
11	11.5
11.5	11.6
11.6	11.62

and: $y^+ = 11.6$

The equation for *buffer zone* may be written as:

$$u^+ = A \ln y^+ + B$$

$$\text{When } y^+ = 5, u^+ = 5$$

$$\text{When } y^+ = 30, u^+ = 2.5 \ln 30 + 5.5$$

Thus: $$5 = A \ln 5 + B$$

$$5.5 + 2.5 \ln 30 = A \ln 30 + B$$

Substracting: $$0.5 + 2.5 \ln 30 = A \ln 6$$

Thus: $A = (0.5 + 2.5 \ln 30)/\ln 6 = 5.02$ and: $B = 5 - A \ln 5 = -3.08$

The difference in the two values of u^+ is a maximum when $y^+ = 11.6$.

From the two-layer theory: $u^+ = 11.6$

From the buffer-layer theory:

$$u^+ = 5.02 \ln +11.6 - 3.08 = 9.2$$

The maximum difference in the two values of u^+ is then: 2.4

PROBLEM 12.22

In the universal velocity profile a "dimensionless" velocity u^+ is plotted against $\ln y^+$, where y^+ is a "dimensionless" distance from the surface. For the region where eddy transport dominates (eddy kinematic viscosity $\gg$ kinematic viscosity), the ratio of the mixing length (λ_E) to the distance (y) from the surface may be taken as approximately constant and equal to 0.4. Obtain an expression for du^+/dy^+ in terms of y^+.

In the buffer zone the ratio of du^+/dy^+ to y^+ is *twice* the value calculated above. Obtain an expression for the eddy kinematic viscosity E in terms of the kinematic viscosity (μ/ρ) and y^+. On the assumption that the eddy thermal diffusivity E_H and the eddy kinematic viscosity E are equal, calculate the value of the temperature gradient in a liquid flowing over the surface at $y^+ = 15$ (which lies within the buffer layer) for a surface heat flux of 1000 W/m^2. The liquid has a Prandtl number of 7 and a thermal conductivity of 0.62 W/m K.

Solution

a) *In the region where eddy effects considerably exceed molecular contributions:*

$$E \gg \mu/\rho$$

Outside the laminar sub-layer and the buffer-layer but still close to the surface:
$R = E\rho(du_x/dy)$ where R is shear stress at surface

Writing E as $\lambda_E u_E$, and approximating, $u_E = \lambda_E |du_x/dy|$ gives:

$$\frac{R}{\rho} = \lambda_E^2 \left(\frac{du_x}{dy}\right)\left(\frac{du_x}{dy}\right)$$

where the modulus sign is dropped as du_x/dy is positive near a surface.

Putting $u^* = \sqrt{R/\rho}$, the shearing stress velocity:

$$u^{*2} = \lambda_E^2 \left(\frac{du_x}{dy}\right)^2$$

$$u^* = \lambda_E \frac{du_x}{dy}$$

Using the Prandtl approximation $\lambda_E = 0.4\, y$ gives:

$$u^* = 0.4\, y \frac{du_x}{dy}$$

$$\therefore \qquad \frac{du_x}{dy} = 2.5 \frac{u^*}{y}$$

Writing $y^+ = yu^*\rho/\mu$ and $u^+ = u_x/u^*$, then:

$$u^* \frac{du^+}{dy^+} = 2.5 \frac{u^*}{y^+}$$

and: $\qquad \dfrac{du^+}{dy^+} = \dfrac{2.5}{y^+}$

b) *In the buffer zone:*

It is given that: $\qquad \dfrac{du^+}{dy^+} = 2\left(\dfrac{2.5}{y^+}\right) = \dfrac{5}{y^+}$

The shear stress is given by:

$$R = (\mu + E\rho)\frac{du_x}{dy}$$

$$\therefore \qquad u^{*2} = \frac{R}{\rho} = \left(\frac{\mu}{\rho} + E\right)\frac{du_x}{dy}$$

and:
$$u^{*2} = \left(\frac{\mu}{\rho} + E\right) \frac{du^+}{dy^+} \frac{u^*}{\mu/\rho u^*}$$

Hence:
$$\frac{\mu}{\rho} = \left(\frac{\mu}{\rho} + E\right) \frac{du^+}{dy^+} = \left(\frac{\mu}{\rho} + E\right) \frac{5}{y^+}$$

When $y^+ = 15$, then: $\mu/\rho = (\mu/\rho + E)/3$

and:
$$E = 2\mu/\rho$$

iii) *For heat transfer in buffer zone:* $q = -(k + C_p \rho E_H) \dfrac{dT}{dy}$

Writing $E_H = E$ gives: $q = -(k + C_p \rho E) \dfrac{dT}{dy}$

When $y^+ = 15$, then putting $E = 2\mu/\rho$ gives:

$$q = -(k + 2C_p \mu) \frac{dT}{dy} = -k(1 + 2Pr) \frac{dT}{dy}$$

Thus:
$$\frac{dT}{dy} = \frac{-q}{k(1 + 2Pr)}$$

Putting: $k = 0.62$ W/m K, $Pr = 7$ and $-q = 1000$ W

then: $dT/dy = 108$ deg K/m or $\underline{\underline{0.108 \text{ deg K/mm}}}$

PROBLEM 12.23

Derive an expression relating the pressure drop for the turbulent flow of a fluid in a pipe to the heat transfer coefficient at the walls on the basis of the simple Reynolds analogy. Indicate the assumptions which are made and the conditions under which it would be expected to apply closely.

Air at 320 K and atmospheric pressure is flowing through a smooth pipe of 50 mm internal diameter and the pressure drop over a 4 m length is found to be 150 mm water gauge. By how much would the air temperature be expected to fall over the first metre if the wall temperature there is 290 K? Viscosity of air $= 0.018$ mN s/m². Specific heat capacity $(C_p) = 1.05$ kJ/kg K. Molecular volume $= 22.4$ m³/kmol at 1 bar and 273 K.

Solution

If a mass of fluid, m, situated at a distance from a surface, is moving parallel to the surface with a velocity of u_s, and it then moves to the surface, where the velocity is zero, it will give up its momentum mu_s in time t. If the temperature difference between the mass of fluid and the surface is θ_s, then the heat transferred to the surface is $(mC_p \theta_s)$ and over a surface of area, A:

$$(mC_p \theta_s)/t = -qA$$

where q is the heat transferred from the surface per unit area per unit time.

If the shear stress at the surface is R_0, the shearing force over area A is the rate of change of momentum or:

$$(mu_s)/t = R_0 A \quad \text{and} \quad C_p \theta_s / u_s = q/R_0$$

Writing $R_0 = -R$, the shear stress acting on the walls and h as the heat transfer coefficient between the fluid and the surface, then:

$$-q/\theta_s = h = -R_0 C_p / u_s = RC_p / u_s \quad \text{or} \quad h/C_p \rho u_s = R/\rho u^2$$

From equation 3.19, the pressure change due to friction is given by:

$$-\Delta P = 4(R/\rho u^2)(l/d)(\rho u^2)$$

and substituting from equation 12.102:

$$-\Delta P = 4(h/C_p \rho u)(l/d)(\rho u^2) = \underline{\underline{4(hu/C_p)(l/d)}}$$

The Reynolds analogy assumes no mixing with adjacent fluid and that turbulence persists right up to the surface. Further it is assumed that thermal and kinematic equilibria are reached when an element of fluid comes into contact with a solid surface. No allowance is made for variations in physical properties of the fluid with temperature.

A further discussion of the Reynolds analogy for heat transfer is presented in Chapter 12.

Density of air at 320 K $= (29/22.4)(273 \times 320) = 1.105$ kg/m^3.

The pressure drop: $-\Delta P = 150$ mm water $= (9.8 \times 150) = 1470$ N/m^2

$$l = 4.0 \text{ m}, \quad d = 0.050 \text{ m}$$

In equation 3.23:

$$-\Delta P d^3 \rho/(4l\mu^2) = (1470 \times 0.050^3 \times 1.105)/[4 \times 4.0(0.018 \times 10^{-3})^2]$$

$$= 3.192 \times 10^7$$

From Fig. 3.8, for a smooth pipe: $Re = 1.25 \times 10^5$

and from Fig. 3.7: $R/\rho u^2 = 0.0021$

The heat transfer coefficient:

$$h = (R/\rho u^2)C_p \rho u = (0.0021 \times 1.05 \times 10^3)\rho u = 2.205\rho u \text{ W/m}^2 \text{ K}$$

Mass flowrate of air, $G = \rho u(\pi/4)0.050^2 = 0.00196\rho u$ kg/s
Area for heat transfer, $A = (\pi \times 0.050 \times 1.0) = 0.157$ m^2.

$$GC_p(T_1 - T_2) = hA(T_m - T_w)$$

where T_1 and T_2 are the inlet and outlet temperatures and T_m the mean value taken as arithmetic over the small length of 1 m.

$\therefore$ $(0.00196\rho u \times 1.050 \times 10^3)(320 - T_2) = (2.205\rho u \times 0.157)(0.5(320 \times T_2) - 290)$

and: $\qquad\qquad\qquad\qquad\qquad\qquad T_2 = 316$ K

The drop in temperature over the first metre is therefore $\underline{\underline{4 \deg \text{K}}}$.

SECTION 13

Humidification and Water Cooling

PROBLEM 13.1

In a process in which benzene is used as a solvent, it is evaporated into dry nitrogen. The resulting mixture at a temperature of 297 K and a pressure of 101.3 kN/m^2 has a relative humidity of 60%. It is required to recover 80% of the benzene present by cooling to 283 K and compressing to a suitable pressure. What must this pressure be? Vapour pressures of benzene: at 297 K $= 12.2$ kN/m^2: at 283 K $= 6.0$ kN/m^2.

Solution

See Volume 1, Example 13.1

PROBLEM 13.2

0.6 m^3/s of gas is to be dried from a dew point of 294 K to a dew point of 277.5 K. How much water must be removed and what will be the volume of the gas after drying? Vapour pressure of water at 294 K $= 2.5$ kN/m^2. Vapour pressure of water at 277.5 K $= 0.85$ kN/m^2.

Solution

When the gas is cooled to 294 K, it will be saturated and $P_{w0} = 2.5$ kN/m^2.
From Section 13.2:

$$\text{mass of vapour} = P_{w0}M_w/\mathbf{R}T = (2.5 \times 18)/(8.314 \times 294) = 0.0184 \text{ kg/m}^3 \text{ gas.}$$

When water has been removed, the gas will be saturated at 277.5 K, and $P_w = 0.85$ kN/m^2.

At this stage, mass of vapour $= (0.85 \times 18)/(8.314 \times 277.5) = 0.0066$ kg/m^3 gas

Hence, water to be removed $= (0.0184 - 0.0066) = 0.0118$ kg/m^3 gas

or: $$(0.0118 \times 0.6) = \underline{\underline{0.00708 \text{ kg/s}}}$$

Assuming the gas flow, 0.6 m^3/s, is referred to 273 K and 101.3 kN/m^2, 0.00708 kg/s of water is equivalent to $(0.00708/18) = 3.933 \times 10^{-4}$ kmol/s.

1 kmol of vapour occupies 22.4 m^3 at STP,

and: volume of water removed $= (3.933 \times 10^{-4} \times 22.4) = 0.00881$ m^3/s

Assuming no volume change on mixing, the gas flow after drying

$$= (0.60 - 0.00881) = \underline{\underline{0.591 \text{ m}^3/\text{s at STP}}}.$$

PROBLEM 13.3

Wet material, containing 70% moisture on a wet basis, is to be dried at the rate of 0.15 kg/s in a counter-current dryer to give a product containing 5% moisture (both on a wet basis). The drying medium consists of air heated to 373 K and containing water vapour with a partial pressure of 1.0 kN/m^2. The air leaves the dryer at 313 K and 70% saturated. Calculate how much air will be required to remove the moisture. The vapour pressure of water at 313 K may be taken as 7.4 kN/m^2.

Solution

The feed is 0.15 kg/s wet material containing 0.70 kg water/kg feed.

Thus water in feed $= (0.15 \times 0.70) = 0.105$ kg/s and dry solids $= (0.15 - 0.105) = 0.045$ kg/s.

The product contains 0.05 kg water/kg product. Thus, if w kg/s is the amount of water in the product, then:

$$w/(w + 0.045) = 0.05 \text{ or } w = 0.00237 \text{ kg/s}$$

and: water to be removed $= (0.105 - 0.00237) = 0.1026$ kg/s.

The inlet air is at 373 K and the partial pressure of the water vapour is 1 kN/m^2.

Assuming a total pressure of 101.3 kN/m^2, the humidity is:

$$\mathscr{H}_1 = [P_w/(P - P_w)](M_w/M_A) \qquad\qquad \text{(equation 13.1)}$$

$$= [1.0/(101.3 - 1.0)](18/29) = 0.0062 \text{ kg/kg dry air}$$

The outlet air is at 313 K and is 70% saturated. Thus, as in Example 13.1, Volume 1:

$$P_w = P_{w0} \times RH/100 = (7.4 \times 70/100) = 5.18 \text{ kN/m}^2$$

and: $\mathscr{H}_2 = [5.18/(101.3 - 5.18)](18/29) = 0.0335$ kg/kg dry air

The increase in humidity is $(0.0335 - 0.0062) = 0.0273$ kg/kg dry air and this must correspond to the water removed, 0.1026 kg/s. Thus if G kg/s is the mass flowrate of dry air, then:

$$0.0273G = 0.1026 \quad \text{and} \quad G = 3.76 \text{ kg/s dry air}$$

In the inlet air, this is associated with 0.0062 kg water vapour, or:

$$(0.0062 \times 3.76) = 0.0233 \text{ kg/s}$$

Hence, the mass of moist air required at the inlet conditions

$$= (3.76 + 0.0233) = 3.783 \text{ kg/s}$$

PROBLEM 13.4

30,000 m^3 of cool gas (measured at 289 K and 101.3 kN/m^2 saturated with water vapour) is compressed to 340 kN/m^2 pressure, cooled to 289 K and the condensed water is drained off. Subsequently the pressure is reduced to 170 kN/m^3 and the gas is distributed at this pressure and 289 K. What is the percentage humidity after this treatment? The vapour pressure of water at 289 K is 1.8 kN/m^2.

Solution

At 289 K and 101.3 kN/m^2, the gas is saturated and $P_{w0} = 1.8$ kN/m^2.

Thus from equation 13.2, $\mathcal{H}_0 = [1.8/(101.3 - 1.8)](18/M_A) = (0.3256/M_A)$ kg/kg dry gas, where M_A is the molecular mass of the gas.

At 289 K and 340 kN/m^2, the gas is in contact with condensed water and therefore still saturated. Thus $P_{w0} = 1.8$ kN/m^2 and:

$$\mathcal{H}_0 = [1.8/(340 - 1.8)](18/M_A) = (0.0958/M_A) \text{ kg/kg dry gas}$$

At 289 K and 170 kN/m^2, the humidity is the same, and in equation 13.2:

$$(0.0958/M_A) = [P_w/(170 - P_w)](18/M_A)$$

or:
$$P_w = 0.90 \text{ kN/m}^2$$

The percentage humidity is then:

$$= [(P - P_{w0})/(P - P_w)](100P_w/P_{w0}) \qquad \text{(equation 13.3)}$$

$$= [(170 - 1.8)/(170 - 0.90)](100 \times 0.90/1.8) = \underline{\underline{49.73\%}}$$

PROBLEM 13.5

A rotary countercurrent dryer is fed with ammonium nitrate containing 5% moisture at the rate of 1.5 kg/s, and discharges the nitrate with 0.2% moisture. The air enters at 405 K and leaves at 355 K; the humidity of the entering air being 0.007 kg moisture/kg dry air. The nitrate enters at 294 K and leaves at 339 K.

Neglecting radiation losses, calculate the mass of dry air passing through the dryer and the humidity of the air leaving the dryer. Latent heat of water at 294 K = 2450 kJ/kg. Specific heat capacity of ammonium nitrate = 1.88 kJ/kg K. Specific heat capacity of dry air = 0.99 kJ/kg K. Specific heat capacity of water vapour = 2.01 kJ/kg K.

Solution

The feed rate of wet nitrate is 1.5 kg/s containing 5.0% moisture or $(1.5 \times 5/100) = 0.075$ kg/s water.

$$\therefore \qquad \text{flow of dry solids} = (1.5 - 0.075) = 1.425 \text{ kg/s}$$

If the product contains w kg/s water, then:

$$w/(w + 1.425) = (0.2/100) \quad \text{or} \quad w = 0.00286 \text{ kg/s}$$

and: the water evaporated $= (0.075 - 0.00286) = 0.07215$ kg/s

The problem now consists of an enthalpy balance around the unit, and for this purpose a datum temperature of 294 K will be chosen. It will be assumed that the flow of dry air into the unit is m kg/s.

Considering the inlet streams:

(i) Nitrate: this enters at the datum of 294 K and hence the enthalpy $= 0$.
(ii) Air: G kg/s of dry air is associated with 0.007 kg moisture/kg dry air.

$$\therefore \qquad \text{enthalpy} = [(G \times 0.99) + (0.007G \times 2.01)](405 - 294) = 111.5G \text{ kW}$$

and the total heat into the system $= 111.5G$ kW.

Considering the outlet streams:

(i) Nitrate: 1.425 kg/s dry nitrate contains 0.00286 kg/s water and leaves the unit at 339 K.

$$\therefore \qquad \text{enthalpy} = [(1.425 \times 1.88) + (0.00286 \times 4.18)](339 - 294) = 120.7 \text{ kW}$$

(ii) Air: the air leaving contains 0.007 G kg/s water from the inlet air plus the water evaporated. It will be assumed that evaporation takes place at 294 K.
Thus:
enthalpy of dry air $= G \times 0.99(355 - 294) = 60.4G$ kW
enthalpy of water from inlet air $= 0.007G \times 2.01(355 - 294) = 0.86G$ kW
enthalpy in the evaporated water $= 0.07215[2450 + 2.01(355 - 294)] = 185.6$ kW
and the total heat out of the system, neglecting losses $= (306.3 + 61.3G)$ kW.

Making a balance:

$$111.5G = (306.3 + 61.3G) \quad \text{or} \quad G = 6.10 \text{ kg/s dry air}$$

Thus, including the moisture in the inlet air, moist air fed to the dryer is:

$$6.10(1 + 0.007) = \underline{\underline{6.15 \text{ kg/s}}}$$

Water entering with the air $= (6.10 \times 0.007) = 0.0427$ kg/s.
Water evaporated $= 0.07215$ kg/s.
Water leaving with the air $= (0.0427 + 0.07215) = 0.1149$ kg/s
Humidity of outlet air $= (0.1149/6.10) = 0.0188$ kg/kg dry air.

PROBLEM 13.6

Material is fed to a dryer at the rate of 0.3 kg/s and the moisture removed is 35% of the wet charge. The stock enters and leaves the dryer at 324 K. The air temperature falls from 341 K to 310 K, its humidity rising from 0.01 to 0.02 kg/kg. Calculate the heat loss to the surroundings. Latent heat of water at 324 K $= 2430$ kJ/kg. Specific heat capacity of dry air $= 0.99$ kJ/kg K. Specific heat capacity of water vapour $= 2.01$ kJ/kg K.

Solution

The wet feed is 0.3 kg/s and the water removed is 35%, or: $(0.3 \times 35/100) = 0.105$ kg/s

If the flowrate of dry air is G kg/s, the increase in humidity $= (0.02 - 0.01) = 0.01$ kg/kg

or: $\quad\quad\quad\quad\quad\quad 0.01G = 0.105 \quad$ and $\quad G = 10.5$ kg/s

This completes the mass balance, and the next step is to make an enthalpy balance along the lines of Problem 13.5. As the stock enters and leaves at 324 K, no heat is transferred from the air and the heat lost by the air must represent the heat used for evaporation plus the heat losses, say L kW.

Thus heat lost by the inlet air and associated moisture is:

$$[(10.5 \times 0.99) + (0.01 \times 10.5 \times 2.01)](341 - 310) = 328.8 \text{ kW}$$

Heat leaving in the evaporated water $= 0.105[2430 + 2.01(310 - 324)] = 252.2$ kW.
Making a balance:

$$328.8 = (252.2 + L) \quad \text{or} \quad \underline{\underline{L = 76.6 \text{ kW}}}$$

PROBLEM 13.7

A rotary dryer is fed with sand at the rate of 1 kg/s. The feed is 50% wet and the sand is discharged with 3% moisture. The entering air is at 380 K and has an absolute humidity of 0.007 kg/kg. The wet sand enters at 294 K and leaves at 309 K and the air leaves at 310 K. Calculate the mass flowrate of air passing through the dryer and the humidity of the air leaving the dryer. Allow for a radiation loss of 25 kJ/kg dry air. Latent heat of water at 294 K $= 2450$ kJ/kg. Specific heat capacity of sand $= 0.88$ kJ/kg K. Specific heat capacity of dry air $= 0.99$ kJ/kg k. Specific heat capacity of vapour $= 2.01$ kg K.

Solution

The feed rate of wet sand is 1 kg/s and it contains 50% moisture or $(1.0 \times 50/100) = 0.50$ kg/s water.

$$\therefore \qquad \text{flow of dry sand} = (1.0 - 0.5) = 0.50 \text{ kg/s}$$

If the dried sand contains w kg/s water, then:

$$w/(w + 0.50) = (3.0/100) \text{ or } w = 0.0155 \text{ kg/s}$$

and: $\qquad$ the water evaporated $= (0.50 - 0.0155) = 0.4845$ kg/s.

Assuming a flowrate of G kg/s dry air, then a heat balance may be made based on a datum temperature of 294 K.

Inlet streams:

(i) Sand: this enters at 294 K and hence the enthalpy $= 0$.
(ii) Air: G kg/s of dry air is associated with 0.007 kg/kg moisture.

$$\therefore \qquad \text{enthalpy} = [(G \times 0.99) + (0.007G \times 2.01)](380 - 294) = 86.4G \text{ kW}$$

and: $\qquad$ the total heat into the system $= 86.4G$ kW.

Outlet streams:

(i) Sand: 0.50 kg/s dry sand contains 0.0155 kg/s water and leaves the unit at 309 K.

$$\therefore \qquad \text{enthalpy} = [(0.5 \times 0.88) + (0.0155 \times 4.18)](309 - 294) = 7.6 \text{ kW}$$

(ii) Air: the air leaving contains $0.07\,G$ kg/s water from the inlet air plus the water evaporated. It will be assumed that evaporation takes place at 294 K. Thus:
enthalpy of dry air $= G \times 0.99(310 - 294) = 15.8m$ kW
enthalpy of water from inlet air $= 0.007G \times 2.01(310 - 294) = 0.23G$ kW
enthalpy in the evaporated water $= 0.4845[2430 + 2.01(310 - 294)] = 1192.9$ kW, a total of $(16.03G + 1192.9)$ kW

(iii) Radiation losses $= 25$ kJ/kg dry air or $25G$ kW and the total heat out $= (41.03G + 1200.5)$ kW.

Mass balance:

$$86.4G = (41.03G + 1200.5) \text{ or } G = 26.5 \text{ kg/s}$$

Thus the flow of dry air through the dryer $= \underline{\underline{26.5 \text{ kg/s}}}$

and the flow of inlet air $= (26.5 \times 1.007) = \underline{\underline{26.7 \text{ kg/s}}}$

As in Problem 13.5, water leaving with the air is: $(26.5 \times 0.007) + 0.4845 = 0.67$ kg/s and humidity of the outlet air $= (0.67/26.5) = \underline{\underline{0.025 \text{ kg/kg}}}$.

PROBLEM 13.8

Water is to be cooled in a packed tower from 330 to 295 K by means of air flowing countercurrently. The liquid flows at the rate of 275 cm^3/m^2 s and the air at 0.7 m^3/m^2 s. The entering air has a temperature of 295 K and a humidity of 20%. Calculate the required height of tower and the condition of the air leaving at the top.

The whole of the resistance to heat and mass transfer can be considered as being within the gas phase and the product of the mass transfer coefficient and the transfer surface per unit volume of column $(h_D a)$ may be taken as 0.2 s^{-1}.

Solution

Assuming, the latent heat of water at 273 K = 2495 kJ/kg
specific heat capacity of dry air   = 1.003 kJ/kg K
specific heat capacity of water vapour   = 2.006 kJ/kg K

then the enthalpy of the inlet air stream is:

$$H_{G1} = 1.003(295 - 273) + \mathscr{H}(2495 + 2.006(295 - 273))$$

From Fig. 13.4, when $\theta = 295$ K, at 20% humidity, $\mathscr{H} = 0.003$ kg/kg, and:

$$H_{G1} = (1.003 \times 22) + 0.003(2495 + (2.006 \times 22)) = 29.68 \text{ kJ/kg}$$

In the inlet air, the humidity is 0.003 kg/kg dry air or $(0.003/18)/(1/29) = 0.005$ kmol/kmol dry air.

Hence the flow of dry air = $(1 - 0.005)0.70 = 0.697$ m^3/m^2 s.

Density of air at 295 K = $(29/22.4)(273/295) = 1.198$ kg/m^3.

and hence the mass flow of dry air = $(0.697 \times 1.198) = 0.835$ kg/m^2 s

and the mass flow of water = 275×10^{-6} m^3/m^2 s or $(275 \times 10^{-6} \times 1000) = 0.275$ kg/m^2 s.

The slope of the operating line, given by equation 13.37 is:

$$LC_L/G = (0.275 \times 4.18/0.835) = 1.38$$

The coordinates of the bottom of the operating line are:

$$\theta_{L1} = 295 \text{ K and } H_{G1} = 29.7 \text{ kJ/kg}$$

Hence, on an enthalpy–temperature diagram (Fig. 13a), the operating line of slope 1.38 is drawn through the point (29.7, 295).

The top point of the operating line is given by $\theta_{L2} = 330$ K, and from Fig. 13a, $H_{G2} = 78.5$ kJ/kg.

From Figs 13.4 and 13.5 the curve representing the enthalpy of saturated air as a function of temperature is obtained and drawn in. This plot may also be obtained by calculation using equation 13.60.

The integral:

$$\int dH_G/(H_f - H_G)$$

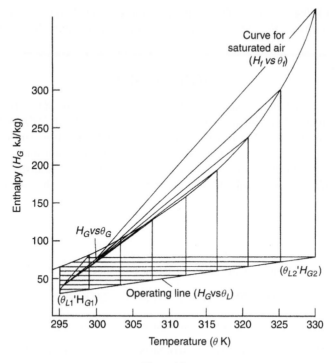

Figure 13a.

is now evaluated between the limits $H_{G1} = 29.68$ kJ/kg and $H_{G2} = 78.5$ kJ/kg, as follows:

H_G	θ	H_f	$(H_f - H_G)$	$1/(H_f - H_G)$
29.7	295	65	35.3	0.0283
40	302	98	58	0.0173
50	309	137	87	0.0115
60	316	190	130	0.0077
70	323	265	195	0.0051
78.5	330	408	329.5	0.0030

From a plot of $1/(H_f - H_G)$ and H_G the area under the curve is 0.573. Thus:

$$\text{height of packing, } z = \int_{H_{G1}}^{H_{G2}} [\mathrm{d}H_G/(H_f - H_G)]G/h_D a \rho \quad \text{(equation 13.53)}$$

$$= (0.573 \times 0.835)/(0.2 \times 1.198)$$

$$= 1.997, \text{ say } \underline{\underline{2.0 \text{ m}}}$$

In Fig. 13a, a plot of H_G and θ_G is obtained using the construction given in Section 13.6.3. and shown in Fig. 13.16. From this plot, the value of θ_{G2} corresponding to $H_{G2} = 78.5$ kJ/kg is 300 K. From Fig. 13.5 the exit air therefore has a humidity of 0.02 kg/kg which from Fig. 13.4 corresponds to a percentage humidity of <u>90%</u>.

PROBLEM 13.9

Water is to be cooled in a small packed column from 330 to 285 K by means of air flowing countercurrently. The rate of flow of liquid is 1400 cm^3/m^{2}s and the flowrate of the air, which enters at 295 K with a humidity of 60% is 3.0 m^3/m^{2}s. Calculate the required height of tower if the whole of the resistance to heat and mass transfer can be considered as being in the gas phase and the product of the mass transfer coefficient and the transfer surface per unit volume of column is 2 s^{-1}. What is the condition of the air which leaves at the top?

Solution

As in Problem 13.8, assuming the relevant latent and specific heat capacities:

$$H_{G1} = 1.003(295 - 273) + \mathcal{H}(2495 + 2.006(295 - 273))$$

From Fig. 13.4, at $\theta = 295$ and 60% humidity, $\mathcal{H} = 0.010$ kg/kg and hence:

$$H_{G1} = (1.003 \times 22) + 0.010(2495 + 44.13) = 47.46 \text{ kJ/kg}$$

In the inlet air, water vapour $= 0.010$ kg/kg dry air or $(0.010/18)/(1/29) = 0.016$ kmol/kmol dry air.

Thus the flow of dry air $= (1 - 0.016)3.0 = 2.952$ m^3/m^{2}s.

Density of air at 295 K $= (29/22.4)(273/293) = 1.198$ kg/m^3.

and mass flow of dry air $= (1.198 \times 2.952) = 3.537$ kg/m^{2}s.

Liquid flow $= 1.4 \times 10^{-3}$ m^3/m^{2}s

and mass flow of liquid $= (1.4 \times 10^{-3} \times 1000) = 1.4$ kg/m^{2}s.

The slope of the operating line is thus: $LC_L/G = (1.40 \times 4.18)/3.537 = 1.66$ and the coordinates of the bottom of the line are:

$$\theta_{L1} = 285 \text{ K}, \quad H_{G1} = 47.46 \text{ kJ/kg}$$

From these data, the operating line may be drawn in as shown in Fig. 13b and the top point of the operating line is:

$$\theta_{L2} = 330 \text{ K}, \quad H_{G2} = 122 \text{ kJ/kg}$$

Again as in Problem 13.8, the relation between enthalpy and temperature at the interface H_f vs. θ_f is drawn in Fig. 13b. It is seen that the operating line cuts the saturation curve, which is clearly an impossible situation and, indeed, it is not possible to cool the water to 285 K under these conditions. As discussed in Section 13.6.1, with mechanical draught towers, it is possible, at the best, to cool the water to within, say, 1 deg K of the wet

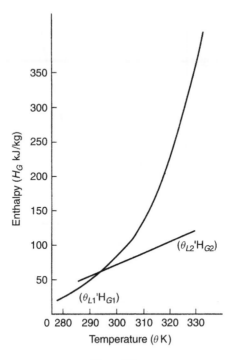

Figure 13b.

bulb temperature. From Fig. 13.4, at 295 K and 60% humidity, the wet-bulb temperature of the inlet air is 290 K and at the best water might be cooled to 291 K. In the present case, therefore, 291 K will be chosen as the water outlet temperature.

Thus an operating line of slope: $LC_L/G = 1.66$ and bottom coordinates: $\theta_{L1} = 291$ K and $H_{G1} = 47.5$ kJ/kg is drawn as shown in Fig. 13c. At the top of the operating line:

$$\theta_{L2} = 330 \text{ K and } H_{G2} = 112.5 \text{ kJ/kg}$$

As an alternative to the method used in Problem 13.8, the approximate method of Carey and Williamson (equation 13.54) is adopted.

At the bottom of the column:

$$H_{G1} = 47.5 \text{ kJ/kg}, \quad H_{f1} = 52.0 \text{ kJ/kg} \quad \therefore \Delta H_1 = 4.5 \text{ kJ/kg}$$

At the top of the column:

$$H_{G2} = 112.5 \text{ kJ/kg}, \quad H_{f2} = 382 \text{ kJ/kg} \quad \therefore \Delta H_2 = 269.5 \text{ kJ/kg}$$

At the mean water temperature of $0.5(330 + 291) = 310.5$ K:

$$H_{Gm} = 82.0 \text{ kJ/kg}, \quad H_{fm} = 152.5 \text{ kJ/kg} \quad \therefore \Delta H_m = 70.5 \text{ kJ/kg}$$

$$\therefore \qquad \Delta H_m/\Delta H_1 = 15.70 \text{ and } \Delta H_m/\Delta H_2 = 0.262$$

and from Fig. 13.17: $f = 0.35$ (extending the scales)

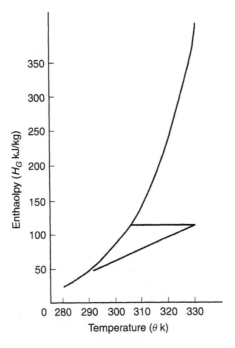

Figure 13c.

Thus:

$$\text{height of packing, } z = \int_{H_{G1}}^{H_{G2}} [dH_G/(H_f - H_G)]G/h_D a\rho \quad \text{(equation 13.53)}$$

$$= (0.35 \times 3.537)/(2.0 \times 1.198) = \underline{0.52 \text{ m}}$$

Due to the close proximity of the operating line to the line of saturation, the gas will be saturated on leaving the column and will therefore be at 100% humidity. From Fig. 13c the exit gas will be at $\underline{306 \text{ K}}$.

PROBLEM 13.10

Air containing 0.005 kg water vapour/kg dry air is heated to 325 K in a dryer and passed to the lower shelves. It leaves these shelves at 60% humidity and is reheated to 325 K and passed over another set of shelves, again leaving with 60% humidity. This is again reheated for the third and fourth sets of shelves after which the air leaves the dryer. On the assumption that the material in each shelf has reached the wet bulb temperature and that heat losses from the dryer can be neglected, determine:

(a) the temperature of the material on each tray,
(b) the rate of water removal if 5 m^3/s of moist air leaves the dryer,

(c) the temperature to which the inlet air would have to be raised to carry out the drying in a single stage.

Solution

See Volume 1, Example 13.4

PROBLEM 13.11

0.08 m^3/s of air at 305 K and 60% humidity is to be cooled to 275 K. Calculate, using a psychrometric chart, the amount of heat to be removed for each 10 deg K interval of the cooling process. What total mass of moisture will be deposited? What is the humid heat of the air at the beginning and end of the process?

Solution

At 305 K and 60% humidity, from Fig. 13.4, the wet-bulb temperature is 299 K and $\mathscr{H} = 0.018$ kg/kg. Thus, as the air is cooled, the per cent humidity will increase until saturation occurs at 299 K and the problem is then one of cooling saturated vapour from 299 K to 275 K.

Considering the cooling in 10 deg K increments, the following data are obtained from Fig. 13.4:

θ (K)	θ_w (K)	% Humidity	$\mathscr{H}$	Humid heat (kJ/kg K)	Latent heat (kJ/kg)
305	299	60	0.018	1.032	2422
299	299	100	0.018	1.032	2435
295	295	100	0.017	1.026	2445
285	285	100	0.009	1.014	2468
275	275	100	0.0045	1.001	2491

At 305 K: the specific volume of dry air $= 0.861$ m^3/kg

the saturated volume $= 0.908$ m^3/kg

and hence the specific volume at 60% humidity $= [0.861 + (0.908 - 0.861)60/100]$

$$= 0.889 \text{ m}^3/\text{kg}$$

Thus: mass flow of moist air $= (0.08/0.889) = 0.090$ kg/s

Thus the flowrate of dry air $= 0.090/(1 + 0.018) = 0.0884$ kg/s.

From Fig. 13.4, specific heat of dry air (at $\mathscr{H} = 0$) $= 0.995$ kJ/kg K.

$\therefore$ enthalpy of moist air $= (0.0884 \times 0.995)(299 - 273) + (0.018 \times 0.0884)$

$\times [4.18(299 - 273) + 2435] + 0.090 \times 1.032(305 - 299) = \underline{\underline{6.89 \text{ kW}}}$

At 295 K: Enthalpy of moist air $= (0.0884 \times 0.995)(295 - 273) + (0.017 \times 0.0884)$

$$\times [4.18(295 - 273) + 2445] = \underline{5.75 \text{ kW}}$$

At 285 K: Enthalpy of moist air $= (0.0884 \times 0.995)(285 - 273) + (0.009 \times 0.0884)$

$$\times [4.18(285 - 273) + 2468] = \underline{3.06 \text{ kW}}$$

At 275 K: Enthalpy of moist air $= (0.0884 \times 0.995)(275 - 273) + (0.0045 \times 0.0884)$

$$\times [4.18(275 - 273) + 2491] = \underline{1.17 \text{ kW}}$$

and hence in cooling from 305 to 295 K, heat to be removed $= (6.89 - 5.75) = 1.14$ kW

in cooling from 295 to 285 K, heat to be removed $= (5.75 - 3.06) = 2.69$ kW

in cooling from 285 to 275 K, heat to be removed $= (3.06 - 1.17) = 1.89$ kW

The mass of water condensed $= 0.0884(0.018 - 0.0045) = \underline{0.0012 \text{ kg/s}}$.

The humid heats at the beginning and end of the process are:

$$\underline{1.082 \text{ and } 1.001 \text{ kJ/kg K}} \text{ respectively.}$$

PROBLEM 13.12

A hydrogen stream at 300 K and atmospheric pressure has a dew point of 275 K. It is to be further humidified by adding to it (through a nozzle) saturated steam at 240 kN/m^2 at the rate of 1 kg steam: 30 kg of hydrogen feed. What will be the temperature and humidity of the resultant stream?

Solution

At 275 K, the vapour pressure of water $= 0.72$ kN/m^2 (from Tables) and the hydrogen is saturated.

The mass of water vapour: $P_{w0}M_w/\mathbf{R}T = (0.72 \times 18)/(8.314 \times 275) = 0.00567 \text{kg/m}^3$ and the mass of hydrogen: $(P - P_{w0})M_A/\mathbf{R}T = (101.3 - 0.72)2/(8.314 \times 275) = 0.0880 \text{ kg/m}^3$

Therefore the humidity at saturation, $\mathscr{H}_0 = (0.00567/0.0880) = 0.0644$ kg/kg dry hydrogen and at 300 K, the humidity will be the same, $\mathscr{H}_1 = 0.0644$ kg/kg.

At 240 kN/m^2 pressure, steam is saturated at 400 K at which temperature the latent heat is 2185 kJ/kg.

The enthalpy of the steam is therefore:

$$H_2 = 4.18(400 - 273) + 2185 = 2715.9 \text{ kJ/kg}$$

Taking the mean specific heat capacity of hydrogen as 14.6 kJ/kg K, the enthalpy in 30 kg moist hydrogen or $30/(1 + 0.0644) = 28.18$ kg dry hydrogen is:

$$(28.18 \times 14.6)(300 - 273) = 11,110 \text{ kJ}$$

The latent heat of water at 275 K is 2490 kJ/kg and, taking the specific heat of water vapour as 2.01 kJ/kg K, the enthalpy of the water vapour is:

$$(28.18 \times 0.0644)(4.18(275 - 273) + 2490 + 2.01(300 - 275)) = 4625 \text{ kJ}$$

Hence the total enthalpy: $H_1 = 15,730 \text{ kJ}$

In mixing the two streams, 28.18 kg dry hydrogen plus $(30 - 28.18) = 1.82$ kg water is mixed with 1 kg steam and hence the final humidity:

$$\mathscr{H} = (1 + 1.82)/28.18 = \underline{\underline{0.100 \text{ kg/kg}}}$$

In the final mixture, 0.1 kg water vapour is associated with 1 kg dry hydrogen or $(0.1/18) = 0.0056$ kmol water is associated with $(1/2) = 0.5$ kmol hydrogen, a total of 0.5056 kmol.

$\therefore$ partial pressure of water vapour $= (0.0056/0.5056)101.3 = 1.11 \text{ kN/m}^2$

Water has a vapour pressure of 1.11 kN/m² at 281 K at which the latent heat is 2477 kJ/kg. Thus if T K is the temperature of the mixture, then:

$$(2716 + 15730) = (28.18 \times 14.6)(T - 273) + 2.82[4.18(281 - 273)$$

$$+ 2447 + 2.01(T - 281)]$$

$$\text{and } \underline{\underline{T = 300.5 \text{ K}}}$$

It may be noted that this relatively low increase in temperature occurs because the latent heat in the steam is not recovered, as would be the case in, say, a shell and tube unit.

PROBLEM 13.13

In a countercurrent packed column, n-butanol flows down at the rate of 0.25 kg/m² s and is cooled from 330 to 295 K. Air at 290 K, initially free of n-butanol vapour, is passed up the column at the rate of 0.7 m³/m² s. Calculate the required height of tower and the condition of the exit air. Data: Mass transfer coefficient per unit volume, $h_D a = 0.1 \text{ s}^{-1}$. Psychrometric ratio, $(h/h_D \rho_A s) = 2.34$. Heat transfer coefficients, $h_L = 3h_G$. Latent heat of vaporisation of n-butanol, $\lambda = 590$ kJ/kg. Specific heat capacity of liquid n-butanol, $C_L = 2.5$ kJ/kg K. Humid heat of gas: $s = 1.05$ kJ/kg K.

Temperature (K)	Vapour pressure of n-butanol (kN/m²)
295	0.59
300	0.86
305	1.27
310	1.75
315	2.48
320	3.32
325	4.49

Temperature (K)	Vapour pressure of n-butanol (kN/m^2)
330	5.99
335	7.89
340	10.36
345	14.97
350	17.50

Solution

See Volume 1, Example 13.10

PROBLEM 13.14

Estimate the height and base diameter of a natural draught hyperbolic cooling tower which will handle 5000 kg/s water entering at 300 K and leaving at 294 K. The dry-bulb air temperature is 287 K and the ambient wet-bulb temperature is 284 K.

Solution

See Volume 1, Example 13.8

Printed and bound by CPI Group (UK) Ltd, Croydon, CR0 4YY

03/10/2024

01040334-0016